Biological roles of copper

*The Ciba Foundation for the promotion of international cooperation in
medical and chemical research is a scientific and educational charity established by
CIBA Limited – now CIBA-GEIGY Limited – of Basle. The Foundation operates
independently in London under English trust law.
Ciba Foundation Symposia are published in collaboration with Excerpta Medica in Amsterdam.*

Excerpta Medica, P.O. Box 211, Amsterdam

Biological Roles of Copper

Ciba Foundation Symposium 79 (new series)

1980

Excerpta Medica

Amsterdam · Oxford · New York

© *Copyright 1980 Excerpta Medica*

All rights reserved. No part of this publication may be reproduced or transmitted in any form or by any means, electronic or mechanical, including photocopying and recording, or by any information storage and retrieval system, without permission in writing from the publishers. However, in countries where specific regulations concerning reproduction of copyrighted matters have been established by law, copies of articles/chapters in this book may be made in accordance with these regulations. This consent is given on the express condition that copies will serve for personal or internal use only and that the copier complies with payment procedures as implemented in the country where the copying is effected.

ISBN Excerpta Medica 90 219 4085 x
ISBN Elsevier/North-Holland 0 444 90177 9

Published in November 1980 by Excerpta Medica, P.O. Box 211, Amsterdam and Elsevier/North-Holland, Inc., 52 Vanderbilt Avenue, New York, N.Y. 10017.

Suggested series entry for library catalogues: Ciba Foundation Symposia.
Suggested publisher's entry for library catalogues: Excerpta Medica.

Ciba Foundation Symposium 79 (new series)

352 pages, 54 figures, 41 tables

Library of Congress Cataloging in Publication Data

Symposium on Biological Roles of Copper, London, 1980.
   Biological roles of copper.

   (Ciba Foundation symposium; 79 (new ser.))
   Bibliography: p.
Includes index.
   1. Copper in the body—Congresses. 2. Copper—Physiological effect—Congresses. I. Title. II. Series: Ciba Foundation. Symposium; new ser., 79. [DNLM: 1. Copper-Metabolism—Congresses. 2. Copper-Deficiency—Congresses. W3 C161F v. 79 1980 / QU130 B6146 1980]
QP535.C9S84   1980   599.01'9214   80-23396
ISBN 90-219-4085-x
ISBN 0-444-90177-9 (Elsevier/North-Holland)

Printed in The Netherlands by Casparie, Amsterdam

Contents

C.F. MILLS Introduction 1

H.T. DELVES Dietary sources of copper 5
Discussion 13

I. BREMNER Absorption, transport and distribution of copper 23
Discussion 36

C.F. MILLS Metabolic interactions of copper with other trace elements 49
Discussion 65

A.E.G. CASS and H.A.O. HILL Copper proteins and copper enzymes 71
Discussion 85

E. FRIEDEN Caeruloplasmin: a multi-functional metalloprotein of vertebrate plasma 93
Discussion 113

H.M. HASSAN Superoxide dismutases 125
Discussion 136

T.L. SOURKES Copper, biogenic amines, and amine oxidases 143
Discussion 154

General discussion Pathology and diagnosis of copper deficiency 157

E.D. HARRIS, J.K. RAYTON, J.E. BALTHROP, R.A. DiSILVESTRO and M. GARCIA-DE-QUEVEDO Copper and the synthesis of elastin and collagen 163
Discussion 177

C.H. McMURRAY Copper deficiency in ruminants 183
Discussion 205

D.M. DANKS Copper deficiency in humans 209
Discussion 221

L.S. HURLEY, C.L. KEEN and B. LÖNNERDAL Copper in fetal and neonatal development   227
Discussion   239

D.M. HUNT   Copper and neurological function   247
Discussion   260

C.A. OWEN JR.   Copper and hepatic function   267
Discussion   277

R. ÖSTERBERG   Therapeutic uses of copper-chelating agents   283
Discussion   293

D.C.H. McBRIEN   Anaerobic potentiation of copper toxicity and some environmental considerations   301
Discussion   315

Final general discussion   319

C.F. MILLS   Chairman's closing remarks   325

Index of contributors   331

Subject index   333

Participants

Symposium on Biological Roles of Copper held at the Ciba Foundation, London, 11–13th March, 1980

C.F. MILLS *(Chairman)* Nutritional Biochemistry Department, The Rowett Research Institute, Greenburn Road, Bucksburn, Aberdeen AB2 9SB, UK

I. BREMNER Nutritional Biochemistry Department, The Rowett Research Institute, Greenburn Road, Bucksburn, Aberdeen AB2 9SB, UK

D.M. DANKS Department of Paediatrics, University of Melbourne, Royal Children's Hospital, Parkville, 3052, Victoria, Australia

H.T. DELVES Department of Chemical Pathology and Human Metabolism, University of Southampton, South Laboratory & Pathology Block, Level D, Southampton General Hospital, Southampton SO9 4XY, UK

T.L. DORMANDY Whittington Hospital, St Mary's Wing, Highgate Hill, London N19 5NF, UK

E. FRIEDEN Department of Chemistry, The Florida State University, Tallahassee, Florida 32306, USA

G.G. GRAHAM Division of Human Nutrition, Department of International Health, School of Hygiene and Public Health, The Johns Hopkins University, 615 North Wolfe Street, Baltimore, Maryland 21205, USA

E.D. HARRIS Department of Biochemistry and Biophysics, Texas A & M University, College of Agriculture, College Station, Texas 77843, USA

H.M. HASSAN Department of Microbiology and Immunology, McGill University, 3755 University Street, Montreal, Quebec, Canada H3A 2B4

H.A.O. HILL Department of Inorganic Chemistry, University of Oxford, South Parks Road, Oxford OX1 3QR, UK

D.M. HUNT School of Biological Sciences, Queen Mary College, University of London, Mile End Road, London E1 4NS, UK

L.S. HURLEY Department of Nutrition, University of California, Davis, California 95616, USA

G. LEWIS Biochemistry Department, Ministry of Agriculture, Fisheries and Food, Central Veterinary Laboratory, New Haw, Weybridge, Surrey KT15 3NB, UK

D.C.H. McBRIEN Department of Biochemistry, School of Biological Sciences, Brunel University, Uxbridge, Middlesex UB8 3PH, UK

C.H. McMURRAY Department of Agriculture, Veterinary Research Laboratories, Stormont, Belfast BT4 3SD, UK

F.R. MANGAN Beecham Pharmaceuticals Research Division, Medicinal Research Centre, Coldharbour Road, The Pinnacles, Harlow, Essex CM19 5AD, UK

R. ÖSTERBERG Department of Medical Biochemistry, University of Gothenburg, Fack, S-400 33 Gothenburg, Sweden

C.A. OWEN JR. Department of Biochemistry, Mayo Clinic, Rochester, Minnesota 55901, USA

J.R. RIORDAN Research Institute, The Hospital for Sick Children, 555 University Avenue, Toronto, Ontario, Canada M5G 1X8

J.C.L. SHAW Paediatric Department, University College Hospital, Huntley Street, London WC1E 6AU, UK

T.L. SOURKES Department of Psychiatry, McGill University, 1033 Pine Avenue West, Montreal, Quebec, Canada H3A 1A1

M.S. TANNER Department of Child Health, Clinical Sciences Building, Leicester Royal Infirmary, P.O. Box 65, Leicester LE2 7LX, UK

S. TERLECKI Ministry of Agriculture, Fisheries and Food, Central Veterinary Laboratory, New Haw, Weybridge, Surrey KT15 3NB, UK

Editors: DAVID EVERED *(Organizer)* and GERALYN LAWRENSON

Chairman's introduction

COLIN F. MILLS

Nutritional Biochemistry Department, Rowett Research Institute, Bucksburn, Aberdeen AB2 9SB, UK

As a participant in a previous symposium organized by the Ciba Foundation (1979), I appreciate that much of the success of the Foundation's meetings is due to the opportunity that they present for discussions to be informal, spontaneous and hard-hitting. One of the particular benefits of fairly small meetings is that the participants can get to know one another and can discuss outstanding and controversial problems.

At this time of rapidly developing studies of the trace elements and their relationship to health and disease, there is a great need for discussion of conflicting issues. We can share pride in some recent achievements, although other aspects are perhaps not quite so satisfactory, either because of questionable standards of experimentation or because of questionable extrapolations of scanty data. This meeting will give us an opportunity to discuss our present knowledge of the biological roles of copper and to assess the inadequacies in our experimental approaches and in their interpretation.

Work on individual trace elements has progressed in a variety of ways. Twenty years ago research on the essentiality of zinc was stimulated primarily by clear evidence of the importance of zinc deficiency diseases in domesticated livestock. Interest in the role of zinc in human nutrition and health was to develop much later, after it had been recognized that the diets of some socially deprived populations, and some therapeutic diets, could effectively be zinc-deficient and that some human subjects had genetic abnormalities in zinc metabolism. Studies of selenium show the same pattern of development. There was great initial interest in the role of selenium in diseases of farm animals but only within the last few years, as firmer evidence has appeared that selenium may be involved in the aetiology of some human diseases, has interest greatly increased in its relevance as an essential nutrient for humans.

The pattern is very different for iron. The relevance of iron to the incidence and consequences of anaemia in humans was the initial stimulus to studies of its role and its metabolism and provides the continuing justification for most current research effort. However, some recent enquiries into the physiological effects of iron deprivation that precede the development of severe anaemia appear to have been initiated more by concern about the ethical acceptability of deliberately depriving meat-producing animals of iron for organoleptic reasons than by anxiety about the relevance of such effects to human health.

The historical development of studies of the role of copper has strongly reflected the importance of copper deficiency as a cause of disease in domesticated animals. The early appreciation of the importance of copper in normal haemopoiesis was succeeded by the conclusion that copper deficiency anaemia was likely to occur only rarely in humans, and few studies were then initiated on copper in human nutrition. In contrast, the first investigations of the roles of copper in nervous tissue metabolism, in maintaining the integrity of connective tissue, in melanogenesis and in body growth all grew directly from the awareness that copper deficiency was a cause of disease in farm livestock.

The present appreciation of the incidence and important economic consequences of copper deficiency in domesticated animals is accompanied, however, by a frequent inability to predict either the development of copper deficiency or the severity of its biochemical and pathological consequences. In humans, interest stimulated by work on copper deficiency of a nutritional origin has been heightened by the recognition, first, of Menkes' disease as a genetic defect in copper metabolism and, second – as with zinc – of the limitations of some therapeutic dietary regimes that have provoked clinical signs of deficiency.

Those faced with the task of detecting or assessing the significance of copper deficiency diseases in both humans and domesticated animals face similar problems: although adequate techniques are available to measure changes in the copper content of tissues that are accessible for sampling they rarely indicate whether such changes have metabolic or pathological significance. For example, we do not yet know the implications of the reductions in plasma copper and caeruloplasmin activity that are early consequences of copper deprivation and that can often precede by weeks or months the first overt indications of deficiency.

Similarly, we are well aware that in the severely copper-deficient animal marked decreases occur in both lysyl oxidase and cytochrome c oxidase activities, yet we know little about the relative sensitivities of target tissues to such changes, or whether the activity of these enzymes becomes rate-limiting at early or late stages of the syndrome. Answers to such biochemical uncertainties will influence not only future pathological studies of the pre-clinical stages of copper deficiency but also, in a wider context, future decisions about the physiological relevance of marginal deficiency or of severe but brief periods of suboptimal copper supply.

For these reasons it may be appropriate to suggest that during this meeting we should pay particular attention to the likely biochemical and pathological effects that arise at early stages of the copper deficiency syndrome in humans and animals. We might attempt to assess whether the measurements commonly used to determine copper status provide a realistic indication of the likelihood of lesions developing when there are no overt signs of deficiency.

Is our understanding of the processes that govern the absorption, transport, storage and subsequent utilization of copper adequate to explain how these processes can be interrupted by the presence of metabolic antagonists in the diet or environment, or to anticipate correctly the consequences of genetic defects in copper metabolism? Without understanding these processes we shall continue to find it difficult to predict conditions that are associated with a high risk or to suggest effective prophylaxis.

I expect that our discussions will reveal many important gaps in our knowledge of the biological roles of copper and of the significance of copper deficiency diseases. The success of the meeting and the value of the published proceedings will directly reflect our willingness to discuss such issues frankly in the congenial environment so willingly provided by the Ciba Foundation.

Reference

Ciba Foundation 1979 Development of mammalian absorptive processes. Excerpta Medica, Amsterdam (Ciba Found Symp 70)

Dietary sources of copper

H.T. DELVES

Chemical Pathology & Human Metabolism Unit, Level D, South Laboratory & Pathology Block, General Hospital, Southampton SO9 4XY, UK

Abstract The dietary intakes of copper by children and adults who consume free diets are often significantly lower than the recommended daily allowances of the National Academy of Sciences (USA) or the World Health Organization. These lower-than-recommended intakes of copper appear to be adequate for healthy individuals since states of copper deficiency have not been observed in the absence of an accompanying metabolic disorder. Copper deficiencies have arisen in preterm infants of low birth weight as a result of breast-milk diets, and in children and adults as a result of fluids that are used for total parenteral nutrition. This paper describes the use of trace-metal balance studies to evaluate the adequacy of copper intake from these sources and from synthetic diets that are used in the treatment of inherited and acquired metabolic disorders.

The richest dietary sources of copper are animal livers, crustacea, shellfish, dried fruit, nuts and chocolate. The copper concentrations in these foods range from 300–800 μmol/kg (20–50 mg/kg) – some 500 times higher than the concentrations in milk and dairy produce, which are the poorest dietary sources of this element (Table 1). The copper content of drinking water can be very high if copper plumbing is used and can provide daily intakes of 12 μmol (0.8 mg).

Adult western-type diets are reported to provide 32–64 μmol (2–4 mg) per day (Underwood 1977) and it is claimed to be difficult to produce a varied diet that will supply less than 32 μmol (2 mg) per day (National Research Council 1977). The validity of this latter statement is questionable when one considers some of the published data on the measured dietary intakes of healthy adults and children (Table 2). Few of the adult diets provide the Recommended Dietary Allowance (RDA) of 32 μmol (2 mg) per day and many contain less than half of this value. Harland et al (1977) have suggested that intakes of less than the RDA may be inadequate but Klevay (1978) has queried the accuracy

TABLE 1

Dietary sources of copper

Food	Copper content	
	$\mu mol\ kg^{-1}$	$mg\ kg^{-1}$
Animal livers, crustacea, shellfish	300 – 800	20 – 50
Poultry	30 – 80	2 – 5
Fish	15 – 30	1 – 2
Cereals	30 – 100	2 – 7
Vegetables		
lima beans, dried split peas;	120 – 140	8 – 9
beets, cauliflower, spinach, mushrooms;	15 – 40	1 – 2.5
cabbage, corn, carrots, soybeans.	8 – 16	0.5 – 1.0
Fruit		
dried figs, apricots, plums, olives, avocados;	50 – 80	3 – 5.0
citrus fruits, berries, apples.	<15	<1
Nuts		
brazil, hazelnut;	170 – 220	11 – 14
walnut, coconut.	50	3
Dairy produce	1.5 – 6.0	0.1 – 0.4
Drinking water	<1.5 – 13.1	<0.05 – 0.84

of some of the higher copper intakes that have been reported. Since copper deficiency has never been reported in adults receiving varied diets, it may be assumed that the intakes in Table 2 were adequate for health even though they were significantly lower than the RDA.

A similar conclusion may be reached from a consideration of the data presented in Table 2 for infants and young children. All the children studied were healthy and none had any overt signs of copper deficiency, yet their copper intakes were less than the suggested daily allowances. These intakes, in terms of body weight, ranged from 1.2 μmol (75 μg) per kg per day at one month of age to 0.7 μmol (43 μg) per kg per day at three months (for the breast-fed infants) and from 0.36 μmol (23 μg) to 0.51 μmol (32 μg) per kg per day (for the infants and young children). These intakes are lower than the suggested values of 1.3 μmol (80 μg) per kg per day for infants and 0.63 μmol (40 μg) per kg per day for older children (World Health Organization 1973,

TABLE 2

Some measured dietary intakes of copper

Subjects	n	Dietary intake of Cu		Reference
		μmol/day	μg/day	
1–3 mth infants on breast milk (Finland)	27	3.9 – 4.9	250 – 310	Vuori 1979
Infants and children 3 mth–8 yr (UK)	11	2.5 – 14.2	156 – 900	Alexander et al 1974
Girls, 12–14 yr (USA)	14	46.3 ± 14.6	2940 ± 930	Greger et al 1978
Children and adults (USA)	20	3.1 – 54.8	200 – 3480[a]	Klevay 1978
Adults, 19–50 yr (New Zealand)	164	23.6 ± 12.6	1500 ± 800[b]	Guthrie et al 1978

[a] Only 2 out of 20 diets provided > 32 μmol (2 mg) Cu per day and 5 out of 20 diets provided < 16 μmol (1 mg) Cu per day
[b] Of 164 diets measured, 33 provided > 32 μmol (2 mg) Cu per day and 38 provided < 16 μmol (1 mg) Cu per day

National Research Council 1974). The lower intake of copper by breast-fed infants agrees with the earlier observations of Picciano & Guthrie (1976). The decrease in copper intake with time of lactation is to be expected from the significant 50% decrease in the copper concentration of human breast milk over the first three months of lactation (Vuori & Kuitunen 1979). Although it may be possible to produce copper deficiency in infants by prolonging breast-feeding this condition is more likely to occur in infants fed cow's-milk diets without supplementation, since cow's milk contains only one-fifth of the copper concentration in human milk.

The measurement of the dietary intakes of copper *per se* does not give sufficient information about the adequacy of the intake. Various dietary constituents such as phytate and fibre have been shown to impair the intestinal absorption of copper (Kelsay et al 1979), so some additional assessment of the copper status of the body is required. Valuable information about the adequacy of natural and synthetic diets can be obtained by the use of metabolic balance techniques in which the total dietary intake and total faecal and urinary excretion of copper (and other elements) are measured and the net absorption and retention are calculated. The net absorption is the difference between the dietary intake and faecal excretion, and the retention is the difference between dietary intake and the sum of the faecal and urinary excretion. Together with serum copper and caeruloplasmin measurements these results provide a good assessment of the body status of copper. Diets that have been evaluated using some of these measurements include: breast-milk diets for preterm infants (Dauncey et al 1977); synthetic oral diets used in the treatment of inherited and acquired metabolic disorders and intolerance to food (Alexander et al 1974, Lawson et al 1977, Thorn et al 1978); and fluids used for parenteral nutrition (Shenkin & Wretlind 1978).

BREAST-MILK DIETS FOR PRETERM INFANTS

Fetal hepatic stores accumulate rapidly in the last three months of pregnancy, presumably to provide for the later needs of the full-term growing infant. Negative balances during the neonatal period have been well documented for full-term and preterm infants (Cavell & Widdowson 1964, Widdowson et al 1974) but it is the preterm infants with their much smaller body stores who are at greater risk of copper deficiency. Dauncey et al (1977) did serial metabolic balance studies of six preterm infants of low birth weight. The observed mean copper intake of 1.3 μmol (85 μg) per kg per day was higher than that reported by Vuori (1979) but the absolute intake of copper was lower, at 1.4–2.6 μmol (91–165 μg), because of the lower body weights of 1.05–1.43 kg. Copper balances were negative until about the 35th day of life and all the infants absorbed inadequate amounts of copper during the 72-day investigation. Three of the six infants experienced a net loss of copper during that period. The net accumulation of body copper was less than 10% of the intrauterine accumulation for an equivalent period. In contrast two light-for-dates but full-term infants were able to absorb significant amounts of copper from breast milk. The authors concluded that fetal immaturity of the alimentary tract was responsible for the impaired absorption of copper by the preterm infants and that such depletion of body copper stores could result in deficiencies in later infancy.

However, Wilson & Lahey (1960) did not detect any difference in indices of body copper status between one group of infants of low birth weight receiving only 0.22 μmol (14 μg) per kg per day for 7–10 weeks and another similar group receiving five to six times this amount. It is difficult to reconcile these observations with those discussed above and more particularly with the well documented cases of severe copper deficiency that have been reported in preterm infants who presented as early as three months of age (al-Rashid & Spangler 1971, Ashkenazi et al 1973).

SYNTHETIC DIETS

Synthetic diets are often used as part of the treatment of inherited and acquired metabolic disorders and of intolerance to foods. Some diets are designed to provide the total daily intake of all essential nutrients whereas others may be used with restricted amounts of natural foods.

A trace-metal supplement, designed for use with the synthetic diets that are used in the treatment of inherited disorders of amino acid metabolism, was originally formulated from the composition of cow's milk (see Table 3). We then evaluated this supplement (M1) by doing metabolic and trace-metal balance studies of children with phenylketonuria or maple-syrup-urine disease. These studies showed that the supplement provided inadequate intakes of copper, zinc, iron and manganese (Alexander et al 1974). A modified mineral and trace-metal supplement was formulated containing increased concentrations of these four elements (see Table 3). This supplement (M2) was later evaluated by Lawson et al (1977), who showed that M2 provided satisfactory amounts of copper, zinc and iron but needed minor changes in its concentration of manganese. The data on copper intake from these studies are summarized in Table 4. All seven patients receiving M2 were in positive copper balance compared with only two of the six patients who received M1. Although none of the patients who received M1 had overt signs of copper deficiency it is clear that M2 is the better dietary source of copper.

Diets based on comminuted chicken as the protein source, with additional fats, vitamins and carbohydrate, but free from milk, lactose, sucrose and gluten, are valuable for feeding children who suffer from malabsorption and intolerance to foods (Larcher et al 1977). The mineral and trace-metal mixtures M1 and M2 were evaluated as supplements to diets based on comminuted chicken in four children aged 2 to 18 months, all of whom had protracted diarrhoea. Three of the children were intolerant of cow's milk; one was also intolerant of soya protein and another of several foods. The fourth patient had malabsorption following a small bowel resection for necrotizing enterocolitis and peritoneal adhesions. Metabolic balance studies were done after the patients had been equilibrated on the diets-plus-mineral mixtures for at least three weeks. When the infants received M1, their mean net absorption

TABLE 3

Iron, copper and zinc content of the trace-metal mixtures in diets M1 and M2

	Diet M1 (per kg)	*Diet M2 (per kg)*
Iron	9.00 mmol (550 mg)	11.2 mmol (628 mg)
Copper	0.99 mmol (63 mg)	2.00 mmol (127 mg)
Zinc	1.84 mmol (120 mg)	7.34 mmol (480 mg)

The diets contained the same amounts of calcium, potassium, sodium, magnesium, iodine, aluminium, cobalt, molybdenum and phosphorus (for further details see Lawson et al 1977).

TABLE 4

Intake, absorption and retention of copper by healthy children on free diets and by patients receiving synthetic diets

Subjects	Copper intake/day	Cu absorbed/day		Cu retained/day	
	Median (obs. range)	Median	% of intake	Median	% of intake
Healthy children (n = 11) on free diets	4.9 (2.5–14.2) μmol 308 (156–900) μg	1.9 μmol 121 μg	+39	1.5 μmol 98 μg	+32
Children (n = 6) on M1[a]	8.9 (5.6–12.8) μmol 566 (355–814) μg	−2.1 μmol −136 μg	−24	−3.7 μmol −238 μg	−42
Children (n = 7) on M2[b]	17.7 (9.1–19.21) μmol 1112 (575–1221) μg	+3.8 μmol +240 μg	+22	+3.2 μmol +198 μg	+18

[a]Six children received synthetic diets with mineral mixture M1 for the inherited metabolic disorders of phenylketonuria (n = 4), or maple syrup urine disease (n = 2).
[b]Seven children with phenylketonuria received synthetic diets with mineral mixture M2.

and retention of dietary copper was 21% and 17% respectively. These values were increased when they were on M2, to 28% and 22% respectively, but the increases were not statistically significant. The serum levels of copper increased in two patients on the M2 diet. It is interesting that the serum copper levels for one of these children increased on the M1 diet from 9.6 to 12.0 μmol/l, and subsequently on the M2 diet to 18.0 μmol/l but they fell to 2.1 μmol/l within three weeks of the child's being regraded onto a cow's milk diet without any supplementation, at the age of 20 months; (the reference range for healthy controls is 12–26 μmol/l). This indicates that the child had very low stores of body copper and emphasizes the inadequacy of a milk-based diet as a source of copper.

FLUIDS FOR PARENTERAL NUTRITION

Long-term total parenteral nutrition (TPN) with infusion fluids that have not been supplemented with copper and other essential trace elements at their optimum concentrations must inevitably lead to deficiency states. Evidence to support this statement may be deduced from the observations of a mean decrease in serum copper concentrations of 1.7 μmol/l (10.8 μg/dl) per week by a group of adults receiving TPN from fluids that contained no detectable amounts of copper (Fleming et al 1976). The detection limit of the method was not quoted but a conservative estimate for flame atomic absorption spectroscopy would be < 1.6 μmol/l (0.1 mg/l), so the patients would probably have received less than 4.7 μmol (300 μg) Cu/day. The lowest serum copper levels observed were about 1.6 μmol/l (10 μg/dl) and when oral feeding was resumed the mean increase was 2.2 μmol/l (14 μg/dl) per week. Severe hypocupraemia is frequently observed in patients on prolonged TPN; the lowest value that we have measured for an adult was < 0.5 μmol/l (3 μg/dl), which is the detection limit of the method.

Severe hypocupraemia, neutropenia and extensive bone changes have been observed in infants and children receiving inadequate amounts of copper when fed intravenously (Karpel & Peden 1972, Heller et al 1978, Yuen et al 1979). In most cases the addition of copper corrected the clinical and biochemical abnormalities. However, the preterm infant of low birth weight investigated by Yuen et al (1979) had, at nine months of age, an apparent bone age of only three months.

Although TPN affords the opportunity for more exact metabolic studies than can be achieved with oral diets, few workers have done trace-metal balance studies. Most appear to rely on measurements of the trace-metal intake together with some additional index of body status, e.g. serum copper or serum caeruloplasmin levels. James & MacMahon (1976) did balance studies for nine elements in preterm infants receiving TPN. Their results suggested that preterm infants could utilize as much as 0.79 μmol (50 μg) per kg per day for normal growth. This is about twice the value recommended by Karpel & Peden (1972) and by Heller et al (1978) for infants on TPN. The tentative recommended daily allowances quoted by Shenkin & Wretlind (1978) agree more closely with these latter values and are 0.3 μmol (19 μg) per kg per day for neonates and infants and 0.07 μmol (4.4 μg) per kg per day as a basal intake for adults, increasing to about 5 and 10 times this value for moderate and high intakes respectively.

As with oral diets, measurements of the total intake of copper alone will not give sufficient information with which to judge the adequacy of the intake. The hyperzincuria obtained when a mixture of D- and L-amino acids was used for TPN (van Rij & McKenzie 1978) would certainly have been accompanied by an increased urinary excretion of copper. There is also the possibility of increased excretion of copper complexes with glucose and amino acids. These complexes form during heat-sterilization of some infusion fluids. Measurements of serum copper and caeruloplasmin levels are important for assessing the body status of copper but a more informative indication of the adequacy of the intake would be obtained from copper balance studies in which the losses of copper in urine or faeces, or from fistulas, are compared with the intake from the infusion fluids.

CONCLUSION

Most adult diets appear to contain adequate amounts of copper to maintain health, even though they provide less than the recommended daily allowance of 32 μmol (2 mg). Milk-based diets have a grossly inadequate copper content, particularly for preterm infants of low birth weight, and non-supplemented infusion fluids that are used for TPN are also inadequately low in copper. This is aptly illustrated in the report by Yuen et al (1979) of severe copper deficiency in a preterm infant of low birth weight who was fed alternately on a milk-based diet and on TPN without copper supplementation.

References

al-Rashid RA, Spangler J 1971 Neonatal copper deficiency. N Engl J Med 285:841-843
Alexander FW, Clayton BE, Delves HT 1974 Mineral and trace-metal balances in children receiving normal and synthetic diets. Q J Med 43:89-111
Ashkenazi A, Levin S, Djaldetti M, Fishel E, Benvenisti D 1973 The syndrome of neonatal copper deficiency. Pediatrics 52:525-533
Cavell PA, Widdowson EM 1964 Intakes and excretions of iron, copper and zinc in the neonatal period. Arch Dis Child 39:496-501
Dauncey MJ, Shaw JCL, Urman J 1977 The absorption and retention of magnesium, zinc and copper by low birth weight infants fed pasteurized human breast milk. Pediatr Res 11:1033-1039
Fleming CR, Hodges RE, Hurley LS 1976 A prospective study of serum copper and zinc levels in patients receiving total parenteral nutrition. Am J Clin Nutr 29:70-77
Greger JL, Bennett OA, Buckley S, Baligar P 1978 Zinc, copper and manganese balance in adolescent girls. In: Kirchgessner M (ed) Trace element metabolism in man and animals. Arbeitskreis für Tierernährungsforschung, Freising-Weihenstephan, vol 3:300-303
Guthrie BE, McKenzie HM, Casey CC 1978 Copper status of New Zealanders. In: Kirchgessner M (ed) Trace element metabolism in man and animals. Arbeitskreis für Tierernährungsforschung, Freising-Weihenstephan, 3:304-306
Harland BF, Prosky L, Vanderveen JE 1978 Nutritional adequacy of current levels of Ca, Cu, Fe, I, Mg, Mn, P, Se and Zn in the American food supply for infants and toddlers. In: Kirchgessner M (ed) Trace element metabolism in man and animals. Arbeitskreis für Tierernährungsforschung, Freising-Weihenstephan, vol 3:311-312
Heller RM, Kirchner SG, O'Neill JA Jr, Hough AJ Jr, Howard L, Kramer SS, Green HL 1978 Skeletal changes of copper deficiency in infants receiving prolonged total parenteral nutrition. J Pediatr 92:947-949
James BE, MacMahon RA 1976 Balance studies of elements during complete intravenous feeding of small pre-term infants. Aust Paediatr J 12:154-162
Karpel JT, Peden VH 1972 Copper deficiency in long-term parenteral nutrition. J Pediatr 80:32-36

Kelsay JL, Jacob RA, Prather S 1979 Effect of fiber from fruits and vegetables on metabolic responses of human subjects. III: Zinc, copper and phosphorus balances. Am J Clin Nutr 32:2307-2311

Klevay LM 1978 Dietary copper and the copper requirement of man. In: Kirchgessner M (ed) Trace element metabolism in man and animals. Arbeitskreis für Tierernährungsforschung, Freising-Weihenstephan, vol 3:307-310

Larcher VF, Shepherd R, Francis DEM, Harries JT 1977 Protracted diarrhoea in infancy. Arch Dis Child 52:597-605

Lawson MS, Clayton BE, Delves HT, Mitchell JD 1977 Evaluation of a new mineral and trace metal supplement for use with synthetic diets. Arch Dis Child 52:62-67

National Research Council, Food and Nutrition Board 1974 Recommended dietary allowances, 8th edn. Washington DC

National Research Council, Committee on medical and biological effects of environmental pollutants 1977 Copper. National Academy of Sciences, Washington DC, p 29

Picciano MF, Guthrie HA 1976 Copper, iron and zinc contents of mature human milk. Am J Clin Nutr 29:242-254

Shenkin A, Wretlind A 1978 Parenteral nutrition. World Rev Nutr Diet 28:1-111

Thorn JM, Aggett PJ, Delves HT, Clayton BE 1978 Mineral and trace metal supplement for use with synthetic diets based on comminuted chicken. Arch Dis Child 53:931-938

Underwood EJ 1977 Trace elements in human and animal nutrition, 4th edn. Academic Press, London

van Rij AM, McKenzie JM 1978 Hyperzincuria and zinc deficiency in total parenteral nutrition. In: Kirchgessner M (ed) Trace element metabolism in man and animals. Arbeitskreis für Tierernährungsforschung, Freising-Weihenstephan, vol 3:288-291

Vuori E 1979 Intake of copper, iron, manganese and zinc by healthy exclusively-breast-fed infants during the first 3 months of life. Br J Nutr 42:407-411

Vuori E, Kuitunen P 1979 The concentrations of copper and zinc in human milk. Acta Paediatr Scand 68:33-37

Widdowson EM, Dauncey J, Shaw JCL 1974 Trace elements in fetal and early postnatal development. Proc Nutr Soc 33:275-284

Wilson JF, Lahey ME 1960 Failure to induce dietary deficiency of copper in premature infants. Pediatrics 25:40-49

World Health Organization 1973 Trace elements in human nutrition. WHO Tech Rep Ser 532

Yuen P, Lin HJ, Hutchison JH 1979 Copper deficiency in a low birth weight infant. Arch Dis Child 54:553-555

Discussion

Mills: I would like to ask how many reports exist of parenteral feeds that have omitted copper and thus, after prolonged usage, have inadvertently led to clinical manifestation of deficiency.

Delves: I don't know the total number but there are probably at least half a dozen reports.

Mills: I believe that such reports continue to appear, some very recently (e.g. Heller et al 1978, Yuen et al 1979). Is the message getting through, to those who prepare and use parenteral and intragastric feeds, that copper is an *essential* element?

Hurley: I think it is getting through.

Mills: But isn't it rather slow?

Graham: Some surgeons are hard to convince!

Frieden: What form of copper is added to parenteral feeds?

Shaw: We add copper sulphate to our solutions. There are now numerous reports of copper deficiency in infants and adults, some of whom receive total parenteral nutrition (Sturgeon & Brubacker 1956, Schubert & Lahey 1959, Griscom et al 1971, al-Rashid & Spangler 1971, Karpel & Peden 1972, Seely et al 1972, Ashkenazi et al 1973, Vilter et al 1974, Dunlap et al 1974, Graham & Cordano 1976, Zidar et al 1977, Heller et al 1978, Sann et al 1978, Yuen et al 1979, Blumenthal et al 1980), and many people believe that we should add both copper and zinc to these feeds. There is, at present, at least one commercially available intravenous additive preparation containing these and other trace elements (Ped-el, KabiVitrum Ltd, London).

Harris: Could copper be deleterious to some of the other components that are present in the intravenous or parenteral preparation? We know that copper is quite an active catalyst chemically. Could it catalyse destruction of some of the vitamins that are present in feeds?

Shaw: I am sure that's a possibility.

Harris: Could that be a reason why copper is left out of feeds?

Shaw: No. I think the main reason is that many hospital pharmacies lack the experience and necessary quality-control facilities to make such additions to feeds. Also, many physicians are uncertain about the point at which the addition of copper becomes essential.

Delves: Intravenous feeding is very complex anyway. To add yet another solution to the different infusion fluids would make the management of patients even more complicated.

Graham: When hospital pharmacists were first made responsible for compounding these solutions, they were afraid to add copper and zinc because they didn't know what effect the elements would have on vitamins and their availability. Copper and zinc are usually added now, but I suspect that although copper deficiency due to hyperalimentation is no longer considered to be rare enough to report, it must still happen quite frequently.

Danks: Surely the amino acid content of most of those mixtures would rapidly mop up the copper that is added, and stop it destroying vitamins or any other components? I would expect the risks from excess copper to be minimal.

Österberg: I would also like to comment on the hyperalimentation mixture used for intravenous feeding. We all know that the amino acids in the mixture will bind copper very strongly. As a result, even if copper is added to the diet, deficiency can arise if the parenteral feeding mixture contains a large excess of amino acids relative to the copper and if the amino acids and their complexes are washed out through the urine. I therefore agree that it's necessary to do strictly balanced studies because of the strong copper complexes formed by parenteral preparations.

I agree that supplementation of copper to the diets would not be a risk because the copper becomes strongly bound in the amino acid complexes.

Frieden: There is an alternative to the parenteral method of administering copper. There are recent reports by Ray Walker and co-workers of the ready absorption of copper salicylate preparations through the skin. This would bypass the need to supplement the intravenous preparations by the direct addition of copper salts.

McMurray: Alternatively, even when parenteral administration is used, continuous infusion of copper may not always be necessary. Parenteral treatment with a bolus of copper boosts the copper reserves in animals, and is used frequently for restoring copper status to normal.

Mangan: At this point I would like to mention some of our observations on anti-inflammatory responses of the rat to copper administered by various routes. When paw oedema induced by carrageenan is used as a model of inflammation, a single subcutaneous dose of a simple copper-containing compound (e.g. copper chloride, copper acetate or copper nitrate) exerts a profound anti-inflammatory effect. However, when these compounds are given orally they have no effect (Boyle et al 1976). Furthermore, complexes of copper with amino acids are also inactive after oral administration whereas, after parenteral administration, the anti-inflammatory activity is directly proportional to the quantity of copper injected (Mangan 1978 cited in Jackson et al 1978). This pattern of response to copper chloride is reflected by the total plasma copper concentrations, which increase only after subcutaneous administration of the compound (F.R. Mangan, unpublished work).

Mills: Dr Delves, you referred several times to cases of copper deficiency associated with diarrhoea. In at least two studies, including George Graham's initial reports from the Anglo-American Hospital at Lima, on copper deficiency in the malnourished child (Graham & Cordano 1969), there was a variable incidence of diarrhoea. When we see copper deficiency in the rat and in ruminants (particularly the bovine but not often the ovine species), there is a variable incidence of diarrhoea that is a consequence rather than a cause of the deficiency (Mills et al 1976). Work with rats and ruminants has indicated that in copper deficiency the cytochrome *c* oxidase activity of the intestinal mucosa is rapidly reduced, while some other tissues do not show such rapid changes (Fell et al 1975). We don't yet know whether this reduction is related to the diarrhoea. How seriously should we therefore consider the possibility that copper deficiency is a *cause* of diarrhoea in the human child?

Delves: We feel that the copper deficiency arises in children *because* of the diarrhoea rather than being a *cause* of diarrhoea, but we don't have any real evidence to support the view.

Mills: What about your studies, Dr Graham?

Graham: Every year in Lima four or five children are admitted to hospital with severe copper deficiency. Almost all of them are between eight and fifteen months of age and have had intermittent diarrhoea since two or three months of age. The diarrhoea may have cleared up completely in these children but they may still be severely copper-deficient. One child from an underdeveloped country was treated for many years in a highly developed country and had copper deficiency for five years (Cordano & Graham 1966). She had intractable diarrhoea and could tolerate nothing except the most elementary of diets. The diarrhoea disappeared like magic when she was given copper.

Mills: Is this the only report of copper-responsive diarrhoea?

Graham: Yes, as far as I know.

Lewis: It is not easy to raise the copper content of cow's milk by supplementation of copper to the diet of cows. In this country, copper deficiency in cattle is widespread; most of our milk is produced from grass-fed cattle and so its copper content is low (Beck 1941a,b). (Previously we used to produce milk from cows fed on concentrated diets during the winter and the copper content of the milk (Beck 1941a,b) might have been higher.) Do you think that widespread copper deficiency in children occurs because the copper content of milk-based diets has gradually decreased over the years, or are the numbers of reports simply related directly to an increased awareness of the deficiency problem?

Graham: All the cases of copper deficiency in full-term babies that I have examined were in babies that had chronic diarrhoea, and therefore severe losses of copper from the body. Only a few tiny premature babies of 800 g weight have developed copper deficiency without diarrhoea (al Rashid & Spangler 1971). By hyperalimentation and more appropriate treatment of severely malnourished infants we are now able to keep alive many who would have died otherwise. As they begin to grow, the copper depletion becomes apparent and milk cannot make up the deficiency.

Sourkes: When we were looking for a mild diet that would produce copper deficiency in rats, we analysed a number of brands of evaporated milk and found that they were sharply divided into two categories. Brands that were very low in copper, containing less than 0.1 mg/l (Missala et al 1967), were from cans with a lacquered interior, while the others were from unvarnished cans. The manufacturers who did not use varnish thought that the infants would dislike the taste that the lacquer would give to the milk, while those who used lacquer claimed it was essential to protect the infants from toxic metals in the can. I mention this because each type of evaporated milk should be analysed separately for dietary balance purposes.

Frieden: There could also be toxic lacquers!

Mills: Are there any comments on the question of the recommended dietary allowances (RDAs) of copper for children?

Shaw: As an aid in determining the requirements of the preterm infant I have used as a reference standard the rate of accumulation of copper by the fetus *in utero*. The accumulation of copper by a human fetus growing along the 50th centile is about 51 μg kg^{-1}day^{-1} between 28 and 36 weeks gestation (Shaw 1980). Pooled breast milk contains about 40 μg copper/100 ml (Dauncey et al 1977), and therefore at a milk intake of 200 ml kg^{-1}day^{-1} copper intake would be 80 μg kg^{-1}day^{-1}. Therefore about 60% of the dietary copper must be absorbed to provide retentions equivalent to those occurring *in utero*. However, copper balances are frequently negative in preterm infants, who seem unable to achieve accumulation rates as high as those *in utero* during the first two months of life (Dauncey et al 1977). In some cases plasma copper falls to low levels, around 40 μg/100 ml, but without any other signs of deficiency such as anaemia, neutropenia or bone changes (Shaw 1980). Therefore, although some infants develop deficiency, the majority seem to manage on much less copper than would be provided *in utero*. During intravenous feeding there may be a problem with the distribution of infused copper. *In utero* and towards term, quite a high proportion of the umbilical venous flow may pass through the liver (Dawes 1968) and may thus provide an opportunity for the formation of the metallothionein-bound copper stores that are evident in the human liver at term (Rydén & Deutsch 1978). If copper is given parenterally the peripheral tissues, but not the liver, will receive the infused copper first and this might lead to an abnormal distribution of copper and possibly to toxic effects. Therefore the optimum intravenous allowance may prove to be less than the *in utero* accumulation rate, while the optimum dietary allowance may prove much higher.

Bremner: There will certainly be differences in the concentrations of copper in tissues after parenteral or oral administration of the metal. However I am not aware of any major differences in the metabolism of copper given by oral or parenteral routes. Concentrations of copper-metallothionein in the liver can be increased in response to both dietary supplementation and injection of copper (Bremner 1979).

Danks: We tend to talk of the copper in the fetal liver as being a copper store. We know that copper does accumulate there, but do we know that it is a functional store, i.e. one that is efficiently used for future requirements? It's possible that most of the copper in the liver may be excreted in the bile, once this excretion starts.

Mills: To return to RDAs, could I ask what percentage absorption of copper is assumed in estimates of RDAs for a child on a liquid milk diet?

Delves: The mean value is about 20% of the ingested copper.

Mills: Is this the same fraction that is absorbed by adults and adolescents on a solid diet?

Bremner: There are major differences in the availability of copper between liquid milk diets and solid diets. For example, in experiments with veal calves fed milk-substitute rations we found that over 50% of the dietary copper was retained in the liver (Bremner & Dalgarno 1973). This compares with a normal availability of only 5–10% of the ingested copper in ruminants that have been weaned onto a solid diet (Suttle 1975).

Hurley: But those experiments reveal the suitability of cow's milk for calves. We don't know how suitable is the copper content of cow's milk for human infants. There may be a difference for human infants between the absorption of copper from cow's milk and that from breast milk.

Owen: I would like to ask a technical question about absorption. I am not really sure how one can measure absorption of copper at all. By that I mean that if I ingest 2 mg copper per day and excrete the same amount, I could be absorbing none of the copper or I could be absorbing 100% of it and be excreting it again through the bile or through other sites. Without the use of radioactive copper and biliary fistulas how can you measure absorption and retention, Dr Delves?

Delves: I think we would define the values we measured as *net absorption* and *net retention*. Retention is the difference between dietary intake and the sum of the faecal and urinary excretion over each three-day period. The net absorption is only an apparent value because we can't take account of the biliary secretion at all – this is one of the big disadvantages of balance studies. I think the more important factor is the amount of copper that is retained.

Owen: Yes. Retention *can* be measured if the balance study is accurate enough. But if net absorption turns out to be zero, then we cannot even guess the amount of copper that has actually been absorbed.

Hill: Dr Delves, you mentioned a patient who developed pneumonia, and you said that the surgeon found osteoporosis. Does it follow that this symptom is associated with copper deficiency?

Delves: I was quoting from a paper by Yuen et al (1979). Osteoporosis does occur with copper deficiency, and these workers attempted to discover why the bones looked so odd. They found from their additional tests that the serum concentration of copper was very low. Before the X-rays were done there was no indication that the child was copper-deficient; it was considered simply to be failing to thrive.

Hill: In what way did the bones look odd?

Delves: On the X-ray (Yuen et al 1979) there was a decreased density in the bones. Fractures can also be seen in the bones of copper-deficient patients.

Danks: I would like to think that any well educated surgeon in a paediatric teaching hospital would regard those observations as symptoms of copper deficiency. When there is osteoporotic bone with expanded bone ends, similar to those observed in scurvy (which would not normally occur at that age), the first thought should be that copper deficiency is the cause.

Hurley: The decreased bone density that Dr Delves has just described is thought to be the result of defective cross-linking of collagen and elastin in the bone (Rucker et al 1975).

Graham: God help the poor child who has to wait until its bones show up the manifestation of deficiency!

Harris: I would like to make a comment about the liquid (i.v.) feeding. I see no problem with the available sources of copper that can be given in the diet – aqueous complexes of copper may be quite adequate for oral administration – but when we introduce copper directly into the bloodstream are we being naive to assume that the ionic form of the metal will suffice? Shouldn't we be more concerned with physiological forms of the metal, e.g. instead of using supplements of copper sulphate we could use a specific concentration of a complex of albumin and copper. Perhaps the body could handle that form much more directly and efficiently than the ionic form.

Delves: Yes, I think it is naive to give ionic copper, and much more work is needed on the correct species of trace metals to be used for i.v. administration. It is said that some of the copper given in combination with amino acids is simply excreted directly, and unchanged, into the urine.

Österberg: There could be some danger in giving copper intravenously. Most of us are familiar with the problems that have arisen with haemodialysis (Matter et al 1969): by accident only a small amount of copper was present in the solution used for haemodialysis and as a result of copper contamination some of the patients died from kidney failure due to haemolysis. Therefore, the form of copper administration is very important.

Graham: In a limited number of studies with Dr Neil Holtzman we gave copper-deficient infants copper acetate intravenously. The severe neutropenia, which is one of the manifestations of copper deficiency, responded promptly to the copper acetate. Administration of caeruloplasmin in sufficient quantity to raise the copper concentration to normal, or administration of enough plasma to raise the caeruloplasmin concentration to normal, produced no neutrophil response (see Graham & Cordano 1976).

Harris: How long did you wait for a response?

Graham: A number of days.

Danks: If one puts together our own experience and that published in numerous articles it seems that the symptoms of Menkes' syndrome that respond to parenteral copper will be corrected promptly by intravenous copper sulphate, copper chloride, copper acetate, or complexes of copper-albumin, copper-histidine, copper-glycine or copper-EDTA (copper-ethylene-diamine-tetraacetic acid). All these substances are approximately equally effective. Copper sulphate given rapidly can produce acute peripheral circulatory failure (Danks et al 1972), and is not to be recommended for rapid administration in large amounts, but all the others seem to be utilized very effectively.

Mills: We now have a rough idea of the variations in copper supply that can arise from a variety of diets but nothing has yet been said about diets in developing countries. From our analyses of diets collected by the MRC Dunn Nutrition Laboratory we now know that the dietary copper content for some West African infants may be as low as 1.5–2.0 mg Cu/kg dry matter (C.F. Mills & N.T. Davies, unpublished). We know nothing of the impact of these diets on such children.

In our discussion we have already mentioned the uncertainty in establishing the most appropriate methods for deciding whether intakes of copper below the recommended dietary allowance are adequate. Perhaps later we should discuss the methods available for detecting copper deficiency in its early stages, before overt symptoms (e.g. skeletal changes) are manifest.

References

al-Rashid RA, Spangler J 1971 Neonatal copper deficiency. N Engl J Med 285:841-843

Ashkenazi A, Levin S, Djaldetti M, Fisher E, Benvenisti D 1973 The syndrome of neonatal copper deficiency. Pediatrics 52:525-533

Beck AB 1941a Studies on the copper content of the milk of sheep and of cows. Aust J Exp Biol Med Sci 19:145-150

Beck AB 1941b Studies on the blood copper of sheep and of cows. Aust J Exp Biol Med Sci 19:249-254

Blumenthal I, Lealman GT, Franklyn PP 1980 Fracture of the femur, fish odour, and copper deficiency in a preterm infant. Arch Dis Child 55:229-231

Boyle E, Freeman PC, Goudie AC, Mangan FR, Thomson M 1976 The role of copper in preventing gastrointestinal damage by acidic anti-inflammatory drugs. J Pharm Pharmacol 28:865-868

Bremner I 1979 Factors influencing the occurrence of copper-thioneins in tissues. In: Kagi JHR, Nordberg M (eds) Metallothionein. Birkhauser, Basel, p 273-280

Bremner I, Dalgarno AC 1973 Iron metabolism in the veal calf. 2: Iron requirements and the effect of copper supplementation. Br J Nutr 30:61-76

Cordano A, Graham GG 1966 Copper deficiency complicating severe chronic intestinal malabsorption. Pediatrics 38:596-604

Danks DM, Stevens BJ, Campbell PE, Gillespie JM, Walker-Smith J, Blomfield J, Turner B 1972 Menkes' kinky-hair syndrome. Lancet 1:1100-1102

Dauncey MJ, Shaw JCL, Urman J 1977 The absorption and retention of magnesium, zinc and copper by low birth weight infants fed pasteurized human breast milk. Pediatr Res 11:1033-1039

Dawes GS 1968 Foetal and neonatal physiology. Year Book Medical Publishers, Chicago

Dunlap WM, James GW, Hume DM 1974 Anaemia and neutropenia caused by copper deficiency. Ann Intern Med 80:470-476

Fell BF, Dinsdale D, Mills CF 1975 Changes in enterocyte mitochondria associated with deficiency of copper in cattle. Res Vet Sci 18:274-281

Graham GG, Cordano A 1969 Copper depletion and deficiency in the malnourished infant. Johns Hopkins Med J 124:139-150

Graham GG, Cordano A 1976 Copper deficiency in human subjects. In: Prasad AS, Oberleas D (eds) Trace elements in human health and disease. Vol 1: Zinc and copper. Academic Press, New York, p 363-372

Griscom NT, Craigh JN, Neuhauser EBD 1971 Systemic bone disease developing in small premature infants. Paediatrics 48:883-895

Heller RM, Kirchner SG, O'Neill JA Jr, Hough AJ Jr, Howard L, Kramer SS, Green HL 1978 Skeletal changes of copper deficiency in infants receiving prolonged total parenteral nutrition. J Pediatr 92:947-949

Jackson GE, May PM, Williams DR 1978 Metal-ligand complexes involved in rheumatoid arthritis. I: Justifications for copper administration. J Inorg Nucl Chem 40:1189-1194

Karpel JT, Peden VH 1972 Copper deficiency in long-term parenteral nutrition. J Pediatr 80:32-36

Matter BJ, Pedersen J, Psimenos G, Lindeman RD 1969 Lethal copper intoxication in hemodialysis. Trans Am Soc Artif Intern Organs 15:309-315

Mills CF, Dalgarno AC, Wenham G 1976 Biochemical and pathological changes in tissues of Friesian cattle during the experimental induction of copper deficiency. Br J Nutr 38:309-331

Missala K, Lloyd K, Gregoriadis G, Sourkes TL 1967 Conversion of ^{14}C-dopamine to cardiac ^{14}C-noradrenaline in the copper-deficient rat. Eur J Pharmacol 1:6-10

Rucker RB, Riggins RS, Laughlin R, Chan MM, Chen M, Tom K 1975 Effects of nutritional copper deficiency on the biomechanical properties of bone and arterial elastin metabolism in the chick. J Nutr 105:1062-1070

Rydén L, Deutsch HF 1978 Preparation and properties of the major copper binding component in human foetal liver. Its identification as metallothionein. J Biol Chem 253:519-524

Sann L, David L, Galy G, Romand-Monier M 1978 Copper deficiency and hypocalcemic rickets in a small-for-date infant. Acta Paediatr Scand 67:303-307

Schubert WK, Lahey ME 1959 Copper and protein depletion complicating hypoferric anaemia in infants. Pediatrics 24:710-733

Seely JR, Humphrey GB, Matler BJ 1972 Copper deficiency in a premature infant fed on iron fortified formula. N Engl J Med 286:109

Shaw JCL 1980 Trace elements in the fetus and young infant. II. Copper, manganese, selenium and chromium. Am J Dis Child 134:74-81

Sturgeon P, Brubacker C 1956 Copper deficiency in infants. A syndrome characterised by hypocupraemia iron deficiency anaemia and hypoproteinaemia. Am J Dis Child 92:254-265

Suttle NF 1975 Change in availability of dietary copper to young lambs associated with age and weaning. J Agric Sci 84:255-261

Vilter RWR, Bozian RC, Hess EV, Zellner DC, Petering HG 1974 Manifestations of copper deficiency in a patient with systemic sclerosis on intravenous hyperalimentation. N Engl J Med 291:188-191

Yuen P, Lin HJ, Hutchison JH 1979 Copper deficiency in a low birth weight infant. Arch Dis Child 54:553-555

Zidar BL, Shadduck RK, Zeigler Z, Winkelstein A 1977 Observations on the anaemia and neutropenia of human copper deficiency. Am J Hematol 3:177-185

Absorption, transport and distribution of copper

IAN BREMNER

Nutritional Biochemistry Department, Rowett Research Institute, Bucksburn, Aberdeen AB2 9SB, UK

Abstract This paper deals with the way animals regulate the supply of copper to sites within the body where the metal exercises its functions. Homeostasis is maintained by the control of both absorption and excretion of the metal, although the efficiency with which this balance is achieved varies between species. These processes are influenced by dietary intake of copper and they also depend on genetic factors, age, dietary composition and the physiological state of the animal. Some of these effects are described, with emphasis on the possible mechanism of absorption of the metal.

The distribution of copper after its absorption from the intestine and, in particular, its uptake by the liver and kidneys are also discussed. Much of the copper removed by these organs is incorporated into, and may induce synthesis of, metallothionein. The precise role of this protein in copper metabolism is still, however, a matter of conjecture. The subsequent binding of hepatic copper to metalloenzymes, its excretion in bile and its incorporation into lysosomes during copper overload are also considered.

This paper is mainly concerned with metallothionein and other proteins that control copper absorption, transport and distribution. However, there has been controversy in recent years about the binding of copper to metallothionein (Bremner 1979) and consequently some preliminary comments on the identification of this protein are merited.

Copper-metallothioneins have been isolated from the liver, kidneys and intestine of several species. They have been characterized by their molecular weight, absorption spectra, metal content and amino acid composition. They have a cysteine content of about 30%, they usually contain about 1% leucine and isoleucine and they do not contain aromatic acids. It has been reported that other copper proteins, of similar molecular weight to metallothionein but with different charge properties and amino acid composition, are also present in these tissues, especially in copper-injected animals. These proteins have cysteine contents of only 2–14% and contain about 7% leucine residues.

Bremner & Young (1976b) suggested that these other proteins may be artifacts, which are derived from copper-metallothionein by unspecified oxidative reactions. Copper-metallothionein is known to be susceptible to oxidation, and in the cases where copper-metallothioneins were successfully isolated from tissues, care was taken either to maintain the proteins under anaerobic conditions and at low temperatures or to include reducing agents in the buffers used. No such precautions have been taken when the other copper proteins were obtained.

I believe, therefore, that metallothionein is the only major copper protein of this molecular weight (6000) present in the liver, kidneys and intestine of animals. However in this paper I shall use the term 'metallothionein' either to describe only the fully characterized protein or only when there is evidence from other published work to support this identification of the protein.

INTESTINAL ABSORPTION OF COPPER

Studies with isolated gut segments suggest that absorption of copper occurs by a process other than simple diffusion and that there are two separate modes of transport from the mucosal to the serosal side of the intestine. The most rapid of these appears to be of low capacity and is inhibited by anoxia and by dinitrophenol. The other mechanism, which is probably of greater importance during high intakes of copper, is less readily saturated, continues to operate for a longer time and is less affected by metabolic inhibitors (Crampton et al 1965).

These conclusions are consistent with the results of Marceau et al (1970) on the rate of entry of an increasing oral dose of ^{64}Cu into specific plasma pools in the rat. The rate of entry of ^{64}Cu into the plasma was linearly related to the amount of ^{64}Cu given by stomach tube only up to doses of 12 μg copper; about 30% of the metal was absorbed under these conditions. At doses between 12 and 36 μg copper, this relationship became non-linear, and the fractional efficiency of absorption decreased to only 13%. Maximum rates of absorption were achieved rapidly, within 30 min, although absorption continued at a lower rate for several hours thereafter. This sequence could reflect the passage of the ^{64}Cu into regions of the small intestine where the absorptive capacity for ^{64}Cu was reduced. Alternatively it could result from the saturation of a rapid transport system and the increased participation of a slower system, which incorporates a 'lag' phase.

Metallothionein and the homeostatic control of copper absorption

Such a lag phase could result from the uptake of the ^{64}Cu by the mucosal cells, and the subsequent slow release of the metal from them (Crampton et al 1965). Evans & Johnson (1978) argued that the binding of much of the copper in mucosal cells to a specific protein, metallothionein, is an important step in the regulation or prevention of copper absorption when dietary intake of the metal is high. By binding ionic copper, metallothionein could protect the cell from possible damaging effects of the metal and could also regulate the passage of copper into the body. In support of this view, Evans & Johnson (1978) reported that synthesis of metallothionein was induced in the rat intestine by oral administration of copper, and that administration of cycloheximide increased the copper absorption, presumably by inhibition of metallothionein synthesis.

Their claim that copper can induce synthesis of intestinal metallothionein was based on the increased incorporation of [^{14}C]lysine into metallothionein after oral administration of copper. However, the same research group reported in an earlier publication (Evans & Le Blanc 1976) that incorporation of [^{3}H]leucine into a similar fraction was also increased after injection of copper. Since intestinal copper-metallothionein does not contain significant amounts of leucine (Evans & Johnson 1978), it is possible that injection of copper merely produces a general stimulation of protein synthesis that is not restricted to metallothionein.

There are also grounds for questioning whether metallothionein normally regulates copper absorption in response to changes in dietary copper intake. There was no increase in the mucosal content of ^{64}Cu in rats when the efficiency of absorption of an oral dose of ^{64}Cu fell from 34% to 12% as the copper content of the diet was increased from 3 to 28 mg/kg (Bremner et al 1979). The ^{64}Cu content of the mucosa in the proximal region of the small intestine 3 hours after the dose was 1.3 ± 0.2 and 0.4 ± 0.1% (mean $\pm$ SE; $n = 6$) of the dose/g at the low and high copper intakes respectively. Although these results were obviously influenced by differences in the specific activity of ^{64}Cu within the alimentary tract, the fractional efficiency of copper absorption nevertheless decreased without a major increase in mucosal content of ^{64}Cu or copper-metallothionein. In addition, the concentrations of non-radioactive copper and of copper-metallothionein in the intestinal mucosa of rats are not greatly affected by oral or parenteral administration of copper (Hall et al 1979). There is therefore a need for a critical assessment of the importance of metallothionein in the homeostatic regulation of copper absorption.

Factors influencing concentrations of intestinal copper-metallothionein

However, the concept of metallothionein acting as a 'mucosal block' to copper absorption, as proposed by Evans & Johnson (1978), may still be valid in other circumstances. For example, when the intestinal absorption of ^{64}Cu in rats was decreased by about 50%, after an increase in the dietary zinc content to 900 mg/kg, there was a major increase in the ^{64}Cu content of the mucosa, with most of the additional ^{64}Cu being bound to metallothionein (Hall et al 1979). Over half of the decrease in ^{64}Cu absorption in the zinc-treated rats could be accounted for by the increased mucosal uptake of the metal by the small intestine (Fig. 1). The basis for the antagonistic effect of zinc on copper absorption therefore seems to be its capacity to increase the concentrations of metallothionein in the mucosa by up to 25-fold. Since copper has a greater affinity than zinc for binding sites on metallothionein (Bremner 1979), it is able to displace some zinc from the protein and thus to assume a form from which it is not readily transported into the blood.

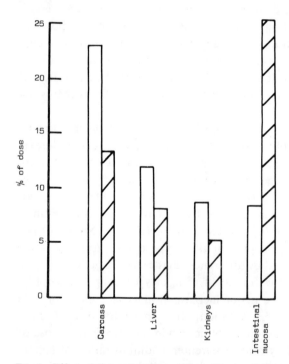

FIG. 1. Effect of dietary supplementation with zinc (900 mg/kg) on the intestinal absorption of ^{64}Cu in the rat. Results are given as the proportion of the administered dose recovered in the 'empty' carcass, in the liver, kidneys and small intestinal mucosa four hours after the dose. Data for both control (☐) and zinc-supplemented rats (▨) are taken from Hall et al (1979).

Similar observations have been made in cadmium-supplemented rats (Davies & Campbell 1977). The decrease in absorption of ^{64}Cu, which occurred when the diet contained 17.6 μg cadmium/g, was associated with an increase in the binding of mucosal ^{64}Cu to a low-molecular-weight protein, which was almost certainly metallothionein. However, there was not in this case a simple inverse relationship between the concentrations of ^{64}Cu-metallothionein and the efficiency of copper absorption at lower intakes of cadmium. Therefore in this case copper absorption might have been affected by some other mechanism. Nevertheless, the relative potency of cadmium and zinc as inhibitors of ^{64}Cu absorption is mirrored to some extent by their relative abilities to induce synthesis of metallothionein in the intestine.

The efficiency of intestinal absorption of copper is also decreased in young brindled mice (Evans & Reis 1978) and in human infants with Menkes' kinky-hair disease (Danks 1977). The development of these congenital copper deficiencies is again associated with increased mucosal concentrations of copper. Moreover fractionation of the mucosa from the mice (on Sephadex G75) has shown that the additional copper occurs principally in a fraction with the same chromatographic properties as metallothionein (Fig. 2).

Similarly Cohen et al (1979) have observed that the changes in ^{64}Cu absorption in tumour-bearing rats and in rats treated with oestrogen are inversely related to the changes in the amount of newly absorbed copper retained by the intestinal mucosal cells. Tumour-bearing rats absorbed more copper and retained less mucosal copper than control rats, whereas the reverse was true of the oestrogen-treated animals. The extent of copper retention by the mucosa was dependent mainly on binding of copper to a cytosolic protein, which had the same apparent molecular weight as metallothionein.

The reason why concentrations of this protein should vary so much in animals of different genetic strain and in different physiological states is still unknown. It could be a consequence of changes in copper flux or of changes in the rate of synthesis or degradation of the protein. If it is the latter, slight variations in the structure or conformation of the protein might influence its susceptibility to proteolytic attack; the biological half-life of hepatic and renal metallothioneins is partly determined by the particular metals bound to the protein (Bremner et al 1978b, Cousins 1979).

Copper absorption by pregnant and suckling animals

The large increase in retention of copper in the body of the maternal rat during pregnancy is partly a consequence of decreased biliary excretion of the metal (Terao & Owen 1977). However, there are also increases in the efficien-

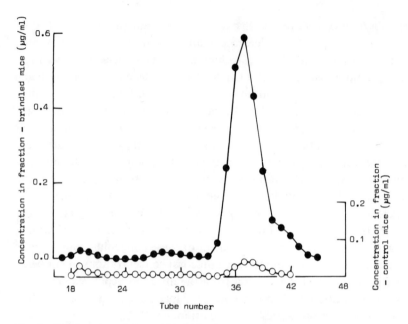

FIG. 2. Fractionation of the small intestine from control and brindled mice on Sephadex G 75. Separations were carried out as described by Hall et al (1979), after homogenization of the intestines in 10 mM-Tris-acetate buffer (pH 7.4). Copper concentrations in the intestine were 15.7 and 3.6 µg/g wet wt for the brindled (•) and control (○) mice respectively. The metallothionein fraction is eluted at around tube 37. (I. Bremner & N.T. Davies, unpublished observations.)

cy of copper absorption; Davies & Williams (1976) found that 54% of a single intragastric dose of ^{64}Cu was absorbed by pregnant rats, compared with only 26% by non-pregnant animals. Because oestrogen administration inhibits copper absorption in rats (Cohen et al 1979), the increased absorption during pregnancy may result from the increased demand for copper rather than from changes in hormonal status.

The increased retention of copper by pregnant animals normally ensures an adequate supply of copper to their offspring, both *in utero* and during the suckling period. However, suckling animals appear to have evolved an additional mechanism for satisfying their copper demands at a time when the concentration of copper in their diets is relatively low, and the availability of copper from milk or from liquid milk substitutes is remarkably high in several species. For example, suckling lambs absorbed 47–71% of the ^{64}Cu added to their diets whereas weaned animals absorbed only 8–10% (Suttle 1975). Similarly 7–10-day-old rats absorbed up to 100% of an intragastric dose of ^{64}Cu, whether it was given as ionic copper, as bile or as caeruloplasmin. The absorption of ^{64}Cu (which included ^{64}Cu bound to the intestinal wall) decreased during the suckling period and decreased even more during weaning (Mistilis & Mearrick 1969). The high uptake of copper was thought to be related to the ability of neonatal animals to absorb intact macromolecules by pinocytosis. In support of this view, cortisone administration, which induces gut closure, caused a reduction in ^{64}Cu absorption in the suckling rats and it prematurely induced a pattern of copper balance found in adult rats.

Most of the copper absorption in the suckling rats occurred in the ileum, in contrast to that in adult rats which absorb copper mainly in the stomach and jejunum. A marked feature of the absorption of copper in the neonatal rats was the high retention of copper in the intestinal wall. Thus, although over 60% of the ^{64}Cu administered in bile to 7–10-day-old rats was apparently absorbed, 38% of the dose was still present in or on the mucosa after 24 hours (Mistilis & Mearrick 1969). Similar observations have been made in neonatal rats dosed with cadmium (Sasser & Jarboe 1977). The intestinal compartment for several metals may therefore be very large and relatively immobile in the first two weeks of life or until the animals are weaned.

TRANSPORT OF COPPER

Most of the copper in plasma is generally present in the form of caeruloplasmin, an α2-globulin which binds 6–8 copper atoms/mole in non-exchangeable forms. However, the principal transport forms of copper are its loosely bound complexes with albumin and, to a lesser extent, with selected amino acids such as histidine, threonine and glutamine. These complexes are probably in equilibrium with each other, perhaps through ternary complexes. Altogether they usually account for 5–10% of the total plasma copper.

Although the forms in which copper passes to the serosal side of the intestinal mucosa have not been established, it is evident that the absorbed metal is rapidly incorporated into complexes with albumin and with amino acids in plasma. These are taken up, again rapidly, by the liver and other tissues. The hepatic copper is then (a) incorporated into caeruloplasmin and released into the plasma or (b) released into the bile or (c) stored temporarily in the liver.

Caeruloplasmin is not involved in the uptake of copper from the intestine but it may function in the transport of copper from the liver and act as a donor of copper for certain copper enzymes. Linder & Moor (1977) showed that appreciable amounts of caeruloplasmin were present not only in the plasma of the rat but also in the heart, liver and brain. Moreover, when [^{3}H]-labelled caeruloplasmin was injected into these animals, 10–20% of the radioactivity was recovered in each organ after 2 hours, which suggests that these organs may be able to take up caeruloplasmin from plasma. The heart, liver and brain are relatively rich in cytochrome *c* oxidase (EC 1.9.3.1) and it is therefore interesting that the activity of this enzyme in tissues from copper-deficient rats was restored more closely to normal levels when the animals were injected with caeruloplasmin than when they received copper-histidine or copper-albumin (Hsieh & Frieden 1975).

These investigations indicate that copper from caeruloplasmin may be incorporated into cytochrome *c* oxidase but it is not clear whether this is a reflection of some specific ability of the plasma protein to act as a copper donor or whether it results merely from variations in tissue distribution of the different forms of injected copper. For example, the copper from injected metallothionein accumulates preferentially in the kidneys and little is taken up by the liver (Bremner et al 1978a).

One of the problems with comparison of the fate of copper from different sources was highlighted by Marceau & Aspin (1973) who injected rats with [^{67}Cu]caeruloplasmin and [^{64}Cu]albumin. The copper from caeruloplasmin was incorporated, in the liver, into a protein thought to be superoxide dismutase (EC 1.15.1.1) and also into cytochrome *c* oxidase. In contrast, much of the copper from albumin was associated with a hepatic protein with a molecular weight of about 10 000. Although these results confirm that major differences occur in the intracellular distribution of copper derived from different sources, it is possible that the results were partly due to differences in the time-scales of the experiment. The distribution of copper from the labelled copper-albumin was determined 10–45 minutes after the injection, whereas the caeruloplasmin-treated rats were not killed until after 20–138 hours (Marceau & Aspin 1973).

HEPATIC ACCUMULATION OF COPPER

The slow development of copper deficiency in animals receiving low-copper diets indicates that many species possess appreciable stores of copper. The main storage organ for copper is the liver, although appreciable amounts of copper also accumulate in other tissues. Whether any specific copper storage protein exists in the liver is still a matter of debate but three different proteins have been considered in this regard, namely hepatocuprein (superoxide dismutase), metallothionein and mitochondrocuprein.

Although superoxide dismutase can account for 20–50% of the total hepatic copper, this occurs only when copper concentrations in the liver are low. As these concentrations increase, the relative contribution made by the protein decreases. In ruminant livers, for example, as little as 4% of the copper may occur as superoxide dismutase. It is difficult to reconcile this observation with the properties that would be expected of a major storage protein.

Natural occurrence of copper metallothionein

When rats are injected with tracer amounts of [^{64}Cu]albumin, much of the ^{64}Cu is rapidly incorporated into a protein of low molecular weight in the liver (Marceau & Aspin 1973). This naturally occurring copper protein in rat liver has never been identified, principally because its concentration is so low. However, hepatic copper concentrations are much higher in copper-supplemented pigs and in newborn calves, and the analogous copper proteins from these animals have been purified (Bremner & Young 1976a, Hartmann & Weser 1977). In both cases, the copper protein had an amino acid composition identical to that of metallothionein, with a cysteine content of about 30%. The metal content of the copper protein is especially noteworthy: up to 10 moles of copper and zinc bound to each mole of protein (whose true molecular weight is approximately 6000). This is equivalent to one metal atom for every two cysteine residues. The copper:zinc ratio in the protein may vary considerably but is generally in the range 4:1 to 1:1. Copper-metallothionein has also been obtained from human fetal liver (Rydén & Deutsch 1978), although the copper and particularly the zinc content of the protein was lower than in the bovine and porcine proteins. It may be relevant that the human protein was isolated by a procedure involving extraction with mercaptoethanol and repeated freezing and thawing of homogenates. This treatment could have released any copper-metallothionein that was originally present, perhaps in polymeric form, in particulate fractions of the liver (see below). This process could explain the high yield of protein, equivalent to 75% of the total copper, from the human fetal livers, compared with only 20-40% in the bovine and porcine livers.

The proportion of hepatic copper present as metallothionein depends on several factors, including hepatic concentration of copper, age, species and the zinc status of the animal (Bremner 1979). The latter effect is particularly striking, as virtually no copper-metallothionein is found in the livers of zinc-deficient animals, regardless of their copper content. Moreover, in ruminant livers the concentration of metallothionein is a function of the hepatic content of zinc, and copper can displace zinc from the protein.

Induction of metallothionein synthesis by copper

The dependence of metallothionein concentration on zinc status can be partly explained by the ability of zinc to induce synthesis of metallothionein by a process that requires synthesis of poly(A)-containing mRNA for the protein (Cousins 1979). However, synthesis of metallothionein can also be induced by administration of copper. Much of the copper that we injected into rats was incorporated into a fraction in the liver with the same molecular weight as metallothionein (Bremner & Davies 1976). We purified this fraction and characterized the three resultant isomers as metallothioneins, on the basis of their amino acid composition, metal content, absorption spectra and electrophoretic behaviour (Bremner & Young 1976b). Since copper injection also increased the incorporation of [^{35}S]cysteine into the proteins (Bremner et al 1978b), and cycloheximide prevented the binding of copper to them (Bremner & Davies 1976) it was apparent that active protein synthesis was involved.

Although these investigations did not show whether synthesis of the copper protein was controlled at the transcription or translation stage, Premakumar et al (1975) have shown that injection of actinomycin D before injection of copper inhibits the incorporation of [^{14}C]lysine into a fraction that has a similar molecular weight to metallothionein. This suggests that induction of the protein synthesis is under transcriptional control. However, Premakumar et al (1975) considered that the protein was copper-chelatin, one of the copper proteins that has a low cysteine content.

Function of copper-metallothionein

Although metallothionein can be one of the major copper-binding proteins in liver, its function, if any, in the control of hepatic copper metabolism is still a matter of conjecture. The rapid incorporation of injected copper into metallothionein could indicate that the protein is involved in the initial hepatic uptake of copper. However, inhibition of metallothionein synthesis by cycloheximide (Bremner & Davies 1976) does not prevent the accumulation of injected copper by the liver, which suggests therefore that metallothionein is not essential for this process.

We provided the alternative hypothesis that the protein functions in the temporary storage or cellular detoxification of copper (Bremner & Davies 1976). Copper may be bound only transiently to metallothionein, and the biological half-life of the copper protein in the rat is much less than that of most other hepatic proteins. The half-life for [^{35}S]-labelled copper-metallothionein from copper-injected rats, as estimated from the rate of removal of either the [^{35}S]cysteine or the metal from the protein, is only 16.9 hours (Bremner et al 1978b). Unfortunately, we have not yet determined the fate of copper after its release from metallothionein.

In considering a possible role for metallothionein in the detoxification of copper, it is noteworthy that the proportion of hepatic copper present in this form in sheep and calves is much less than that in pigs (Bremner 1979). The greater susceptibility of sheep and calves to chronic copper poisoning could be related to their inability to accumulate large amounts of copper as monomeric metallothionein in their livers. This might occur if the degree of liver damage that develops in copper-loaded animals is influenced by the intracellular distribution of copper, as has been suggested for Wilson's disease (Goldfischer & Sternlieb 1968). Alternatively, a particular pattern of copper distribution may favour the excretion rather than the accumulation of copper. This may explain why hepatic concentrations of copper in the pig fall rapidly when the animals are transferred from high-copper to low-copper rations, whereas excretion of copper from ovine liver occurs more slowly and is relatively insensitive to changes in hepatic copper concentration. Since copper homeostasis is achieved primarily by control of the biliary excretion of copper, these differences in response are obviously of considerable importance.

Subcellular distribution of copper in liver

It has generally been assumed that the disappearance of copper from hepatic metallothionein in copper-injected rats is associated with the degradation of the protein (Bremner et al 1978b). However, it is also possible that oxidation and aggregation of the protein may occur under some circumstances. Porter (1974) has suggested that much of the copper in neonatal bovine liver is present in the lysosomes, in a polymeric form of metallothionein. A significant amount of copper in the liver of copper-loaded animals may also occur in the lysosomes (Goldfischer & Sternlieb 1968, Evans 1973). For example, when we examined the liver from copper-poisoned sheep by electron-microscopy, we found that copper had accumulated in residual bodies, which were rich in acid phosphatase (EC 3.1.3.2) and were derived from lysosomes (King & Bremner 1979). Since we also detected sulphur within the residual bodies, the copper may have originated from metallothionein, which is rich in sulphur. In addition, these residual bodies were generally close to the portal tracts within the liver and they did not appear to be excreted in the bile.

Using subcellular fractionation of the liver by differential centrifugation we have shown that the distribution of copper is greatly influenced by hepatic concentrations of copper. Fig. 3 shows that for sheep liver, the capacity of the heavy mitochondrial fraction, rich in cytochrome c oxidase, to bind copper appears to be saturated at a hepatic copper content of about 125 μg/g. Sur-

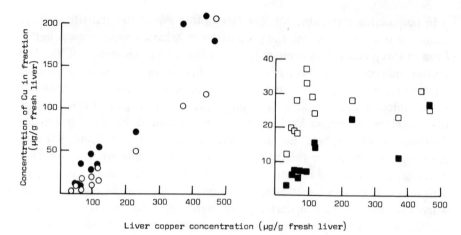

FIG. 3. Subcellular fractionation of normal and copper-loaded sheep livers. Livers were obtained from sheep receiving copper-supplemented rations, homogenized in 0.25 M sucrose and fractionated by standard differential centrifugation techniques, to give nuclear (●), heavy mitochondrial (□), light mitochondrial (■), microsomal (not shown) and supernatant (○) fractions. (I. Bremner and E. Philip, unpublished observations.)

prisingly, there was no significant accumulation of copper in the light mitochondrial fraction, which is generally rich in lysosomes. It seems likely, however, that the copper which was detected in lysosomes by electron-microscopy was sufficiently dense to sediment with the nuclear fraction. This frequently accounts for a high proportion of the copper in copper-loaded livers, although little of the copper is actually associated with the nuclei or with the plasma membranes (Corbett et al 1978).

RENAL ACCUMULATION OF COPPER

Although copper is rapidly taken up by the kidneys after its absorption from the gut or after intravenous injection of the metal, renal concentrations of copper in animals are usually quite low. However, increased renal concentrations have been reported in copper-poisoned sheep (Bremner 1979), in brindled and blotchy mice (Danks 1977, Prins & Van den Hamer 1979) and in mature rats (Bremner 1979). In all these animals, the additional copper was present principally as metallothionein or in an uncharacterized protein with the same molecular weight as metallothionein.

The accumulation of copper-metallothionein in rat kidneys appears to be influenced by hormonal factors, because the concentrations of the complex are greater in female than in male animals. Moreover, ovariectomy and postsurgical stress decrease the concentrations of this protein whereas progesterone administration increases them (Bremner 1979). It is not known whether these effects arise from a direct hormonal influence on synthesis of the protein or whether they are secondary to other changes in copper metabolism.

Unfortunately the function of this renal protein has not been established. Its turnover is rapid (Bremner et al 1978b) and it may therefore have some regulatory function, perhaps in the control of tubular reabsorption and urinary excretion of copper.

References

Bremner I 1979 Factors influencing the occurrence of copper-thioneins in tissues. In: Kagi JHR, Nordberg M (eds) Metallothionein. Birkhauser, Basel, p 273-280

Bremner I, Davies NT 1976 Studies on the appearance of a hepatic copper-binding protein in normal and zinc-deficient rats. Br J Nutr 36:101-112

Bremner I, Young BW 1976a Isolation of (copper, zinc)-thioneins from pig liver. Biochem J 155:631-635

Bremner I, Young BW 1976b Isolation of (copper, zinc)-thioneins from the livers of copper-injected rats. Biochem J 157:517-520

Bremner I, Hoekstra WG, Davies NT, Young BW 1978a Metabolism of ^{35}S-labelled copper-, zinc- and cadmium-thionein in the rat. Chem-Biol Interact 23:355-367

Bremner I, Hoekstra WG, Davies NT, Young BW 1978b Effect of zinc status of rats on the synthesis and degradation of copper-induced metallothioneins. Biochem J 174:883-892

Bremner I, Young BW, Mills CF 1979 The effect of ammonium tetrathiotungstate on the absorption and distribution of copper in the rat. Biochem Soc Trans 7:677-678

Cohen DI, Illowsky B, Linder MC 1979 Altered copper absorption in tumor-bearing and estrogen-treated rats. Am J Physiol 236:E309-E315

Corbett WS, Saylor WW, Long TA, Leach RM 1978 Intracellular distribution of hepatic copper in normal and copper-loaded sheep. J Anim Sci 47:1174-1179

Cousins RJ 1979 Synthesis and degradation of liver metallothionein. In: Kagi JHR, Nordberg M (eds) Metallothionein. Birkhauser, Basel, p 293-301

Crampton RF, Matthews DM, Poisner R 1965 Observations on the mechanism of absorption of copper by the small intestine. J Physiol (Lond) 178:111-126

Danks DM 1977 Copper transport and utilisation in Menkes' syndrome. Inorg Perspect Biol Med 1:73-100

Davies NT, Campbell JK 1977 The effect of cadmium on intestinal copper absorption and binding in the rat. Life Sci 20:958-960

Davies NT, Williams RB 1976 The effects of pregnancy on uptake and distribution of copper in the rat. Proc Nutr Soc 35:4A

Evans GW 1973 Copper homeostasis in the mammalian system. Physiol Rev 53:535-570

Evans GW, Johnson PE 1978 Copper- and zinc-binding ligands in the intestinal mucosa. In: Kirchgessner M (ed) Trace element metabolism in man and animals. Arbeitskreis für Tierernährungsforschung, Freising-Weihenstephan, vol 3:98-105

Evans GW, Le Blanc FN 1976 Copper-binding protein in rat intestine: amino acid composition and function. Nutr Rep Int 14:281-288

Evans GW, Reis BO 1978 Impaired copper homeostasis in neonatal male and adult female brindled mice. J Nutr 108:554-560

Goldfischer S, Sternlieb E 1968 Changes in the distribution of hepatic copper in relation to the progression of Wilson's disease (hepatolenticular degeneration). Am J Pathol 53:883-894

Hall AC, Young BW, Bremner I 1979 Intestinal metallothionein and the mutual antagonism between copper and zinc in the rat. J Inorg Biochem 11:57-66

Hartmann H-J, Weser U 1977 Copper-thionein from fetal bovine liver. Biochim Biophys Acta 491:211-222

Hsieh HS, Frieden E 1975 Evidence for caeruloplasmin as a copper transport protein. Biochem Biophys Res Commun 67:1326-1331

King TP, Bremner I 1979 Autophagy and apoptosis in liver during the prehaemolytic phase of chronic copper poisoning in sheep. J Comp Pathol 89:515-530

Linder MC, Moor JR 1977 Plasma ceruloplasmin. Evidence for its presence in and uptake by heart and other organs of the rat. Biochim Biophys Acta 499:329-336

Marceau N, Aspin N 1973 The intracellular distribution of radio copper derived from ceruloplasmin and from albumin. Biochim Biophys Acta 293:338-350

Marceau N, Aspin N, Sass-Kortsak A 1970 Absorption of copper 64 from gastrointestinal tract of the rat. Am J Physiol 218:377-383

Mistilis SP, Mearrick PT 1969 The absorption of ionic, biliary, and plasma radiocopper in neonatal rats. Scand J Gastroenterol 4:691-696

Porter H 1974 The particulate half-cystine-rich copper protein of newborn liver. Biochem Biophys Res Commun 56:661-668

Premakumar R, Winge DR, Wiley RD, Rajagopalan KV 1975 Copper-induced synthesis of copper-chelatin in rat liver. Arch Biochem Biophys 170:267-277

Prins HW, Van den Hamer CJA 1979 Primary biochemical defect in copper metabolism in mice with a recessive X-linked mutation analogous to Menkes' disease in man. J Inorg Biochem 10:19-27

Rydén L, Deutsch HF 1978 Preparation and properties of the major copper-binding component in human fetal liver. J Biol Chem 253:519-524

Sasser LB, Jarboe GE 1977 Intestinal absorption and retention of cadmium in neonatal rat. Toxicol Appl Pharmacol 41:423-431

Suttle NF 1975 Change in availability of dietary copper to young lambs associated with age and weaning. J Agric Sci 84:255-261

Terao T, Owen CA 1977 Copper metabolism in pregnant and postpartum rat and pups. Am J Physiol 232:E172-E179

Discussion

Dormandy: We have tried to raise serum copper concentrations in normal adults by giving them enormous doses of copper in many different forms and this had no effect whatever on the serum copper concentrations, as measured by atomic absorption spectroscopy. The only way that we could raise the concentrations slightly was by giving the copper in combination with alcohol. Are there great species differences in copper absorption?

Bremner: It is more accurate to say that there are major species differences in overall copper balance and in other aspects of copper metabolism. We cannot claim that these differences necessarily reflect differences in the efficiency of copper absorption.

Dormandy: Yes, but would you not expect the serum copper concentration to reflect intake of copper to some extent?

Bremner: It's actually quite difficult to increase serum copper concentrations even when there is a high intake of the element. Even though copper-poisoned sheep build up massive concentrations of copper in their livers, there is hardly any increase in the serum copper levels until the terminal stages of the disease when there is a sudden outflow of copper from the liver.

McMurray: Dr Dormandy, what concentration of serum copper were you trying to achieve in your human subjects? In sheep and cattle, it is not normally possible to raise the concentration of serum copper much above 0.7–0.8 mg/l. Higher levels are associated with acute-phase reactions or with copper poisoning in conjunction with the haemolytic crisis.

Dormandy: Our aim was really to see if we could raise serum caeruloplasmin by giving copper by mouth instead of by inducing an 'acute-phase' reaction. But we succeeded only in inducing a certain amount of nausea.

I take Dr Bremner's point that serum copper is not a reflection of copper absorption but I would have expected a slight rise in both serum copper and serum caeruloplasmin in response to relatively large oral doses of copper.

McBrien: I believe that women on the contraceptive pill have considerably increased serum copper concentrations compared to control women (Carruthers et al 1966). Did you try using hormones to increase serum copper?

Dormandy: Women on the pill certainly have a high serum copper because they have a high serum caeruloplasmin (Cranfield et al 1979): this is also true for pregnant women and for men who are on oestrogen treatment for prostatic cancer. In all these states the rise in serum copper and caeruloplasmin is part of an 'acute-phase-protein' response.

Frieden: In fact, there are numerous reports of increases in serum copper and caeruloplasmin levels after administration of oestrogens.

I would like to ask Dr Bremner some further questions about the storage forms of copper. It seems to me that these fit into three categories: (1) the metallothioneins; (2) the so-called hepatocupreins, which include superoxide dismutase; and (3) caeruloplasmin itself. How would you judge the potential of these substances as storage forms of copper? Do you distinguish between the storage forms of metallothionein and the metallothioneins that are induced by toxic concentrations of copper?

Bremner: Many research groups have reported that in the livers of copper-injected rats there is synthesis of a variety of copper proteins with apparent molecular weights of 8000–12 000 (e.g. Winge et al 1975, Irons & Smith 1977, Riordan & Gower 1975, Evans et al 1975). However, I believe that these proteins are not separate species and that they are actually artifacts that are derived from metallothionein. Their formation is probably a consequence of the inadequate exclusion of oxygen during their purification. If anaerobic conditions and low temperatures are maintained (Bremner & Young 1976), or if 2-mercaptoethanol is included in the buffers (Rydén & Deutsch 1978), then only copper-metallothionein is obtained. Indeed, Jack Riordan (personal communication) and Evans (Evans & Johnson 1978) now also use anaerobic conditions to purify the protein and they agree that metallothionein is produced. I believe, therefore, that only one type of copper protein with this molecular weight is induced by copper, and that is metallothonein.

I don't think there is any major difference between the biological properties of copper-metallothioneins that occur naturally and those that are induced by oral or parenteral administration of toxic concentrations of copper.

Frieden: Is metallothionein a very mobile form of copper?

Bremner: If the metallothionein is present in monomeric form it is certainly quite mobile. For example, in our studies on the turnover rate of copper-metallothionein in copper-injected rats, we estimated that the half-life was only 17 h. Clearly the copper is removed from the protein very quickly.

However this may not happen if the copper-metallothionein is present in polymeric form. Porter (1974) has suggested that the copper protein called mitochondrocuprein, present in the mitochondria or lysosomes of newborn animals, is actually a polymeric form of metallothionein. I suspect that this protein takes longer to be utilized. Certainly lysosomal copper in sheep liver persists for a long time before it is eventually excreted (Gopinath & Howell 1975, King & Bremner 1979).

To come back to your other point about hepatocuprein (or superoxide dismutase) acting as a storage form of copper, I think this is most unlikely. Even though a high proportion of the total hepatic copper may be present in this form in rats, less is present in other species that have a greater ability to increase hepatic copper concentrations. We found that in ruminants, for example, only about 5% of the hepatic copper was present in the fraction corresponding to hepatocuprein when the cytosol was subjected to gel filtration (Bremner & Marshall 1974).

Hill: You said that metallothionein was induced. Have you or others gone any further in studying *thionein?* Is it metallothionein that is induced or is it actually thionein; and does thionein exist independently of the metal? From your electronmicroscopy and microprobe analysis studies, was it possible to account for the sulphur separately from the metal?

Bremner: No one has ever found the apoprotein (thionein), probably because it is so unstable.

Hill: But it is thionein that polymerizes so readily. When you mentioned polymerization I wondered if you were referring to thionein.

Mills: Is there any evidence that thionein polymerizes *in vivo?*

Hill: In vitro at pH 7 thionein readily forms a polymer under aerobic conditions.

Riordan: Most other workers don't observe that, because they do not handle the thionein.

Hill: Under anaerobic conditions it *is* possible to handle the thionein at pH 7.

Riordan: I want to comment on the storage forms of copper and their possible relevance to excretion. In human fetal liver the major form of copper-thionein is the insoluble form, described by Dr Bremner, which may or may not be associated with the lysosome. The soluble form, however, is a zinc-thionein and is seemingly the same one that is found in adult liver. We have recently subfractionated metallothionein (Riordan & Richards 1980); we purified a zinc-enriched form from the soluble fraction of human fetal liver and a copper-enriched form from the insoluble fraction. So this protein seems to be present in the normal physiological state, and not just when animals are given a high dose of copper.

Mills: Does the soluble form of the metallothionein contain only zinc, or is it a zinc–copper (i.e. a mixed-metal) soluble form?

Riordan: Both the soluble and insoluble forms contain mixed metals but the soluble form predominantly contains zinc whereas the insoluble form predominantly contains copper.

Mills: I am thinking of the problems that paediatricians face because of perinatal accumulation of copper. Most of the copper is, presumably, in the insoluble polymeric form. This balance between the soluble and the insoluble forms may be determined in some way by zinc status. There must be marked differences in turnover between the soluble and insoluble forms.

In the same context of copper storage in the liver in the form of copper metallothionein, could I ask Dr Tanner about the question of copper accumulation in infants? Perhaps he would comment briefly on the problem of Indian childhood cirrhosis, with its high accumulation of hepatic copper?

Tanner: The electronmicroscopic appearance of liver from patients with Indian childhood cirrhosis is strikingly similar to the picture that Dr Bremner showed of the copper-poisoned sheep, in that electron-dense residual bodies are prominent (M.S. Tanner, D. Pallot, S. Bulman, K. Ibe & T. King, unpublished work). X-ray dispersive microprobe analysis showed us that copper and sulphur were localized to these residual bodies, which presumably are responsible for the granular orcein staining seen in paraffin sections (Portmann et al 1978). Our problem with Indian childhood cirrhosis is that patients present at a late stage of the disease. We do not know what happens in the early stages, nor whether copper accumulation is a primary or a secondary phenomenon — though we strongly suspect the former.

Mills: How high a concentration of hepatic copper is reached in this disease?

Tanner: All cases exceed 1000 µg/g dry weight and may be as high as 2500 µg/g (Tanner et al 1979).

Danks: I'd like to return to discussion of the infantile polymeric form of copper and to discuss whether this is really a storage form or a pre-excretory form of copper.

Harris: May I add the further point that these residual bodies that you described, Dr Bremner, may not be polymerized forms of metallothionein, but rather lysosomal-engulfed proteins in the cytosol that are in the process of turning over?

Bremner: The possibility that metallothionein is taken up by the lysosomes and subsequently degraded by the hydrolases present there has been considered by several research groups. Feldman et al (1978) have shown, for example, that cadmium- and especially zinc-thioneins are readily degraded by lysosomal extracts. The relative stability of the proteins under these conditions compares well with what we know about their biological stability *in vivo*.

Harris: Can we then consider copper storage as an inability to turn over the storage protein?

Bremner: This is an attractive possibility. However our investigations on the turnover rate of copper-metallothionein (Bremner et al 1978) suggested that it was less stable biologically than the zinc and cadmium proteins. The half-lives for these different forms (Cu, Zn and Cd proteins) are about 17 h, 19 h and 3–4 days respectively. Presumably the nature of the metal bound to the protein affects the ease with which the protein is degraded. Incidentally, the biological stabilities of the different forms of the protein are not related simply to the strength of the metal–protein bond, since the copper protein is the most stable chemically.

These observations indicate that the copper protein would break down readily in the lysosomes, which is really the opposite of what you suggested. However, there is some recent evidence (W.G. Hoekstra, personal communication) which suggests that copper-metallothionein may be more resistant to hydrolysis than we thought. Bill Hoekstra could apparently detect no signs of degradation of copper-metallothionein when it was incubated *in vitro* with lysosomal extracts. There is obviously a need for further examination of this whole system. I suspect that we may have been wrong to attribute the disappearance of [^{35}S]cysteine from metallothionein in the liver of copper-injected rats to the degradation of the protein (Bremner et al 1978). Conceivably, we were actually looking at the consequences of aggregation of the protein, which would be consistent with what we know of the tendency of copper-metallothionein to form polymeric species. We are currently examining this possibility.

Harris: Is the metallothionein an obligatory intermediate in intracellular metabolism, i.e. does the metabolic route of copper pass through metallothionein and does the subsequent fate of copper depend upon binding to this protein?

Bremner: I think it may do, although we still know very little about the physiological role of the protein. Dr Owen's group (Hazelrig et al 1966) suggested that there were three separate pools of copper in the liver: one was used for caeruloplasmin synthesis, one was the source of biliary copper, and the other was a temporary storage pool. It is possible that metallothionein could constitute that temporary storage pool.

Owen: We generally direct all our attention to copper-binding proteins with molecular weights of about 10 000; I wonder whether we're missing something. Dr Bremner showed that with separation by Sephadex G 75, some of the copper binds to a large-molecular-weight form in the void volume. We use Sephadex G 100, and the protein is still in the void volume. At first we thought this was an artifact because most of the cytosolic protein comes through in the void volume and therefore we thought it was just carrying copper with it. However, if radiolabelled copper is given to a normal rat then, after half an hour, 20% of the labelled copper in the liver is in this large void-volume fraction; after 12 h there is 10% and after 24 h there is virtually nothing. It is difficult to believe that this evolution is simply an artifact and we wondered whether there is a second copper-binding protein, and if so, whether it is carrying copper into a pathway entirely different from the metallothionein pathway.

McBrien: I believe that when sheep are kept indoors, even on a normal low-copper diet, they accumulate much more copper in their livers than if they are left outdoors over the winter.

Bremner: In many cases the development of copper poisoning is really a reflection of increased copper intake. When animals are fed indoors, on diets based on concentrates, it is quite common for the diet to contain around 20 mg copper/kg, whereas in the pasture the normal copper intake may be closer to 5 mg/kg diet. However, it is true that the availability of the dietary copper will be higher in some types of diets.

McBrien: What finally causes the sudden release of copper from the liver?

Bremner: The short answer is that we don't know. The onset of the terminal haemolytic episode is often associated with some type of stress, but the specific cause is not known. It can occur after a change of diet or if the sheep is unfortunate enough to be chased by a dog! I should emphasize that the release of copper from the liver represents merely the terminal stages of the disease; considerable liver damage occurs before this point is reached.

Danks: I'd like to go back to proteins that have similar molecular weights but have been claimed to differ from metallothionein. Perhaps David Hunt could defend his observations on the material that he and his colleagues extract from mottled mice (Port & Hunt 1979). Is it really a degraded metallothionein?

Hunt: I cannot defend our observations too strongly because the purification of the copper-binding proteins was not carried out under anaerobic conditions (Port & Hunt 1979). Nevertheless, the proteins that we found in neonatal mice are dissimilar in their amino acid composition to metallothionein and they resemble more closely the protein isolated by Riordan & Gower (1975). Perhaps there is a different form of copper-binding protein in young animals from that in the adult. We have previously found that in liver and kidney tissue of young mice, the copper-binding protein is synthesized at a high rate and this rate is not further increased by the administration of cupric chloride (Hunt & Port 1979). This is unlike the adult, in which the protein is highly inducible.

Danks: Another protein that has been claimed to be different from metallothionein has been called copper-chelatin (Winge et al 1975). Do you have any comments on that?

Bremner: I partly covered this point in my earlier comments to Earl Frieden. I find it difficult to understand why anyone who is working with a protein rich in thiol groups and in copper should choose to subject that protein to a heat-coagulation step under aerobic conditions. Since these conditions were used to isolate copper-chelatin, I think it is not unreasonable to suggest that chelatin could be an artifact.

Sourkes: A number of years ago Gregoriadis made an observation in my lab which may be pertinent here (Gregoriadis & Sourkes 1968). We loaded rats with copper by i.p. injections of copper sulphate, and the liver content of copper increased to high levels. We then followed the disappearance of the copper over a period of a few days. If these rats were treated with an inhibitor of protein synthesis, like cycloheximide, there was a great delay in the loss of copper from the liver. This observation fits in with Dr Owen's previous comment about the complexity of copper removal. We did not measure metallothionein at the time, so we don't know which protein might be involved.

Frieden: I have a question about the excretion of copper. Is copper excretion similar to iron excretion, for which there does not appear to be an endogenous mechanism?

Bremner: It is generally accepted that there is an endogenous mechanism for copper excretion in the bile. Indeed this is probably the most important way in which copper homeostasis is achieved in most species. However, I believe that there are possible exceptions to this and that the susceptibility of sheep to copper poisoning is a reflection of some limitation in the sheep's ability to excrete copper in the bile.

Frieden: Is it possible that copper isn't completely absorbed?

Bremner: I don't think so. Even though copper homeostasis is usually achieved by control of excretion, there is of course some variation in the efficiency of copper absorption in response to changes in dietary copper intake and possibly also in demand for copper. I described some of these effects in my paper.

Mills: We have found that in calves, endogenous losses of copper can range from 2–90 µg/kg body weight but the losses are closely related to hepatic copper content, so there is an effective regulatory mechanism there.

Frieden: Is this related to bile formation?

Mills: We do not know.

Bremner: There's some evidence in ruminants for another route of copper excretion, perhaps in the abomasum. This evidence has been obtained by comparing the concentration of copper in the digestive tract with that of a non-digestible marker. It is a relatively crude technique but the findings have been confirmed in several studies. I do not know whether the same applies to non-ruminant (monogastric) species.

Danks: To return to the subject of factors affecting the absorbability of copper, could anyone comment on the absorbability of that form of copper (whatever it is) that is excreted in the bile? Does an enterohepatic circulation of that form of copper exist?

Owen: If rats are given radioactive-labelled copper and the first couple of hours of bile collection is ignored, the bile that comes out thereafter contains labelled copper that is virtually unabsorbable when placed in the intestinal tract of another rat (Owen 1964). However, if the copper-labelled bile is put into the intestinal tract of a newborn animal then it *is* absorbed (Mistilis & Mearrick 1969). Whether this observation reflects a different absorptive mechanism or a more primitive gut in the newborn animal, I do not know. It is, in fact, fortunate that in the adult rat the 'late bile' is unabsorbable, because if it *were* absorbable there would be no net loss during enterohepatic circulation. It has so far proved impossible to identify the nature of the copper complex — it is not a protein-bound compound and every bioassay that I could think of has failed to identify it. All that I can say is that the compound probably has a molecular weight smaller than 4000.

Danks: It seems necessary that the compound shouldn't be reabsorbable.

Mills: I believe that in the neonatal rat absorption of copper, cadmium and other elements from a liquid diet is inhibited by cortisone administration, which suggests that at this early stage of postulated development, absorption may be by a pinocytotic mechanism that is closed down by cortisone.

Bremner: You referred just now, Dr Owen, to absorption of the 'late bile'. What about the 'early bile' — could copper be available for absorption in that?

Owen: About 20–30% of labelled copper in early bile (i.e. bile that is excreted *within* the first few hours of the dose of labelled copper) is absorbable. In the late bile (i.e. that collected *after* the first 2–4 hours) less than 10% of the labelled copper can be absorbed.

Bremner: Is the copper present in different chemical forms in the early bile and in the late bile?

Owen: I don't know. I can only quote our numerical results, but I can't explain them. In fact, if copper is added to bile in a test tube, 70% of the complexed copper becomes instantly unabsorbable *in vivo*.

Hurley: One of the most important factors that determine the absorbability of copper is the form in which the copper is present. We have some indirect information on absorbability from some experiments in which we gave copper to lactating rats in combination with nitrilotriacetic acid (NTA) (Keen et al 1980). We added copper-NTA to the drinking water and then we milked the rats and found that the copper concentration in the milk was significantly increased. We also gave iron-NTA, zinc-NTA, NTA alone and no supplements. The only group of rats that had increased levels of copper in the liver was the group receiving copper-NTA. This result was found in both the dams and the pups. In the kidney and spleen there was an increase in copper only in the dams, and in the brain there was no increase in copper content in either dams or pups. Various published reports (for refs. see Keen et al 1980) suggest that additional copper supplements in the maternal diet do not increase the concentration of copper in her milk, whereas our results show not only that absorption of copper was increased, in both the dams and the pups, by supplementation with copper-NTA, but also that copper concentration in milk can be increased by appropriate supplements. This search for appropriate supplementation is relevant to the copper deficiency in infants which we were just discussing.

Mills: Are there any naturally occurring materials that might increase the copper content of milk or, by passage into milk, modify the availability of the copper it contains?

Hurley: I don't know of any substances but this area has not been studied very much.

Frieden: Citrate or ascorbate are likely to increase the availability of copper in milk.

Mills: In the Indian childhood cirrhosis that you have studied, Dr Tanner, there are indications that the disease starts before the child is weaned (Tanner et al 1979, Parekh & Patel 1972). Should we therefore invoke an increase in the availability of copper in milk to explain this?

Tanner: That is possible, but all of the patients that I have studied in detail have actually drunk water that was stored in copper pots, or cow or buffalo milk that was diluted and heated in copper or brass pots.

Frieden: Have you any idea of the form of the copper that is present in non-supplemented milk, Professor Hurley?

Hurley: Our studies show that in human milk, copper is not bound to a low-molecular-weight component (Lönnerdal et al 1980a), although some *zinc* is bound as zinc citrate, which does have a low molecular weight (Lönnerdal et al 1980b). We have found that both copper and zinc are bound to the same high-molecular-weight protein. We are trying to isolate the protein and so far we have been unable to separate the zinc from the copper, so we think they must be attached to the same compound.

Hill: Has there been much consideration of natural ligands in the diet, other than phytic acid (Oberleas 1973)? Years ago I was interested in the possible fate of the products from glucosinolates, which give rise to nitriles and sulphur-containing materials.

Bremner: There was some evidence from Kirchgessner & Grassmann (1970) indicating that amino acids increased copper absorption. They examined a range of amino acids and small peptides and monitored the hepatic uptake of copper in rats. There was an increase in hepatic uptake with some copper complexes, but the effects were not very significant.

Graham: We have speculated that an infant can reabsorb copper from bile. If a full-term baby is fed nothing but cow's milk for a year, it will certainly not become copper-deficient because it has enough copper in the liver to last that long. However, if the same baby develops diarrhoea for three or four months it will be severely depleted of copper. As a result, we have postulated, perhaps naively, that such a baby does not reabsorb its own copper but loses it by some unknown route.

Mills: How satisfactory is the growth of a child who is maintained entirely on cow's milk?

Graham: There is excellent growth, with normal copper status maintained.

Hurley: But surely the child's copper comes from its hepatic stores? Couldn't you explain the depletion of copper in a child with severe diarrhoea by invoking decreased absorption of copper?

Graham: Well, we have actually postulated decreased *re*absorption; the child's diet is practically devoid of copper.

Hurley: But cow's milk, although it contains only *small* amounts of copper, does actually contain *some*.

Graham: The infants we have studied, who have had diarrhoea for weeks and months, have been drinking rice water, weak tea and broth – a diet that is practically devoid of copper.

Mills: And in the full-term babies without diarrhoea, what criteria are you using to say that their copper status is normal?

Graham: We measure their serum concentrations of caeruloplasmin and copper and their haematological status.

McMurray: Is the causal agent for diarrhoea relevant, and does it matter whether the upper or lower gastrointestinal tract is involved?

Graham: No, most of the chronic diarrhoea is no longer related to a specific agent, but just to starvation. The infants have malabsorption of practically all dietary constituents, and they are severely copper-deficient even when they are only 3 to 4 kg in weight – one doesn't have to wait until they are at the 10 kg weight of a one year-old, at which point, if they had been on a copper-free or copper-poor diet, they might have exhausted their hepatic stores.

Tanner: What is the effect of iron deficiency on copper absorption?

Bremner: There have been a few studies on this, mainly with the everted sac technique, and all the evidence suggests that iron deficiency has no effect on copper absorption (Forth 1970, El-Shobaki & Rummel 1979). But what *is* found consistently (and we ourselves found this) is an increase in hepatic copper stores during iron deficiency (e.g. Sourkes et al 1968). We have just completed a preliminary study on copper absorption in iron-deficient rats, using a whole-body counting technique, and we have an indication that there might be an *increase* in copper absorption, which is contrary to what others have found before. Did you find an increase in copper absorption, Professor Sourkes?

Sourkes: Not in absorption, but in *content* of copper. We have found an inverse relationship between iron and copper in the liver of rats (Sourkes et al 1968).

Mills: I believe it is also true that if rat pups are born from iron-deficient dams, there is an increased transfer of copper to the fetal liver (Sherman & Tschiember 1979).

References

Bremner I, Marshall RB 1974 Hepatic copper- and zinc-binding proteins in ruminants. 1: Distribution of Cu and Zn among soluble proteins of livers of varying Cu and Zn content. Br J Nutr 32:283-291

Bremner I, Young BW 1976 Isolation of (copper,zinc)-thioneins from the livers of copper-injected rats. Biochem J 157:517-520

Bremner I, Hoekstra WG, Davies NT, Young BW 1978 Effect of zinc status of rats on the synthesis and degradation of copper-induced metallothioneins. Biochem J 174:883-892

Carruthers ME, Hobbs CB, Warren RL 1966 Raised serum copper and caeruloplasmin levels in subjects taking oral contraceptives. J Clin Pathol 19:498-501

Cranfield LM, Gollan JL, White AG, Dormandy TL 1979 Serum antioxidant activity in normal and abnormal subjects. Ann Clin Biochem 16:299-306

El-Shobaki FA, Rummel W 1979 Binding of copper to mucosal transferrin and inhibition of intestinal iron absorption in rats. Res Exp Med 174:187-195

Evans GW, Johnson PE 1978 Copper- and zinc-binding ligands in the intestinal mucosa. In: Kirchgessner M (ed) Trace element metabolism in man and animals. Arbeitskreis für Tierernährungsforschung, Freising-Weihenstephan, vol 3:98-105

Evans GW, Wolenetz ML, Grace CI 1975 Copper-binding proteins in the neonatal and adult rat liver soluble fraction. Nutr Rep Int 12:261-269

Feldman SL, Failla ML, Cousins RJ 1978 Degradation of rat liver metallothioneins in vitro. Biochim Biophys Acta 544:638-646

Forth W 1970 Absorption of iron and chemically related metals in vitro and in vivo; the specificity of an iron-binding system in the intestinal mucosa of the rat. In: Mills CF (ed) Trace element metabolism in animals. E & S Livingstone, Edinburgh, p 298-310

Gopinath, C, Howell JMcC 1975 Experimental chronic copper toxicity in sheep: changes that follow the cessation of dosing at the onset of hemolysis. Res Vet Sci 19:35-43

Gregoriadis G, Sourkes TL 1968 Role of protein in removal of copper from the liver. Nature (Lond) 218:290-291

Hazelrig JB, Owen CA, Ackerman E 1966 A mathematical model for copper metabolism and its relation to Wilson's disease. Am J Physiol 211:1075-1080

Hunt DM, Port AE 1979 Trace element binding in the copper deficient mottled mutants in the mouse. Life Sci 24:1453-1466

Irons RD, Smith JC 1977 Isolation of a non-thionein copper-binding protein from liver of copper-injected rats. Chem-Biol Interact 18:83-89

Keen CL, Lönnerdal B, Sloan MV, Hurley LS 1980 Effect of dietary iron, copper and zinc chelates of nitrilotriacetic acid (NTA) on trace metal concentrations in rat milk and maternal and pup tissues. J Nutr 110:897-906

King TP, Bremner I 1979 Autophagy and apoptosis in liver during the prehaemolytic phase of chronic copper poisoning in sheep. J Comp Pathol 89:515-530

Kirchgessner M, Grassmann E 1970 The dynamics of copper absorption. In: Mills CF (ed) Trace element metabolism in animals. E & S Livingstone, Edinburgh, p 277-285

Lönnerdal B, Keen CL, Hoffman B, Hurley LS 1980a Copper ligands in human milk – a vehicle for copper supplementation in the treatment of Menkes' disease. Am J Dis Child, in press

Lönnerdal B, Stanislowski AG, Hurley LS 1980b Isolation of a low molecular weight zinc binding ligand from human milk. J Inorg Biochem 12:71-78

Mistilis SP, Mearrick PT 1969 The absorption of ionic, biliary, and plasma radiocopper in neonatal rats. Scand J Gastroenterol 4:691-696

Oberleas D 1973 Phytates. In: Toxicants occurring naturally in foods. (Food and Nutrition Board). National Academy of Sciences, Washington DC, p 363-371

Owen CA Jr 1964 Absorption and excretion of Cu^{64}-labeled copper by the rat. Am J Physiol 207:1203-1206

Parekh SR, Patel BD 1972 Epidemiological survey of Indian childhood cirrhosis. Indian Pediatr 9:431-439

Port AE, Hunt DM 1979 A study of the copper-binding proteins in liver and kidney tissue of neonatal normal and mottled mutant mice. Biochem J 183:721-730

Porter H 1974 The particulate half-cystine-rich copper protein of newborn liver. Biochem Biophys Res Commun 56:661-668

Portmann B, Tanner MS, Mowat AP, Williams R 1978 Orcein positive liver deposits in Indian childhood cirrhosis. Lancet 1:1338-1340

Riordan JR, Gower I 1975 Small copper-binding proteins from normal and copper-loaded liver. Biochim Biophys Acta 411:393-398

Riordan JR, Richards V 1980 Human fetal liver contains both zinc- and copper-rich forms of metallothionein. J Biol Chem 255:5380-5383

Rydén L, Deutsch HF 1978 Preparation and properties of the major copper-binding component in fetal liver. J Biol Chem 253:519-524

Sherman AR, Tschiember N 1979 Tissue copper in offspring of iron deficient rats. Fed Proc 38:453 (abstr)

Sourkes TL, Lloyd K, Birnbaum H 1968 Inverse relationship of hepatic copper and iron concentrations in rats fed deficient diets. Can J Biochem 46:267-271

Tanner MS, Portmann B, Mowat AP, Williams R, Pandit AN, Mills CF, Bremner I 1979 Increased hepatic copper concentration in Indian childhood cirrhosis. Lancet 1:1203-1205

Winge DR, Premakumar R, Wiley RD, Rajagopalan KV 1975 Copper chelatin: purification and properties of a copper binding protein from rat liver. Arch Biochem Biophys 170:253-266

Metabolic interactions of copper with other trace elements

COLIN F. MILLS

Nutritional Biochemistry Department, Rowett Research Institute, Bucksburn, Aberdeen AB2 9SB, UK

Abstract Metabolic interactions between copper and other trace elements influence not only the susceptibility of animals and humans to deficiency or toxicity of copper but also the biochemical and pathological consequences of these states. Antagonistic trace elements affect the utilization of copper by reducing its solubility within the intestinal lumen, by competing with copper during its absorption or transport, or by modifying its distribution between receptors in body tissues.

The derangement of iron-dependent functions during the development of copper deficiency provides a further example of such interactions which, in this instance, arise from the involvement of copper in processes that apparently regulate the flux of iron between plasma and cellular pools.

This paper deals with recent studies of the mechanisms of the interactions between copper and zinc, cadmium, molybdenum and iron.

Susceptibility of animals and humans to deficiency or to excess of many of the essential trace elements is often increased by the presence of inorganic or organic antagonists of these elements in the diet or environment. At least 25 such interactions are known to be involved either in the aetiology or in the modification of the effects of trace element deficiency or excess.

Two principal interactions that affect the outcome of dietary deficiency or dietary excess of copper are now recognized. First, although the steps in copper absorption and in its transport to functional sites in tissues are poorly defined, many of the various stages are susceptible to interference from the simultaneous presence of ions that either reduce the solubility of copper within the intestinal lumen or modify its fate after absorption. Such 'competitive' interactions may increase either the adverse effects of a low copper supply or the tolerance of an excess.

A second type of interaction is seen when the metabolism of one element is influenced by an enzyme-catalysed step for which a second element is essential. In the present context, the most obvious example is the participation of copper in iron metabolism.

Many experiments in which such interactions have been studied have used extremely high ratios of antagonist to 'target' element. Although such work has revealed the development of metabolic defects which can be corrected by restoring the balance between elements, the extreme and acute challenges imposed by these high ratios are such that their relevance to the aetiology of human or animal disease is questionable. This paper deals mostly with less extreme conditions which may be more relevant to the incidence or pathological manifestations of trace element diseases attributable to defects in copper metabolism.

CADMIUM AND ZINC AS COPPER ANTAGONISTS

Metabolic interrelationships between copper, cadmium and zinc significantly modify both the biochemical and pathological manifestations of copper, cadmium or zinc intoxication and the tolerance of animals to excess of any of these three elements. Such interactions largely escaped attention until Van Reen (1953) showed that high levels of dietary zinc inhibited copper utilization by rats sufficiently to reduce the activity of hepatic cytochrome c oxidase (EC 1.9.3.1). During a study of the metabolic effects of copper poisoning in pigs, Suttle & Mills (1966) found biochemical and clinical evidence of an induced deficiency of zinc. Nevertheless, high dietary copper always increased hepatic retention of zinc. Suspicions that much of this was physiologically unavailable were confirmed by the apparent prophylactic effectiveness of additional dietary zinc for the control of copper poisoning.

A rational approach to the investigation of such interactions was subsequently proposed by Hill & Matrone (1970), who suggested that elements with similar valence-shell hybrid orbitals might compete for specific binding sites on proteins involved in their absorption and perhaps also during the *de novo* synthesis of metalloenzymes. Although steric considerations sometimes preclude such substitutions, the concept has proved useful in the search for metabolic antagonists of copper and, conversely, in investigations of its toxic effects. In our studies of such interactions we have used less extreme imbalances than those chosen by Hill & Matrone (1970) to demonstrate the validity of their ideas. We have paid particular attention to the consequences of an imposed imbalance when the dietary supply of copper or zinc is just adequate to meet normal requirements but is insufficient to permit deposition of tissue reserves that could reduce or nullify the effects of a potential antagonist.

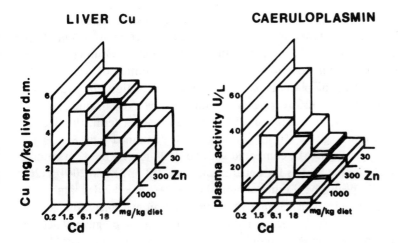

FIG. 1. The influence of dietary cadmium and zinc content on hepatic copper and on caeruloplasmin activity in growing rats. The basal diet provided 2.6 mg Cu/kg D.M. (dry matter). Treatments were maintained for 63 days before these measurements were taken.

Fig. 1 illustrates the effects of a range of dietary concentrations of cadmium and zinc, separately or together, on the hepatic copper content and on the plasma caeruloplasmin activity of rats (Campbell & Mills 1974, Campbell et al 1977). When either cadmium or zinc was given to the rats, at concentrations within the range commonly found in crops growing in areas adjacent to some industrial complexes (Mills & Dalgarno 1972), there was a reduction in caeruloplasmin activity, in hepatic copper retention, in plasma copper and in the thickness of cortical bone in the femurs. In contrast to previous studies which showed that zinc alleviates the teratogenic effects (Ferm & Carpenter 1967) and testicular damage (Parizek 1957) caused by acute exposure to cadmium, we found that chronic and simultaneous exposure to zinc and cadmium had an additive inhibitory effect on copper metabolism. Although this may suggest that these two antagonists of copper metabolism act identically, subsequent work has suggested several notable differences.

When sheep are given diets containing up to 12 mg cadmium/kg dry matter (D.M.) during pregnancy, although the placenta effectively discriminates against the transfer of cadmium to the fetus, this transfer appears to be achieved at the expense of copper transfer, as indicated by the copper content of the fetal liver at birth (Mills & Dalgarno 1972). In a similar study with rats, Choudhury et al (1977) also found that cadmium (17 mg/l in drinking water) reduced the total copper content of the fetus by 40%, even though no cadmium was transferred to the offspring. These findings are consistent with evidence that cadmium, acting as a copper antagonist, is involved in the aetiology of the demyelinating disease, enzootic ataxia, in newborn lambs in areas adjacent to industrial complexes that emit cadmium (Grun et al 1977).

Although high dietary zinc also reduces the tissue content of copper in the fetus and newborn, the placenta fails to discriminate against zinc transport, and a decrease in fetal tissue copper is accompanied by a rise in zinc content. Thus, an experiment by Campbell (1979) showed that, at term, the liver of lambs born to ewes given 750 mg zinc/kg D.M. contained only 8.4 ± 1.7 (SEM) mg copper/kg D.M., compared with 80 ± 34 mg/kg D.M. in control lambs; corresponding figures for hepatic zinc were 252 ± 21 and 2044 ± 260 mg/kg D.M. respectively. Perinatal mortality in these zinc-loaded lambs was high. Despite ample evidence that the lambs had an extremely low copper status at birth, a five-fold increase in dietary copper during pregnancy neither restored fetal hepatic copper content to normal nor prevented perinatal losses of copper when a high zinc intake was also maintained during pregnancy.

Competitive interactions between zinc and copper can also occur in the mammary gland. Thus, changes in the zinc content of bovine milk induced by zinc depletion and repletion are accompanied by reciprocal changes in copper content. It is not yet known whether low concentrations of cadmium in the diet of the lactating female have a similar competitive interaction with copper. However, the presence of adventitious cadmium in the diet of young animals that subsist primarily on a liquid diet has a strong inhibitory effect on tissue retention of copper. Our study with the suckling lambs which had access to the cadmium-containing diets given to their dams (3.5 mg cadmium/kg D.M.) showed that although hepatic cadmium increased by only about 1 mg/kg D.M. between birth and eight weeks of age, hepatic copper content decreased by 75% compared with that of control lambs that did not have access to a cadmium-containing diet (Mills & Dalgarno 1972). The sensitivity of the newborn lamb to cadmium was confirmed in a later study (Campbell 1979) which provided additional evidence that the inhibitory effect of cadmium upon growth was abolished if dietary copper was increased from 4.5 to 15 mg/kg D.M.

The competitive interaction between zinc and copper has been exploited in the control of copper toxicity in copper-intolerant species such as the sheep. Bremner et al (1976) found that zinc given in the diet, at 220 or 440 mg/kg D.M., effectively reduced both the rate of hepatic copper retention and the rise in plasma arginase (EC 3.5.3.1) activity that indicates liver damage. In addition, the zinc protected against the haemolytic crisis that terminates in death when lambs are given diets high in copper.

MECHANISMS OF THE ANTAGONISM BETWEEN CADMIUM, ZINC AND COPPER

The mechanisms whereby cadmium or zinc modify copper transport through the placenta and mammary gland are not known. In the weaned animal, the antagonism of cadmium or zinc is partly attributable to a decrease in the efficiency of copper absorption and partly to changes in the distribution of copper between body tissues, e.g. between cytosolic proteins in the liver and in the kidney. Evidence of these changes in distribution is based (1) on the increased retention of copper in the kidney, even when hepatic and total body copper is declining and (2) on increases in the proportion of total copper that associates with metallothionein, a cysteine-rich protein of low molecular weight, in the cytosol of hepatic, renal and intestinal mucosal cells (Bremner 1974, Mills 1974).

Excess cadmium or zinc induces metallothionein synthesis in a wide variety of tissues. As the tissue content of metallothionein increases, the proportion of total cytosolic copper bound to this fraction also increases, either because copper is incorporated during *de novo* synthesis of the protein or because it displaces zinc or cadmium that was previously bound. When the absorption of copper is inhibited in rats by a diet high in zinc, up to 60% of the total (non-intestinal) tissue deficit of copper can be accounted for by copper bound to intestinal mucosal metallothionein (Fig. 2). After an oral dose of ^{64}Cu, the incorporation of copper into metallothionein increased six-fold (Hall et al 1979).

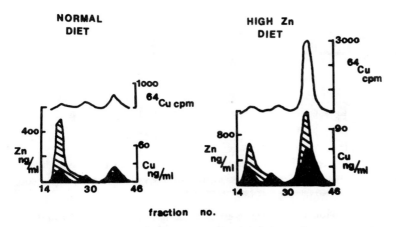

FIG. 2. The influence of adequate (30 mg/kg D.M.) and high (900 mg/kg D.M.) dietary zinc on the distribution of copper (■), zinc (▨) and orally administered ^{64}Cu in the cytosol of the proximal small intestine of rats. Rats were killed 4 h after ^{64}Cu administration. Sephadex G75 gel filtration and fraction volumes of 3.5 ml were used. (Hall et al 1979.)

The biological half-life of copper bound to metallothionein may be as long as 17 h in the liver and 19 h in the kidney (Bremner et al 1978). If the copper- and zinc-containing metallothionein of enterocytes has a similar half-life, it can be appreciated why increases in the proportion of enterocytic copper associated with metallothionein may account for a decreased absorption of copper. This effect could arise from the appearance of this relatively immobile copper-containing pool in enterocytes, which have only a transient existence (estimated half-life 15–25 h) before they desquamate from intestinal villi, with the copper they have retained (see Fig. 3).

Although the retention of ^{64}Cu by the intestinal mucosa during a period of four hours increased, similarly, from 11% to 20% of an orally administered dose in rats given 4 or more mg cadmium/kg D.M. (see Table 1), it is less clear than for zinc that this effect, and the subsequently reduced transport of ^{64}Cu, was attributable primarily to 'trapping' of copper by the metallothionein fraction of mucosal cells (Campbell et al 1977, Davies & Campbell 1977).

These and many other studies make it clear that 'environmentally relevant' concentrations of cadmium or zinc can so impair utilization of copper, in animals whose diets are only marginally adequate in copper, that defects attributable to a conditioned copper deficiency are induced. The evidence for this, in addition to the observed changes in tissue content of copper, includes depression of caeruloplasmin and cytochrome c oxidase activity in some tissues, impaired melanogenesis and defective osteogenesis in rats and defec-

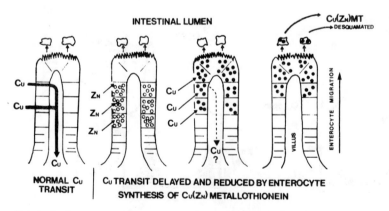

FIG. 3. Probable mechanism of the inhibitory effect of high dietary zinc on copper absorption, accounting for (1) the initially increased mucosal retention of copper and (2) the subsequent decrease in copper flux into other body tissues. Key: ▨ metallothionein synthesis induced by high luminal concentrations of zinc; ⋮⋮ subsequent incorporation of copper to form Cu-Zn-metallothionein which is mostly retained in the enterocyte until desquamation takes it into the lumen.

TABLE 1

Effects of dietary cadmium and zinc on absorption and retention of orally administered ^{64}Cu in intestinal mucosa of rats

	Expt. 1				Expt. 2			
Dietary Cd (mg/kg D.M.):	< 0.1	< 0.1	< 0.1	< 0.1	< 0.1	4.4	8.8	17.6
Zn (mg/kg D.M.):	30	150	450	900	30	30	30	30
Fraction (%) of dietary ^{64}Cu absorbed into carcass[a]	43	39	43	27	39	40	33	22
^{64}Cu retained by intestinal mucosa, relative to low Cd, low Zn group (100)	100	101	113	301	100	180	172	184
^{64}Cu associated with metallothionein in mucosal cytosol, relative to low Cd, low Zn group (100)	100	113	168	627	100	455	270	181

[a]Absorption estimated from retention of ^{64}Cu by carcass after removal of gastrointestinal tract. Treatments were maintained for 7 days before oral administration of ^{64}Cu. Rats were killed 4 h after ^{64}Cu administration. Note that the decrease in efficiency of ^{64}Cu absorption induced by zinc appears to be more closely related than that induced by cadmium to the mucosal retention of ^{64}Cu in the metallothionein fraction. (Calculated from data of Campbell et al 1977, Davies & Campbell 1977, Hall et al 1979.)

tive keratinization of wool in sheep. With the exception of the adverse effects of a high intake of zinc on perinatal survival, all these effects are preventable by an increase in copper supply in the diet. Thus, future estimates of the tolerable upper threshold concentrations of cadmium or zinc must allow for any imbalances that arise when copper supply is only just adequate to meet the body's requirements.

EFFECTS OF MOLYBDENUM AS A COPPER ANTAGONIST

The development of a low redox potential and of near-neutral or alkaline conditions in molybdenum-containing soils frequently induces a sufficient increase in the molybdenum content of forages to provoke clinical copper deficiency in herbivores. This deficiency arises because molybdenum strongly inhibits copper utilization, an effect that is potentiated by increases in the dietary content of inorganic sources of sulphur or sulphur amino acids. Molybdenum acts as a copper antagonist when the molybdenum content of the diet exceeds about 1 mg/kg D.M. and thus it causes copper deficiency in ruminants in many parts of the world.

Studies of the antagonism between molybdenum and copper in ruminants have shown that when dietary molybdenum content increases, the nature of its inhibitory action upon copper metabolism also changes. At relatively low dietary contents of molybdenum (1 to 10 mg/kg D.M.), its effects are attributable to a marked decrease in the efficiency with which dietary copper is absorbed. The resulting decline in tissue content of copper occurs with no marked accompanying increase in tissue content of molybdenum. Although copper absorption remains reduced when molybdenum intake is higher than 10 mg/kg D.M. systemic defects in copper metabolism also develop. Such effects include a marked redistribution of copper between plasma proteins. The enzyme activity and the copper content of the caeruloplasmin fraction fall and marked increases occur in both the copper and molybdenum content of the albumin fraction. In the kidney, there is evidence of a decrease in the proportion of copper in the cytosol and in the fractions that normally contain copper metallothionein or account for superoxide dismutase activity (EC 1.15.1.1) (Mills et al 1977, Mills & Bremner 1980).

The potency of molybdate as a copper antagonist is much greater in ruminants than in monogastric species. Also, the synergistic effect of dietary sources of sulphur in ruminants depends in some way on the reaction of molybdenum with sulphide generated either from bacterial reduction of inorganic sulphate or from the degradation of sulphur amino acids within the digestive tract. There is evidence that the initial sequence of this reaction involves progressive substitution of the oxygen in the MoO_4^{2-} ion to yield either tetrathiomolybdate (MoS_4^{2-}) or a series of oxythiomolybdates:

$$MoO_4^{2-} \rightarrow MoO_3S^{2-} \rightarrow MoO_2S_2^{2-} \rightarrow MoOS_3^{2-} \rightarrow MoS_4^{2-}$$

We have investigated the possibility that one of these derivatives may be the active agent that exerts an inhibitory effect on copper metabolism in rats.

Apart from some exceptions described later, molybdenum given as MoO_4^{2-} is remarkably well tolerated by non-ruminant species. Furthermore, when molybdenum in this form is present in the diet of rats at high concentrations (e.g. > 250 mg/kg D.M.), there is little clinical or biochemical evidence that the observed adverse effects on growth are primarily attributable to an induced copper deficiency. In contrast, defects in copper metabolism became apparent when 2 or more mg molybdenum/kg D.M. were given as MoS_4^{2-}. The effects of this ion (which are not reproduced by equivalent dosage with free S^{2-}, in the form of CaS) include inhibition of caeruloplasmin synthesis and of melanogenesis in hair (Mills & Bremner 1980), arrested osteogenesis and cartilage dysplasia (Spence et al 1980) and the development of an anaemia largely attributable to an abrupt arrest in erythrogenesis. All these effects were prevented by oral or intraperitoneal administration of copper supplements.

Resolution of the copper-containing proteins of rat plasma and kidney cytosol by Sephadex gel filtration showed that MpS_4^{2-} induced changes in copper distribution (Fig. 4) that were virtually identical to those found in ruminants given diets that included inorganic sulphate and high concentrations (> 15 mg/kg D.M.) of molybdenum in the form of molybdate. A further similarity between the responses of rats and ruminants to such treatment was the appearance of copper deficiency well before plasma copper had declined to the concentration usually associated with deficiency (< 0.5 mg/l). An abnormally high proportion of plasma copper occurred in the albumin fraction which also retained copper and molybdenum in forms that, in contrast to normal plasma, were not released by treatment with 5% trichloroacetic acid.

Although this evidence indicated that thiomolybdate or its metabolic derivatives induced systemic defects in the metabolism of tissue copper, this form of molybdenum also markedly inhibited the absorption of dietary cop-

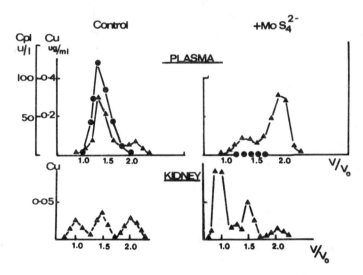

FIG. 4. The influence of dietary MoS_4^{2-} on the distribution of copper between kidney cytosol and plasma proteins and on caeruloplasmin (Cpl) activity in rats. Gel filtration was performed on Sephadex G100 (plasma) or G75 (kidney cytosol). Basal diet for both groups provided 3 mg Cu/kg D.M.; MoS_4^{2-} provided 6 mg Mo/kg D.M. Key: ▲—▲, Cu content of fraction; ●—●, caeruloplasmin activity. MoS_4^{2-} increased the copper content of the plasma albumin fraction ($V/V_o \approx 1.8$) and decreased that of the metallothionein (V/V_o 2.0) and superoxide dismutase-containing fraction (V/V_o 1.5) of kidney cytosol (from data of Mills & Bremner 1980, El-Gallad 1979). V/V_o = total eluate volume before emergence of fraction/void volume of column.

per. Studies of the fate of orally administered ^{64}Cu in rats given tetrathiomolybdate (Table 2) confirmed the existence of an absorptive defect and illustrated that this species both prolongs the retention of ^{64}Cu in plasma and inhibits its incorporation into the liver and kidney. Our studies with the thio- and oxythio-derivatives of molybdenum and tungsten show that only the tetrathio-derivatives inhibit copper absorption and induce clinical changes that are preventable by increases in dietary copper content.

This evidence from experiments on rats suggests that the ultimate development of copper deficiency in ruminants whose diets contain high concentrations of molybdate (MoO_4^{2-}) may depend on the synthesis of MoS_4^{2-} or of structurally related derivatives that are much more potent antagonists of copper metabolism. Whenever the tetrathiomolybdate ion is given, however, it invariably induces systemic defects in copper utilization that are not always evident when a copper deficiency is induced in ruminants by relatively low dietary concentrations of molybdenum (1 to 15 mg/kg D.M.). Since the systemic effects appear to be related to the facility with which the MoS_4^{2-} ion is absorbed, we are now examining the effect of polymeric derivatives which, while they retain a strong affinity for copper, would act solely within the gastrointestinal tract. Present views on the mode of action of thiomolybdates and their analogues as copper antagonists are summarized in Fig. 5.

It has been widely assumed that the action of molybdenum as a copper antagonist is largely irrelevant to the aetiology of copper deficiency in humans and other monogastric species, but this view is questionable. The molybdate

TABLE 2

Contrasting effects of dithiomolybdate [$MoO_2S_2^{2-}$] and tetrathiomolybdate [MoS_4^{2-}] on ^{64}Cu absorption and tissue distribution in rats

Treatment	Control[a]	$MoO_2S_2^{2-}$ [b]	MoS_4^{2-} [b]
Absorption of orally administered ^{64}Cu			
(% of dose)	47	42	9.5
Fraction of *absorbed* dose (%) retained by			
Liver	28	18	9.0
Kidney	19	11	3.5
Blood	8	24	57

[a] Without molybdenum
[b] These diets provided 6 mg Mo/kg D.M. Treatments were maintained for seven days before oral administration of ^{64}Cu. Rats were killed five hours after ^{64}Cu administration. Other studies (Mills & Bremner 1980) have shown that MoO_4^{2-} at the same dietary concentration of molybdenum neither inhibits ^{64}Cu absorption nor influences its distribution between blood and tissues.

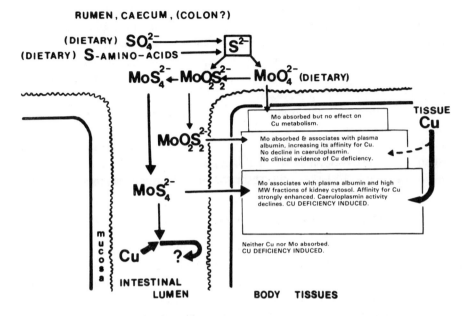

FIG. 5. A summary of recent observations on the probable action of thio- and oxythiomolybdates (and their tungsten analogues) as metabolic antagonists of copper within the intestinal lumen and body tissues (from data of Mills & Bremner 1980, El-Gallad 1979).

ion is usually well tolerated by rats and, in contrast to ruminants, any adverse effects in rats are abolished if molybdenum absorption is inhibited competitively by the SO_4^{2-} ion (Cardin & Mason 1976). However, when the sulphate intake of a rat is sufficiently high for some of it to escape absorption and to penetrate to the lower gut (subsequently to be reduced to sulphide by microflora), systemic changes in copper metabolism appear which are identical to those produced by tetrathiomolybdate. These effects can also be reproduced by traces of free sulphide in a diet to which molybdate has been added (El-Gallad 1979). These findings explain the apparently discordant observation of Gray & Daniel (1964) that SO_4^{2-} (added at the unusually high concentration of 27 g/kg D.M.) provoked a copper-responsive syndrome in rats given 10 mg molybdenum/kg D.M. as MoO_4^{2-}; they also illustrate the circumstances under which an antagonism between molybdenum, sulphur and copper may become established in monogastric species. In some geographical areas, it is not uncommon to encounter a high molybdenum content in legume crops consumed by humans. Simultaneous consumption of these crops and a diet high in inorganic sulphate or rich in sulphur amino acids, which could provoke sulphide generation within the lower bowel, could produce appropriate conditions for the synthesis of MoS_4^{2-} or its analogues, and could initiate the metabolic and pathological lesions described above.

The rapidity of the effect and the low doses of tetrathiomolybdate required to inhibit copper metabolism make this ion probably the most effective copper antagonist yet discovered. However, because it is easily absorbed and produces dramatic pathological changes in the skeleton, connective tissue and gastrointestinal mucosa (Fell et al 1979), investigation of its value as a therapeutic agent for use in disorders such as Wilson's disease should not be attempted. The development of analogues or derivatives of higher molecular weight that would be less readily absorbed should circumvent this difficulty.

METABOLIC INTERRELATIONSHIPS BETWEEN COPPER AND IRON

Interest in the role of copper in iron metabolism has centred primarily on its relationship to the development of anaemia in copper deficiency syndromes. However it is now evident that defects in iron metabolism arise during copper deficiency long before overt signs of the deficiency appear. Increases in tissue retention of iron have been noted in copper-deficient rats, sheep and calves before growth fails and anaemia develops, and these increases are often detectable at early stages of the syndrome when effects such as cardiac hypertrophy, mitochondrial hyperplasia, inhibited melanogenesis and defects in development of skeletal and connective tissue are appearing. While defects in the biosynthesis of elastin and melanin are attributable to changes in the activity of the specific copper-dependent enzymes, lysyl oxidase and monophenol monooxygenase (EC 1.14.18.1), other metabolic lesions of copper deficiency may not have such a direct origin. Interest is therefore growing in the wider implications of deranged iron metabolism during copper deficiency.

Although there is good evidence that the copper enzyme, caeruloplasmin, helps to mobilize cellular iron before its incorporation into transferrin, several effects of copper deficiency on iron metabolism cannot be explained by this finding alone, not least because there are marked species differences in response. In copper-deficient pigs, the observed decline in hepatic iron and in total body iron probably results from inhibited absorption of copper. There is a concurrent increase in histochemically demonstrable iron-containing particulates in mucosal cells and a marked reduction in the turnover rate of iron in enterocytes (Lee et al 1976). Although copper deficiency also decreases the efficiency of iron absorption by rats and ruminants, the most significant change in tissue distribution of iron in these species is a marked increase in the deposition of ferritin in liver, kidney, spleen and bone marrow (e.g. Nozu & Kitamura 1977). These effects are strongly enhanced by increases in dietary iron suggesting that, in these species, a less effective block to iron absorption has developed. The origins of such species differences are not yet known.

The pattern of changes in tissue retention of iron in rats and ruminants during copper depletion suggests that more than one copper-dependent process may be involved. Typical changes in both species are early but moderate increases in hepatic iron, followed by a period when little further change occurs. At about the time when gross clinical signs of deficiency and retardation of growth occur, a further rapid increase in hepatic iron takes place. Typical data for the changes caused by copper depletion in calves are given in Fig. 6 (Mills et al 1976). It is noteworthy that the late increases in hepatic iron are not usually accompanied by any further decrease in caeruloplasmin activity and that they frequently precede the appearance of even a mild anaemia in both rats and cattle. It is difficult to reconcile such evidence of a biphasic change in iron metabolism with the concept that the only relevant lesion is inhibited oxidation of Fe^{2+} to Fe^{3+}, arising from a failure of caeruloplasmin

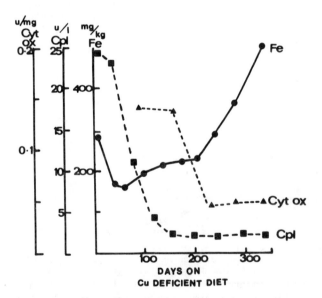

FIG. 6. Changes in bovine hepatic iron (mg/kg D.M.), cytochrome c oxidase activity (U/mg protein) and plasma caeruloplasmin activity (U/l) during copper depletion. Pooled data from Mills et al (1976) and C.F. Mills & A.C. Dalgarno (unpublished). Diets provided < 1 mg Cu and 40–60 mg Fe/kg D.M. Note: the decline in hepatic iron between days 0 and 50 is an age-dependent effect and is not related to the experimental treatment. (Units of enzyme activity as defined by IUPAC/IUB.)

synthesis. Thus, interest is growing in recent evidence that changes in the activity of terminal respiratory enzymes, or of cytochrome c oxidase in particular, may influence the rate of entry and fate of mitochondrial iron. The indirect involvement of copper in these processes is suggested by the following observations.

(1) The work of Ponka et al (1977) indicates that the mitochondrial pool of haem regulates, by negative feedback, the rate of Fe^{3+} release from transferrin that is in the cytosol or in close association with mitochondrial membranes. Thus a decline in haem content would be expected to increase Fe^{3+} flux into mitochondria.

(2) Evidence for the accumulation of a non-haem pool of Fe^{3+}, with a concurrent decrease in the haem content of copper-deficient mitochondria, has recently been obtained by Williams et al (1976, 1978). Their data also suggest that the reduction of Fe^{3+} to Fe^{2+}, which is obligatory for the incorporation of iron into haem, may require electron transfer from reduced cytochrome c oxidase.

(3) Work on several species has produced convincing evidence that decreases in cytochrome c oxidase activity and in haem A content occur in copper-depleted tissues. It is generally presumed (without direct evidence) that the loss of cytochrome c oxidase activity is always attributable solely to a reduction in mitochondrial copper content to levels that are inadequate for

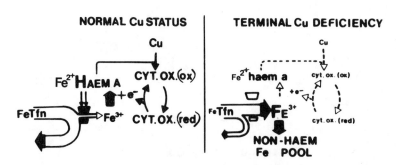

FIG. 7. Processes probably involved in the excessive accumulation of non-haem Fe^{3+} during terminal stages of copper deficiency. Lower case symbols indicate decreased pool size or decreased flux. The suggested sequence of events in terminal copper deficiency is as follows: (1) a decline in cytochrome c oxidase activity initiated solely by copper deficiency, (2) a decrease in the rate of reduction of Fe^{3+} to Fe^{2+}, obligatory for haem A synthesis (Williams et al 1976), (3) a decrease in haem A pool size, (4) loss of inhibitory feedback control of Fe^{3+} entry (Ponka et al 1977), (5) further decline in cytochrome c oxidase activity resulting from decreased synthesis of the haem A moiety of the enzyme. Implicit in the argument is that loss of regulation of step (4) more than compensates for the lower plasma flux of Fe transferrin [FeTfn] that arises from any earlier decline in caeruloplasmin (ferroxidase I) activity.

synthesis of the enzyme. However, the results of Williams et al (1976, 1978) suggest that at later stages of copper deficiency the failure of synthesis of the haem A moiety of cytochrome c oxidase may also contribute to this loss of activity.

If cytochrome c oxidase therefore maintains the balance between haem and non-haem iron while haem regulates the entry of Fe^{3+} into the mitochondrion, changes in cytochrome c oxidase activity, mediated by copper supply, could sensitively control the tissue accumulation of iron. This sequence (Fig. 7) is derived from several independent studies and requires validation from a detailed investigation of sequential changes in mitochondrial flux of iron and in cytochrome c oxidase activity in copper-deficient animals. Nevertheless, it may explain why the observed changes in the terminal accretion of iron appear to be more closely related to changes in cytochrome c oxidase activity than to changes in caeruloplasmin activity. The results of Williams et al (1976, 1978) also emphasize the wider implications of the decline in terminal respiratory enzyme activity that is a consistent feature of most copper deficiency syndromes.

CONCLUSION

Trace element interactions can be significant in the aetiology and consequences of copper deficiency. Quantitative aspects of the interactions have not been adequately investigated, and therefore the minimal copper requirements of many species are poorly defined. Metabolic interactions between cadmium, zinc and copper are now known also to influence both the nature and the severity of the pathological changes that ensue from excessive exposure to these trace elements, and therefore much information on the tolerance of potentially toxic dietary and environmental concentrations of Cd, Zn and Cu requires re-evaluation. Variations in dietary composition must be expected to influence the biochemical and pathological consequences of trace element deficiency or excess.

ACKNOWLEDGEMENTS

I am indebted to my many colleagues in Nutritional Biochemistry Department, Rowett Research Institute, for the many stimulating, critical and fruitful discussions on this topic that have punctuated our mutual efforts in this field of research. The financial support granted for some of these studies by Rio Tinto Zinc Services Limited, the International Copper Research Organisation and the International Lead-Zinc Research Organisation is also acknowledged with gratitude.

References

Bremner I 1974 Heavy metal toxicities. Q Rev Biophys 7:75-124
Bremner I, Young BW, Mills CF 1976 Protective effect of zinc against copper toxicosis in sheep. Br J Nutr 36:551-561
Bremner I, Hoekstra WG, Davies NT, Young BW 1978 Effect of Zn status of rats on the synthesis and degradation of Cu-induced metallothionein. Biochem J 174:883-892
Campbell JK 1979 Growth, reproduction and copper metabolism in animals exposed to elevated dietary levels of cadmium and zinc. PhD thesis, University of Aberdeen
Campbell JK, Mills CF 1974 Effects of dietary cadmium and zinc on rats maintained on diets low in copper. Proc Nutr Soc 33:15A-16A
Campbell JK, Davies NT, Mills CF 1978 Interactions of Cd, Cu and Zn in animals chronically exposed to low levels of dietary Cd. In: Kirchgessner M (ed) Trace element metabolism in man and animals. Arbeitskreis für Tierernährungsforschung, Freising-Weihenstephan vol 3:553-556
Cardin CJ, Mason J 1976 Molybdate and tungstate transfer by rat ileum; competitive inhibition by inorganic sulphates. Biochim Biophys Acta 455:937-946
Choudhury H, Hastings L, Menden E, Brockman D, Copper GP, Petering HG 1978 In: Kirchgessner M (ed) Trace element metabolism in man and animals. Arbeitskreis für Tierernährungsforschung, Freising-Weihenstephan vol 3:549-552
Davies NT, Campbell JK 1977 The effect of cadmium on copper absorption and binding in the rat. Life Sci 20:955-960
El-Gallad TT 1979 The antagonistic effect of Mo and S compounds upon Cu metabolism in animals. PhD thesis, University of Aberdeen
Fell BF, Dinsdale D, El-Gallad TT 1979 Gut pathology of rats dosed with tetrathiomolybdate. J Comp Pathol 89:495-514
Ferm VH, Carpenter SJ 1967 Teratogenic effect of cadmium and its prevention by zinc. Nature (Lond) 216:1123
Gray LF, Daniel LJ 1964 Effect of the copper status of the rat on the copper-molybdenum-sulfate interaction. J Nutr 84:31-37
Grun M, Anke M, Partschfeld M 1977 The Cd-exposure of cattle and sheep in the GDR. In: Bolk F (ed) Kadmium-Symposium. Friedrich Schiller University, Jena, p 253-257
Hall AC, Young BL, Bremner I 1979 Intestinal metallothionein and the mutual antagonism between copper and zinc in the rat. Inorg Biochem 11:57-66
Hill CH, Matrone G 1970 Chemical parameters in the study of in vivo and in vitro interactions of transition elements. Fed Proc 29:1474-1481
Lee GR, Williams DM, Cartwright GE 1976 Role of copper in iron metabolism and heme biosynthesis. In: Prasad AS, Oberleas D (eds) Trace elements in human health and disease. Vol 1: Zinc and copper. Academic Press, New York, p 373-390
Mills CF 1974 Trace element interactions: effects of dietary composition on the development of imbalance and toxicity. In: Hoekstra WG et al (eds) Trace element metabolism in animals. University Park Press, Baltimore, vol 2:79-90
Mills CF, Bremner I 1980 Nutritional aspects of molybdenum in animals. In: Coughlan M (ed) Molybdenum and molybdenum-containing enzymes. Pergamon, Oxford, p 517-542
Mills CF, Dalgarno AC 1972 Copper and zinc status of ewes and lambs receiving increased dietary concentrations of cadmium. Nature (Lond) 239:171-173
Mills CF, Dalgarno AC, Wenham G 1976 Biochemical and pathological changes in tissues of Friesian cattle during the experimental induction of copper deficiency. Br J Nutr 35:309-331
Mills CF, Bremner I, El-Gallad TT, Dalgarno AC, Young BL 1978 Mechanisms of the molybdenum/sulphur antagonism of copper utilisation. In: Kirchgessner M (ed) Trace element metabolism in man and animals. Arbeitskreis für Tierernährungsforschung, Freising-Weihenstephan, vol 3:150-158
Nozu T, Kitamura K 1977 Iron metabolism in hypocaeruloplasminaemic rats with special reference to the reticuloendothelial system. Showa Igakkai Zasshi (Showa Med Assoc J) 37: 525-533

Parizek J 1957 The destructive effect of cadmium ion on testicular tissue and its prevention by zinc. J Endocrinol 15:56-63
Ponka P, Neuwirt J, Borova J, Fuchs O 1977 Control of iron delivery to haemoglobin in erythroid cells. In: Iron metabolism. Excerpta Medica, Amsterdam (Ciba Found Symp 51), p 167-187
Suttle NF, Mills CF 1966 Studies of the toxicity of copper to pigs. Br J Nutr 20:135-148
Spence JA, Suttle NF, Wenham G, El-Gallad TT, Bremner I 1980 A sequential study of the skeletal abnormalities which develop in rats given ammonium tetrathiomolybdate. J Comp Pathol 90:139-153
Van Reen R 1953 Effects of excessive dietary zinc in the rat and the interrelationship with copper. Arch Biochem Biophys 46:337-344
Williams DM, Loukopoulos D, Lee GR, Cartwright DE 1976 Role of copper in mitochondrial iron metabolism. Blood 48:77-85
Williams DM, Barbuto AJ, Atkin CL, Lee GR 1978 Evidence for an iron carrier substance in copper deficient mitochondria. Prog Clin Biol Res 21:539-549

Discussion

Frieden: I would like to comment about the decrease in caeruloplasmin concentrations that can be produced by the simultaneous administration of various heavy metal ions other than copper. Whanger & Weswig (1970) have studied the effect on plasma caeruloplasmin concentrations of feeding zinc, silver, and cadmium. Molybdenum was not particularly effective in reducing plasma caeruloplasmin in their experiments, and I suspect that they found the same as you have, Dr Mills — that they had used a relatively ineffective form of molybdenum. Their experiments produced results similar to yours — there was a marked interference with the biosynthesis of caeruloplasmin when Ag^+, Cd^{2+} and Zn^{2+} were fed. Subsequently, they observed symptoms of anaemia in one form or another.

I also have a comment about the biphasic changes in iron which you mentioned. For the control of iron mobilization there must be a large excess of plasma caeruloplasmin; this has been confirmed by experiments *in vitro* and *in vivo*. This excess would occur even when plasma caeruloplasmin was reduced beyond the levels necessary to decrease cytochrome *c* oxidase activity. So I'm not too surprised that you obtained complex biphasic responses. I would expect to see differential effects on iron-responsive systems: on haemoglobin biosynthesis, on cytochrome *c* oxidase concentrations and on superoxide dismutase concentrations. In the light of your work, Dr Mills, I think we should begin to look more closely for those effects. Ultimately these differences in response to iron during copper depletion could also have an impact on the systems for iron transfer and iron release.

Mills: The terminal hepatic accumulation of iron would be easy to explain if we had evidence that it was accompanied by the development of anaemia, but it isn't.

Frieden: There is, in fact, evidence for that in the rat, from work by Evans & Abraham (1973).

Mills: But the fall in haemoglobin concentration that we often observe is only from 12.5 to 11.5 g/100 ml when the rapid accumulation of iron is taking place. So the source of the accumulated iron cannot be the degraded haemoglobin.

Frieden: In the experiments of Evans & Abraham (1973) the changes in haemoglobin concentration were much greater.

Mills: People have become obsessed with the idea that defects in iron metabolism arising during copper deficiency are secondary consequences of a failure in haemoglobin synthesis, but there is much evidence that the major changes taking place in iron metabolism in copper-deficient animals are not initiated by the appearance of anaemia (e.g. Grassman & Kirchgessner 1973, Evans & Abraham 1973, Mills et al 1976, Nozu & Kitamura 1977).

Österberg: I would like to comment on the interaction of copper with cadmium and zinc. Some years ago we did some calculations on these three metal ions (see e.g. Österberg 1974, 1978) and we were able to correlate the calculations with experiments of the type that you have done, Dr Mills. The amino acids may be considered as some kind of discriminating system. *In vivo,* the amino acid concentration is very high compared to the concentrations of labile copper and zinc ions; and, at a relatively harmless cadmium intoxication, the cadmium concentration is lower still. Now if we consider protein sites where copper, zinc or cadmium can bind, and if we consider the binding in the presence of the *in vivo* concentration of amino acids, then Cu(II) ions should have difficulty in binding when there is a large concentration of amino acids. On the other hand, cadmium ions, which do not bind very strongly to amino acids, should bind more easily to the protein sites. These results are in excellent agreement with your results on animal diets whose low content of cadmium leads to simultaneous reduction in the concentration of the copper proteins.

Mills: We repeatedly observe that cadmium is at least 20–30 times more potent than zinc as an antagonist of copper (e.g. Campbell & Mills 1974).

Frieden: What about the silver ion?

Österberg: Silver, of course, binds very poorly to amino acids because it coordinates only the amino group of the amino acid, and therefore behaves similarly to Cu(I). A consequence of this poor binding is that silver should bind easily and rapidly to metallothionein, perhaps like cadmium.

I would also like to comment about the interaction of copper with molybdenum and sulphide. Some years ago we did some calculations on redox potentials and pH (Österberg 1975). I believe that molybdenum interacts unfavourably with copper when it is present in the grass that ruminants eat. However, we could not find a reducing agent strong enough to reduce sulphate to sulphide in non-ruminant mammals. Therefore it is most interesting to see that if you feed tetrathiomolybdate (MoS_4^{2-}) to the non-ruminant mammal then the results are much the same as those obtained with ruminants.

Is it correct that the molybdenum and sulphur of MoS_4^{2-}, together with the associated copper, go into the albumin fraction of plasma?

Mills: We know that the molybdenum is there but, although we strongly suspect it, we cannot yet be sure that the sulphur is there. When molybdate is the source of molybdenum, accumulation of copper and molybdenum on the albumin fraction of plasma or in high-molecular-weight proteins of kidney cytosol is not as great. The problem is to detect the MoS_4^{2-} ion at the low concentrations that may be present. Nor have we yet succeeded in preparing sufficient active $Mo^{35}S_4^{2-}$ to allow us to follow the fate of the sulphur. However, the copper that becomes associated with the albumin is no longer released by 5% trichloracetic acid (TCA), as it would normally be. If a solution of bovine serum albumin is gassed briefly with sulphide, and if free sulphide is dialysed away under gaseous nitrogen until no trace of sulphide remains, then any copper that is added to the preparation becomes TCA-insoluble, as it does if albumin is treated with MoS_4^{2-} before dialysis. In this instance, the molybdenum is also retained in the TCA-insoluble fraction, as it is in plasma albumin from animals given MoS_4^{2-} orally or by intravenous infusion. So our evidence is indirect. We also have evidence that all the four sulphurs of tetrathiomolybdate (MoS_4^{2-}) are needed to induce clinical signs of copper deficiency and that tetrathiotungstate behaves similarly (Bremner et al 1979). Dithiomolybdate ($MoO_2S_2^{2-}$) does not inhibit copper absorption when given orally (Mills & Bremner 1980).

Österberg: I wonder if the binding of MoS_4^{2-} to albumin is similar to that for ferridoxin models (McCarthy & Lovenberg 1968)? Iron cysteinyl sulphide seems to bind quite well to albumin.

Mills: Tetrathiomolybdate may be a low-molecular-weight analogue of the forms of molybdenum that cause copper deficiency in ruminants. It is absorbed readily as the MoS_4^{2-} ion, perhaps too readily to account for those instances in which copper deficiency develops with little or no increase in tissue content of molybdenum. In such cases an analogue or derivative of MoS_4^{2-} with a higher molecular weight may be the species responsible for blocking copper absorption. The monomeric ion, however, satisfactorily reproduces the effect of high intakes of molybdenum in ruminants in which systemic defects in copper metabolism are evident. Molybdenum may also be linked to copper deficiency in other species. When molybdate (MoO_4^{2-}) is given to rats receiving diets with a high sulphate content, changes in plasma copper content and distribution occur which strongly resemble those caused by MoS_4^{2-}. It seems likely that reduction of sulphate to sulphide in the caecum or lower digestive tract may lead to synthesis of tetrathiomolybdates or oxythiomolybdates which are then absorbed. The claim of Gray & Daniel (1964) that *sulphate* potentiated the effect of molybdate as a copper antagonist in rats, just as it does in ruminants, was based on the use of diets that were very high in sulphate. Their results conflicted with many others which indicated that sulphate ameliorated the effects of molybdate in rats and the discrepancy is likely to be attributable to differences in sulphate intake. Low doses of sulphate prevent absorption of molybdate while high doses favour endogenous generation of sulphide and the synthesis of thiomolybdates.

McMurray: Is all the molybdenum in herbage in the form of molybdate, or is it converted?

Mills: Plants obviously contain a variety of chemical species of molybdenum. Some molybdenum is contained in nitrate reductase, but the form in which the remaining large amount is present is unknown.

McMurray: What are the equilibrium and kinetic constants between sulphur complexes and molybdenum–sulphur complexes that may exist in the digestive tract?

Mills: The kinetics of the reactions with copper have not been studied in sufficient detail. Insoluble derivatives are present at the end of the reaction sequence. When copper is in the cupric form, the end products are CuS and MoS_3; when it is in the cuprous form, a cuprous thiomolybdate ion is probably produced as an intermediate. Which of these sequences occurs generally *in vivo* is not known.

Frieden: There are finite limits to the absorption of iron, even from foodstuffs that are rich in iron. Does copper absorption present the same problem – is copper absorption as incomplete as iron absorption?

Mills: As long ago as 1962, Davis et al did some work on phytic acid and copper availability. They examined how the presence of soya bean protein affected the utilization of dietary zinc, copper and manganese in chicks. They found that the availability of copper for haemoglobin synthesis was reduced by the phytic acid present in the soya bean protein. This work has recently been repeated in Aberdeen in studies of net copper balance and retention in rats, and phytic acid was found to be almost as effective an inhibitor of copper uptake as it is of iron uptake.

Frieden: Does molybdenum affect the absorption of other metals?

Mills: I am not aware of any work on that.

References

Bremner I, Young BW, Mills CF 1979 Effect of ammonium tetrathiotungstate on the absorption and distribution of copper in the rat. Trans Biochem Soc 7:667-668

Campbell JK, Mills CF 1974 Effects of dietary cadmium and zinc on rats maintained on diets low in copper. Proc Nutr Soc 33:15A-16A

Davis PN, Norris LC, Kratzer FH 1962 Interference of soyabean proteins with the utilisation of trace minerals. J Nutr 77:217-223

Evans JL, Abraham PA 1973 Anaemia, iron storage and ceruloplasmin in copper nutrition in the growing rat. J Nutr 103:196-201

Grassman E, Kirchgessner M 1973 Zur Fe- und Cu-Verfügbarkeit im Stoffwechsel bei unterschiedlicher Fe- und Cu-Versorgung. Z Tierphysiol Tierernähr Futtermittelkd 31:113-120

Gray LF, Daniel LJ 1964 Effect of the copper status of the rat on the copper-molybdenum-sulfate interaction. J Nutr 84:31-37

McCarthy K, Lovenberg W 1968 Optical properties of artificial non-heme iron proteins derived from bovine serum albumin. J Biol Chem 243:6436-6441

Mills CF, Bremner I 1980 Nutritional aspects of molybdenum in animals. In: Coughlan M (ed) Molybdenum and molybdenum-containing enzymes. Pergamon, Oxford, p 517-542

Mills CF, Dalgarno AC, Wenham G 1976 Biochemical and pathological changes in tissues of Friesian cattle during the experimental induction of copper deficiency. Br J Nutr 35:309-331

Nozu T, Kitamura K 1977 Iron metabolism in hypocaeruloplasminaemic rats with special reference to the reticulo endothelial system. Showa Igakkai Zasshi 37:525-533

Österberg R 1974 Models for copper protein interaction based on solution and crystal structure studies. Coord Chem Rev 12:309-347

Österberg R 1975 A critical review of copper in medicine. International Copper Research Association, New York (INCRA report 234), p 1-70

Österberg R 1978 Från grundämnen till liv. In: The year book of the Swedish Natural Science Research Council, p 165-174

Whanger PD, Weswig PH 1970 Effect of some copper antagonists on induction of ceruloplasmin in the rat. J Nutr 100:341-348

Copper proteins and copper enzymes

A.E.G. CASS and H.A.O. HILL

Inorganic Chemistry Laboratory, South Parks Road, Oxford OX1 3QR, UK

Abstract The copper proteins that function in homeostasis, electron transport, dioxygen transport and oxidation are discussed. Particular emphasis is placed on the role of the ligands, their type and disposition which, in conjunction with other residues in the active site, determine the role of the copper ion. It is proposed that copper proteins can be considered in four groups. Those in Group I contain a single copper ion in an approximately tetrahedral environment with nitrogen and sulphur-containing ligands. Group II proteins have a single copper ion in a square-planar-like arrangement. Group III proteins have two copper ions in close proximity. Group IV consists of multi-copper proteins, composed of sites representative of the other three groups.

> *'Gold is for the mistress — silver for the maid —*
> *Copper for the craftsman, cunning at his trade.'*

But how well crafted are these biological copper vessels? Although we can barely recognize the outline of those unearthed, and although excavations still continue, we are here interested in how well fitted they are to their function. In addition to the capture and transport of copper, the principal roles of copper proteins concern the transfer of electrons and molecular oxygen (or dioxygen) and the catalysis of combined reactions of these two in oxidases. Those proteins most studied are described briefly in Table 1.

COPPER ACQUISITION AND TRANSPORT

The least-studied aspects of the multifarious roles of inorganic elements in biological systems are the mechanisms by which they are adsorbed, by which their levels are 'controlled' and by which they are transported and inserted into their appropriate sites. The first of these, adsorption, is 'poorly under-

TABLE 1

Copper proteins

Function	Examples
Acquisition Storage Transport	Metallothionein Caeruloplasmin
Dioxygen transport	Haemocyanin
Electron transfer	Azurin Plastocyanin Rusticyanin Stellacyanin
Monooxygenation	Monophenol monooxygenase (EC 1.14.18.1) Dopamine-β-monooxygenase (EC 1.14.17.1) p-Coumarate 3-monooxygenase (EC 1.14.17.2)
Dioxygenation	Quercetin 2,3-dioxygenase (EC 1.13.11.24) Indole 2,3-dioxygenase (EC 1.13.11.17) 2,3-Dihydroxybenzoate-2,3-dioxygenase (EC 1.13.11.28)
Superoxide dismutation	Superoxide dismutase (EC 1.15.1.1)
Substrate oxidation with $O_2 \rightarrow H_2O_2$	Galactose oxidase (EC 1.1.3.9) Amine oxidases (EC 1.4.3.6; 1.4.3.4)
Terminal oxidase	Laccase (EC 1.10.3.2) Caeruloplasmin (EC 1.16.3.1) Ascorbate oxidase (EC 1.10.3.3)

stood', a euphemism which barely disguises our ignorance. Homeostasis of copper is a role which some favour for metallothionein although toxicologists have recently claimed that it may function in metal ion detoxification (Kagi & Nordberg 1979). Whether or not it has a role in the detoxification of heavy metal ions, there is good evidence (Bremner 1980) that metallothionein is induced in response to the intake of divalent metal ions, particularly zinc(II), cadmium(II) and copper(II) although little is known about concentrations of thionein – the metal-free protein – before, during or after the introduction of the metal ion. Metallothionein is cysteine-rich, lacks aromatic amino acid residues and is remarkably well conserved. Indeed a similar small protein isolated from *Neurospora crassa* has a closely related amino acid sequence (Lerch 1979) although it is less than half the size of metallothioneins isolated from higher organisms. Several spectroscopic methods have been used to elucidate the structure of the metal-binding sites in metallothionein. The preponderance of cysteinyl residues invites comparison with the rubredoxins and the ferredoxins. The ratio of cysteinyl residues to zinc ions is 3:1 and seven zinc ions per molecule are commonly found; the available evidence does not implicate another ligand. Possible molecular clusters for zinc-metallothioneins are shown in Fig. 1(a). For the copper-metallothioneins the ratio of cysteinyl residues to copper ions is closer to 2:1 and a likely structure for these proteins is two interlocking tetrahedra, analogous to the four-iron ferredoxins (Fig. 1b). Such a structure could act as a transient store for copper; it would be relatively unreactive and not predisposed to catalysis, and the copper ions would be released, perhaps in concert, by a reduction in the local pH. However, much work remains to be done particularly on the induction of the protein. Of course we must always feel a little uneasy in consigning a protein to a transport role. To adapt an old adage, those who can, study function; those who can't, study structure; and those who can't do either, study transport.

ELECTRON TRANSPORT

We are on much more secure ground here since we have recently benefited from the X-ray diffraction studies of two copper redox proteins, poplar plastocyanin (Colman et al 1978) and azurin from *Pseudomonas aeruginosa* (Adman et al 1978). This is not the place to review the role of the blue copper

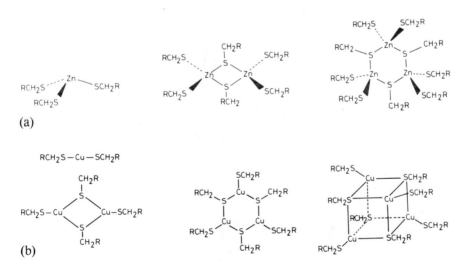

FIG. 1. (a) Possible structures of the metal-binding sites in zinc-metallothioneins. (b) Possible structures of the metal-binding sites in copper-metallothioneins.

proteins as a siren luring biochemists, spectroscopists and inorganic chemists into hazardous speculative shoals. Few got the answer completely wrong. However, not only is the structure worth seeing (Fig. 2); it is worth going to see (Freeman 1979). A reasonable hypothesis is that for fast electron transfer to occur, either there should be little associated movement of the nuclei at the redox centre, or electron transfer should be coupled to a vibration that links the two redox centres, or both. At the resolution available in X-ray diffraction it appears that the structure around the copper is a flattened tetrahedron which can be considered to be intermediate between the 'preferred' geometries of copper(II) and copper(I). Similarly the ligands can be considered to be a compromise that determines the redox potential, which influences the rate of

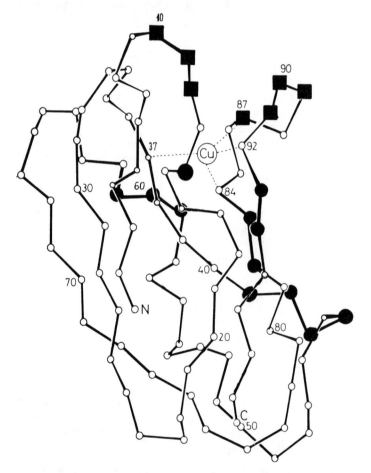

FIG. 2. The X-ray diffraction crystal structure of poplar plastocyanin (2.7 Å resolution).

electron transfer. Modification of the distortion, particularly of the ligands, will lead to a change in the redox potential. Not all blue copper proteins are alike. It may be that, once Nature had found a winning hand, only a shuffling among members of the same suit became necessary. Thus in stellacyanin, with $E_o = +184$ mV, the ligands may be two cysteinyl and two histidinyl residues (Hill & Lee 1979) whereas in rusticyanin ($E_o = +680$ mV) they may be two methioninyl and two histidinyl residues. A further refinement would include the coordination of the cysteine as the thiol rather than as the thiolate. In this way the standard electrode potential, which is the key variable, could be readily altered to fit the functional requirements (Fig. 3).

In common with other redox proteins for which reliable structural information exists, the blue copper proteins have the redox centre *at*, but not *on* their surfaces. Two surface sites have been identified for the binding of redox reagents to plastocyanin (Cookson et al 1980, Handford et al 1980): positively charged reagents bind in the vicinity of tyrosine 83 (to the lower right of the copper atom in Fig. 2) and reagents that are negatively charged bind at sites that lie over and above the copper. The relevance of these results to the biological transfer of electrons has yet to be established but they indicate the possibility of both inward and outward pathways for electron transfer to and from the protein's biological partners.

DIOXYGEN TRANSPORT

More than one solution exists to the problem of transport of dioxygen in biological systems. Of the proteins that have been characterized, the most

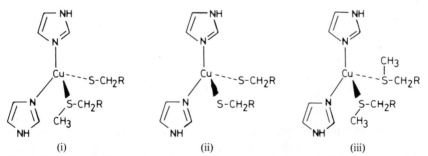

FIG. 3. Variations on the structural theme of electron transfer centres in 'blue' copper proteins. (i) Copper site as revealed by X-ray crystallography of azurin and plastocyanin; $E_o \sim 300$ mV. (ii) Possible structure of stellacyanin centre; $E_o = 184$ mV. (iii) Possible structure of rusticyanin centre; $E_o = 680$ mV.

studied are the globins, in which the single atom of iron, in conjunction with its immediate environment, is sufficient to bind reversibly one molecule of dioxygen. In each molecule of haemerythrin, the ratio of binding is one molecule of dioxygen to two atoms of iron (Okamura & Klotz 1973). Each of these transport systems probably involves electron transfer from the metal ion to dioxygen, to yield something akin to a superoxy complex for haemoglobin, and a peroxy complex for haemerythrin. Haemocyanin is found in invertebrates that belong to the phyla Mollusca and Arthropoda and it is the second most widely distributed pigment in the animal kingdom (Brunori et al 1979). A question often asked is: why is haemocyanin used physiologically? Though we are not here concerned with teleological questions it is a puzzle nonetheless, especially as myoglobin is also present (in the radular muscles) in most molluscs that contain haemocyanin. The structures of the haemocyanins are remarkably complex. Irrespective of their source and complexity, the ratio of dioxygen bound to copper is always one molecule O_2 : two atoms copper. The most straightforward interpretation of this ratio is that the binding is related to that in haemerythrin. To complete the analogy to iron would require that another copper protein exists in which the ratio of bound dioxygen and copper was 1 : 1.

OXIDATIVE COPPER PROTEINS

In addition to those copper proteins involved in electron transfer or in dioxygen transfer there is a class of proteins that couples together these two fundamental processes in substrate oxidation. Enzyme-catalysed oxidative reactions are of three general types; those catalysed by monooxygenases, by dioxygenases and by oxidases, shown schematically below (Hayaishi 1974).

$$SH + O_2 + 2H^+ + 2e^- \rightarrow SOH + H_2O \quad \text{Monooxygenation}$$

$$\text{or} \begin{cases} S + O_2 + 2H^+ + 2e^- & \rightarrow S(OH)_2 \\ SH_2 + O_2 & \rightarrow S(OH)_2 \end{cases} \text{Dioxygenation}$$

$$\text{or} \begin{cases} 2SH_2 + O_2 & \rightarrow 2S + 2H_2O \\ SH_2 + O_2 & \rightarrow S + H_2O_2 \end{cases} \text{Oxidation}$$

Copper-containing enzymes are represented in all three classes and may contain either a single copper site or multiple copper sites. A member of the former class of simple copper proteins has attracted much attention, especially since its mundane role as the transport protein, erythrocuprein (or hepatocuprein), has been replaced by its newer role as the enzyme superoxide dismutase (EC 1.15.1.1) (Fridovich 1975, Hassan 1980). This enzyme catalyses most effectively the disproportionation of the superoxide radical by a mechanism that is presumed to involve the alternate reduction and oxidation of the copper ion at the active site.

$$Cu(II) + O_2^- \longrightarrow Cu(I) + O_2$$

$$Cu(I) + O_2^- \xrightarrow{2H^+} Cu(II) + H_2O_2$$

$$\text{net effect: } 2O_2^- \xrightarrow{2H^+} O_2 + H_2O_2$$

Rate constants for each step are close to the diffusion control limit. At present superoxide dismutase is the only copper enzyme whose crystal structure is known (Richardson et al 1975). Fig. 4 shows the structure at 3 Å resolution; the immediate environment of the copper is shown in Fig. 5. One of the interesting features of this enzyme is that despite the complex structure of the active site and the high degree of conservation of this structure it is no better a catalyst for superoxide dismutation than the aquated copper(II) ion. The reasons why such an enzyme is used for this apparently simple reaction have recently been discussed (Cass & Hill 1980). Only a macromolecule can combine under physiological conditions the dual, and, to some degree, the mutually exclusive properties of tight binding of the metal ion and free access of substrate. In bovine superoxide dismutase this may be enhanced by the inclusion of an arginine at the active site, thereby increasing the rate of the catalysed reaction, which is otherwise slowed by chelation of the metal ion in the protein. This is a good example of a metalloenzyme in which the active site is more than just the metal ion and its ligands.

The basic structure of the copper site in superoxide dismutase can be considered to be an archetype of the metal site provided by the protein in the copper-containing oxygenases. Thus, while the copper site retains a symmetry related to a square-plane, variations in the ligands, their type and charge, can modify the oxidation–reduction potential of the protein and the thermodynamics and kinetics of ligand binding and substrate binding. Additional selectivity can be introduced by nearby active site residues.

In using copper as the active centre of mono- or dioxygenases, Nature has been extremely frugal, in contrast to her lavish employment of either iron- or flavin-containing enzymes. Indeed the characterized copper-containing oxygenases are distinguished not only by their small number but also by the disparity of the reactions that they catalyse. The extracellular enzyme

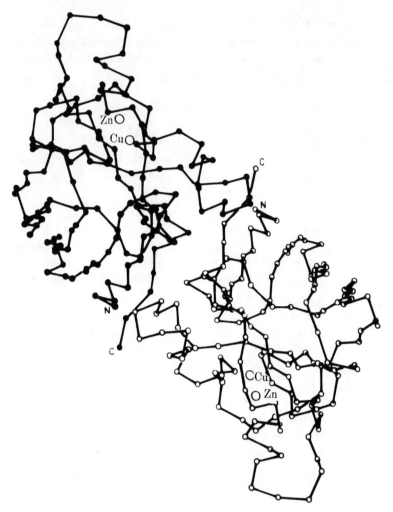

FIG. 4. The X-ray diffraction crystal structure of bovine erythrocyte superoxide dismutase (3 Å resolution).

quercetin 2,3-dioxygenase catalyses the dioxygenation of quercetin, forming carbon monoxide as one of its products. No reducing agents are necessary. In contrast to this the mammalian enzyme dopamine-β-monooxygenase (dopamine-β-hydroxylase) requires the addition of ascorbate as a source of reducing equivalents. The third enzyme of this group, monophenol monooxygenase (or tyrosinase), derives the required reducing equivalents from oxidation of the product of the monooxygenation (Vanneste & Zuberbühler 1974). These three reactions are shown in Fig. 6.

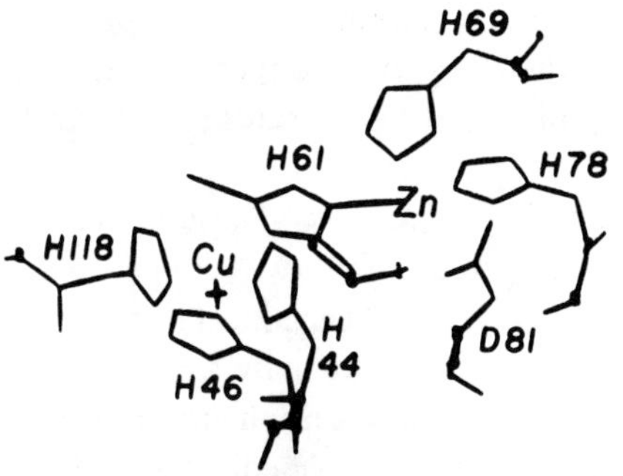

FIG. 5. Detail of the metal-binding sites in copper–zinc-superoxide dismutase.

Very little is known about quercetin 2,3-dioxygenase and we shall not discuss it further except to mention that carbon monoxide is a good ligand for copper(I) but not for copper(II). Therefore breakdown of the proposed intermediate, a cyclic peroxide, in the reaction catalysed by quercetin 2,3-dioxygenase, would be expected to occur when the copper centres are in

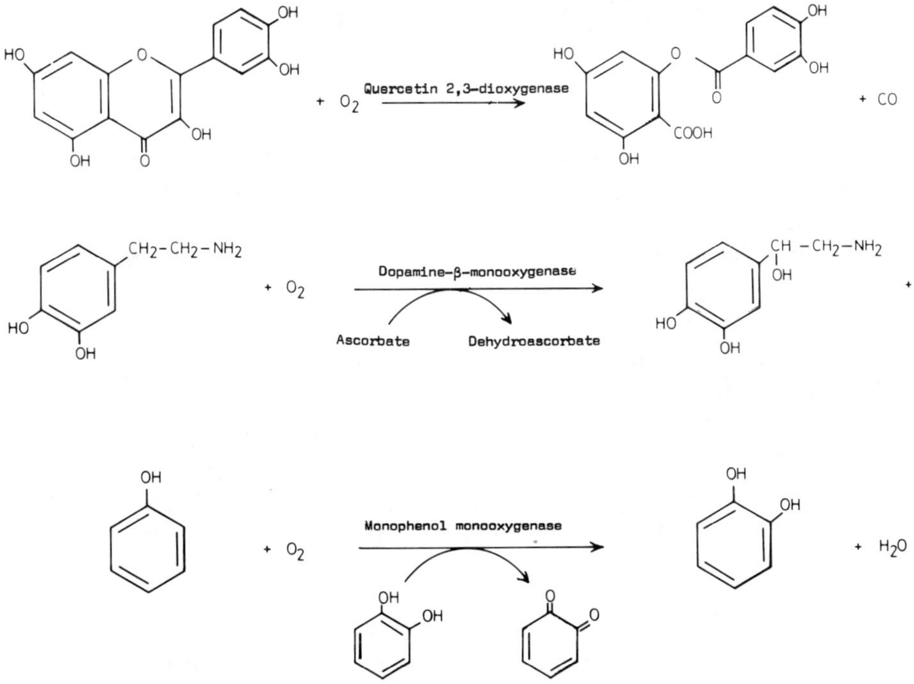

FIG. 6. Reactions catalysed by the copper-containing oxygenases.

the Cu^{2+} state, to avoid possible auto-inhibition. The state of the copper ions in this enzyme is not well characterized; there are two copper(II) ions per molecule and they are not reduced during binding of substrates (Vanneste & Zuberbühler 1974),

Monophenol monooxygenase and dopamine-β-monooxygenase appear to have a similar function and an active site containing two copper ions, although for monophenol monooxygenase this is not yet certain. The reaction scheme seems to include the reduction of the copper either by ascorbate (in dopamine-β-monooxygenase) or by an o-diphenol (in monophenol monooxygenase). The reduction is followed by reaction of the intermediate with dioxygen to yield a complex that is analogous to oxyhaemocyanin, although the complex with a 1:1 ratio of copper to dioxygen, referred to earlier, may also be formed. With haemocyanin the dioxy complex is stable; with the monooxygenases it is broken down by the substrate to water and product.

The copper-containing oxidases that reduce dioxygen to hydrogen peroxide are probably related to superoxide dismutase by the nature of their active sites. The reaction catalysed by amine oxidases is an oxidative deamination (Vanneste & Zuberbühler 1974),

$$RCH_2NH_2 + O_2 + H_2O \longrightarrow RCHO + NH_3 + H_2O_2$$
(R = alkyl or aryl group)

and the enzymes contain pyridoxal phosphate as well as copper. The role of the metal ion in these enzymes is not well defined. The observations that the amine–enzyme complex still contains copper(II) under anaerobic conditions and the ping-pong kinetics, which reveal release of product before binding of dioxygen, suggest that the metal ion functions in coenzyme reoxidation.

Galactose oxidase converts the *C*-6 hydroxyl group on galactose to the aldehyde (Bereman et al 1977). There is considerable controversy about the state of the copper in the mechanism of action of galactose oxidase; indeed every permutation from a constant redox state to an interconversion between Cu(I) and Cu(III) has been postulated. The resting form of the enzyme has an e.p.r. spectrum that is consistent with the expected structure of this type of copper: two nitrogen and two oxygen atoms in an approximately square-planar arrangement around the copper ion (Bereman et al 1977).

MULTI-COPPER PROTEINS

The most striking feature of the two most thoroughly studied multi-copper proteins, laccase (Malmström 1978) and cytochrome *c* oxidase (Nicholls &

Chance 1974) is the elegant mechanism by which four one-electron transfer reactions are converted into one four-electron transfer reaction by the device of four redox centres in the proteins – four copper ions in laccase and two copper ions and two haem moieties in cytochrome c oxidase.

Caeruloplasmin may also fall into this category. At first it was considered simply as a copper transport protein and, more recently, as an oxidase with perhaps iron(II) as the physiological substrate. This protein is discussed in greater detail elsewhere (Frieden 1980, this volume) and although some consider that the protein is much too complex for the apparently simple task of oxidation of iron(I), its complexity may be associated more with its other substrate, dioxygen. The simple one-electron reduction of O_2 (by e.g. Fe(II)) can lead, via the formation of the superoxide ion (O_2^-) to the highly reactive and damaging hydroxyl radical (Halliwell & Foyer 1976). Therefore the complexity of caeruloplasmin may be required to ensure the direct reduction of O_2 to H_2O,

$$4Fe^{2+} + O_2 \xrightarrow{4H^+} 4Fe^{3+} + 2H_2O$$

without the release of noxious intermediates. Laccase and cytochrome c oxidase also avoid the release of oxygen-derived intermediates. A simple explanation for the mechanism of these enzymes is that one copper site with properties not unlike those of azurin or plastocyanin (though the E_o of this centre is closer to that of rusticyanin; see Fee 1975, Cox et al 1978) acts as a conduit for electrons to the other site, perhaps somewhat removed, at which dioxygen binds. Presumably two of the copper atoms at this site are so close that their magnetic properties are those of an antiferromagnetic pair, and therefore they may resemble the copper–copper pair in haemocyanin. There are obviously many ways in which the environment provided by the protein could perturb this structure so as to make the reverse reaction (the release of dioxygen) very slow. One method would be to employ the fourth copper atom to act both as the final link in this electron-transfer chain and to stabilize further reduction products and transition states leading to the peroxy intermediate (Reinhammar 1979). It takes a certain amount of *chutzpah* even to comment on the vast amount of work and varied opinions on cytochrome c oxidase! This enzyme has some features in common with laccase; it contains one copper site that functions in electron transfer to a complex dioxygen-binding site, which comprises the second copper in close proximity to one of the haem moieties (Tweedle et al 1978).

CONCLUSIONS

Copper proteins have previously been discussed as Type 1, Type 2 and Type 3. This classification is based on the physical properties, particularly spectroscopic, of the copper ions (Fee 1975). These physical properties are determined by the structures of the copper ions as influenced by their environments – principally, but not exclusively, local ones. As our knowledge of the copper proteins increases the classification will be revised to one based on molecular structure. From the evidence now available the proteins can be classed into four main groups with a number of sub-groups.

Group I proteins have a single copper ion in a structure with a distorted tetrahedral symmetry. The ligands provided by the protein are histidine (probably two residues) and one or two sulphur-containing groups, cysteine or methionine, or both. A number of combinations are possible, especially if the thiol form of cysteine can act as a ligand. This site is designed for rapid electron transfer.

Group II comprises proteins in which the structure around a single copper has a symmetry related to a square-plane. A large number of sub-groups is possible, depending not only on the number and type of ligands but also on the non-ligand environment provided by the protein. The ligand permutations will allow a variety of functions in which copper can act not only as a redox centre but also as a Lewis acid. The ability to bind ligands, whether these are derived from dioxygen or organic substrates, will depend very much on the nature of the other ligands to the copper. The spectroscopic properties will also reflect these structural details but will be similar to those of simple copper(II) complexes since these often have a symmetry related to that of a square-planar structure.

Group III contains proteins that have a single copper–copper pair. Again a wide variety of sub-groups is possible, depending on the number and nature of the ligands, the number of bridging groups and the presence of other charged groups in the active site. The function of this site is to bind dioxygen; the reversibility of the reaction can be modulated by the particular structure and local environment.

The Super-Group IV contains proteins that have one or more of their sites corresponding to those in Groups I–III. The combination of an efficient electron-transfer site (Group I) with a dioxygen binding site (Group III) in the vicinity of combined electron-transfer or Lewis acid site (Group II) allows for the catalysis of more complex reactions. Given the number of sub-groups in (I), (II) and (III), many such combinations are possible and it will be interesting to learn how many of them are used physiologically.

Copper enzymes do not consist simply of the metal ion with a few amino acid side-chains making up its coordination sphere. The protein not only provides the metal ligands but also controls the nature of the microenvironment around the metal ion and its immediate coordination sphere. Functions of this type include modification of the polarity of the active site, control of the hydration and dehydration of the substrate and provision of charged groups as general acid or base catalysts or as agents to neutralize or promote charge separation during the course of the reaction.

For most functions undertaken by a copper protein there exists a counterpart in the biochemistry of iron (Table 2). Although iron is used more widely than copper, it is in conjunction with non-proteinaceous ligands, the porphyrins, that iron is able to express fully its inherent properties.

TABLE 2

Biological roles of iron and copper proteins

Function	Cu-Protein	Haem Fe-Protein	Non-Haem Fe-Protein
Acquisition Storage Transport	Caeruloplasmin Metallothionein	–	Ferritin Transferrin
Dioxygen transport	Haemocyanin	Haemoglobin Myoglobin	Haemerythrin
Electron transfer	Azurin Plastocyanin	Cytochromes	Ferredoxins
Monooxygenation	Monophenol monooxygenase Dopamine-β- monooxygenase	Cytochrome P-450	Benzoate-4-mono- oxygenase (EC 1.14.13.12)
Dioxygenation	Quercetin 2,3-dioxygenase	Indole 2,3-dioxygenase (EC 1.13.11.17)	Lipoxygenase (EC 1.13.11.12)
Superoxide dismutation	Superoxide dismutase	–	Superoxide dismutase
Oxidation of substrate with $O_2 \rightarrow H_2O_2$	Galactose oxidase Amine oxidases	–	–
Terminal oxidase	Laccase Caeruloplasmin Ascorbate oxidase	Cytochrome c oxidase (EC 1.9.3.1)	–

> *'Gold is for the mistress — silver for the maid —*
> *Copper for the craftsman, cunning at his trade.'*
> *'Good!' said the Baron, sitting in his hall,*
> *'But Iron — Cold Iron — is master of them all.'*
>
> Rudyard Kipling, *Cold Iron*

ACKNOWLEDGEMENTS

This is a contribution from the Oxford Enzyme Group, of which one of us, H.A.O.H., is a member. We thank the Science Research Council, the Medical Research Council, The British Heart Foundation and the Wellcome Trust for support.

References

Adman ET, Stenkamp RE, Sieker LC, Jensen LH 1978 A crystallographic model for azurin at 3 Å resolution. J Mol Biol 123:35-47

Bereman RD, Ettinger MJ, Kosman DJ, Kurland RJ 1977 The copper(II) site in galactose oxidase. Adv Chem Ser 162:263-280

Bremner I 1980 Absorption, transport and distribution of copper. In: Biological roles of copper. Excerpta Medica, Amsterdam (Ciba Found Symp 79), p 23-48

Brunori M, Girdina B, Bannister JV 1979 Respiratory proteins. In: Hill HAO (senior reporter) Inorganic biochemistry: a specialist periodical report. The Chemical Society, London, vol 1:159-209

Cass AEG, Hill HAO 1980 Anion binding to copper(I) superoxide dismutase: a high resolution ^{1}H nuclear magnetic resonance spectroscopic study. In: Bannister JV, Hill HAO (eds) Chemical and biochemical aspects of superoxide and superoxide dismutase. Elsevier/North-Holland, Amsterdam (Developments in biochemistry series, vol 11A), p. 290-298

Colman PM, Freeman HC, Guss JM, Murata M, Norris VA, Ramshaw JAM, Venkatappa MP 1978 X-ray crystal structure analysis of plastocyanin at 2.7 Å resolution. Nature (Lond) 272:319-324

Cookson DJ, Hayes MT, Wright PE 1980 Electron transfer reagent binding sites on plastocyanin. Nature (Lond) 283:682-683

Cox JC, Aasa R, Malmström BG 1978 EPR studies on the blue copper protein rusticyanin: a protein involved in Fe^{2+} oxidation at pH 2.0 in *Thiobacillus ferro-oxidans*. FEBS (Fed Eur Biochem Soc) Lett 93:157-160

Fee JA 1975 Copper proteins — systems containing the 'blue' copper centre. Struct Bonding 23:1-60

Freeman HC 1979 Elegance in molecular design: The copper site of a photosynthetic electron-transfer protein. J Proc R Soc NSW 112:45-62

Frieden E 1980 Caeruloplasmin: a multi-functional metalloprotein of vertebrate plasma. In: Biological roles of copper. Excerpta Medica, Amsterdam (Ciba Found Symp 79), p 93-124

Fridovich I 1975 Superoxide dismutases. Annu Rev Biochem 44:147-159

Halliwell B, Foyer CH 1976 Ascorbic acid, metal ions and the superoxide radical. Biochem J 155:697-700

Handford PM, Hill HAO, Lee W-K, Henderson RA, Sykes AG 1980 Investigation of the binding of inorganic complexes to blue copper by ^{1}H NMR spectroscopy I. The interaction between the $Cr(phen)^{3+}$ and $Cr(CN)_6^{3-}$ ions and the Cu(I) form of parsley plastocyanin. J Inorg Biochem 13:83-88

Hassan HM 1980 Superoxide dismutase. In: Biological roles of copper. Excerpta Medica, Amsterdam (Ciba Found Symp 79), p 125-142
Hayaishi O 1974 (ed) Molecular mechanisms of oxygen activation. Academic Press, New York
Hill HAO, Lee W-K 1979 Investigation of the structure of the blue copper protein from *Rhus vernicifera* stellacyanin by ^{1}H nuclear magnetic resonance spectroscopy. J Inorg Biochem 11:101-113
Kägi JHR, Nordberg M 1979 (eds) Metallothionein. Birkhäuser Verlag, Basel (Exper Suppl (Basel) 34)
Lerch K 1979 Amino acid sequence of copper metallothionein from *Neurospora crassa*. In: Kägi JHR, Nordberg M (eds) Metallothionein. Birkhäuser Verlag, Basel (Exper Suppl (Basel) 34), p. 173-179
Malmström BG 1978 Copper containing oxidases. In: da Silva JRRF, Williams RJP (eds) New trends in bioinorganic chemistry. Academic Press, London
Nicholls P, Chance B 1974 Cytochrome *c* oxidase. In: Hayaishi O (ed) Molecular mechanisms of oxygen activation. Academic Press, New York, p 479-534
Okamura MY, Klotz IM 1973 Hemerythrin. In: Eichhorn GL (ed) Inorganic biochemistry. Elsevier, Amsterdam, vol 1:320-343
Reinhammar B 1979 Laccase. In: Eichhorn GL, Marzilli LG (eds) Advances in inorganic biochemistry. Elsevier/North-Holland, New York, vol 1:91-118
Richardson JS, Thomas KA, Rubin BH, Richardson DC 1975 Crystal structure of bovine Cu, Zn superoxide dismutase at 3 Å resolution – chain tracing and metal ligands. Proc Natl Acad Sci USA 72:1349-1353
Tweedle MF, Wilson LJ, Garcia-Iniguez L, Babcock GT, Palmer G 1978 Electronic state of heme in cytochrome oxidase III. J Biol Chem 253:8065-8071
Vanneste WH, Zuberbühler A 1974 Copper-containing oxygenases. In: Hayaishi O (ed) Molecular mechanisms of oxygen activation. Academic Press, New York, p 371-404

Discussion

Dormandy: You mentioned the role of copper in electron transport and in electron transfer, Dr Hill. Perhaps one also ought to mention its role in electron trapping. I think one of the main functions of copper proteins could be to protect against 'loose' electrons (i.e. free radicals) which are potentially harmful. This role is essentially an electron-trapping function rather than a transport function (Dormandy 1980). Copper is generally useful biologically when it is combined with protein. What about smaller complexes of copper with amino acids or peptides? Do you think they have any role?

Hill: I have been considering only the catalytic role of the complexes. If I could just emphasize your first point – colloquially one can describe copper redox proteins, e.g. azurin or plastocyanin as transient resting places for electrons. You were correct to refer to radicals as rather dangerous entities; the availability of redox traps may therefore be quite important.

Österberg: I would like to ask Dr Hill a question about electron transport. During the reaction of oxygen with cytochrome *c* oxidase, equal numbers of protons and electrons become bound to oxygen. Do you have a model for the addition of protons similar to those you showed for the addition of electrons?

Hill: No, I don't. I think that if we consider the terminal oxidase as a capacitor, it can probably be discharged only by the protons. The protein cytochrome *c* oxidase can be reduced fully but it, or rather the complex formed with dioxygen (O_2), will discharge (in the sense of adding four electrons to oxygen) only in the presence of protons. This is because after the first electron transfer to oxygen, subsequent electron transfers require very high energy, except in the presence of protons or metal ions.

Österberg: Water is generally the protein solvent and so one can normally think of the protons as coming from the solvent (and it is presumably possible to check whether the protons come from the solvent or from the protein). Do you think the protons that are added to the oxygen come from the protein, so that there is a specific role for the protein in the addition of both protons and electrons to the oxygen, or do the proteins provide electrons, while the protons come from the solvent?

Hill: I don't know the answer to that. The properties of water at an active site are not necessarily related to the properties of bulk water. For example, in many active sites there are relatively small numbers of water molecules, which usually bind by some kind of hydrogen bonding to the protein surface of the active site. It has been speculated (see Cass & Hill 1980) that when proton transfer is involved (as it is in the activity of the dismutase), the water is organized and must be considered as an integral part of the active site. This would enable rapid proton transfer to occur from the bulk solvent. We are forced to consider this possibility, particularly with an enzyme like superoxide dismutase (SOD) which is a very effective catalyst for this reaction. The rate of the reaction is close to the diffusion limit and therefore no single step can be far from the diffusion rate.

Frieden: In the development of iron in biological systems, the extreme insolubility of ferric iron may have 'forced' the protoporphyrin mould on to iron in many reactions (Frieden 1974) whereas the problem has not been quite as severe with copper.

Hill: The problem is there with copper, but as you say it is not too severe. Whether your explanation is the reason or not may be impossible to confirm. The porphyrin ligand allows Nature to do with iron what she can do quite easily with copper and a protein alone.

Harris: There is no question that copper has to be permanently bound to a protein structure in order to perform some of its biological functions. But in the exchange, transport or delivery of copper, we are possibly dealing with ligand exchange reactions. What types of site would favour the ready exchange of copper from one protein to another?

Hill: If the number of ligands were low and if they were uncharged, then rapid exchange of copper would be favoured, whereas charged ligands and a high coordination number would tend to slow down the process. When one uses models of SOD *in vivo,* e.g. small copper complexes like those with aspirin, tyrosine and penicillamine, the problem is that some of the complexes will rapidly lose their copper. Indeed, Sorenson (1976) has speculated that the ability of these small complexes to act as dismutases depends on this loss of copper, because the free copper, rather than the complex, could act as the catalyst. However, copper is not lost from the dismutase until about pH 4.5, so at pH 7 the exchange rate is very slow.

Harris: What type of energy barriers have to be overcome in order to get copper into the proper binding site on the protein; can this be done by a simple mixing of the protein with the metal?

Hill: We have found (Cass et al 1979) that when zinc and copper are removed from SOD, only the immediate environment of the copper is affected; the β-barrel structure remains unperturbed. After zinc is bound to the enzyme, the structure of the protein is essentially that of the native protein. In other words, the copper site is preformed by the zinc.

Mills: Are you suggesting, Dr Hill, that when the system is trying to synthesize SOD in the presence of a low concentration of zinc, steric considerations would preclude the incorporation of copper, even if it were present?

Hill: No; loose or randomly distributed copper could be present. We believe that one of the reasons why there are two subunits in SOD is that there are also metal binding sites at the subunit interface, and when zinc binds, these sites are shuffled and blocked off.

Frieden: But surely no other metal will replace copper in that system?

Hill: No other metal will replace copper whilst allowing the enzyme to retain its activity (Beem et al 1974), but other metals will certainly bind to the enzyme.

Mills: Is there any evidence that SOD activity is influenced by zinc status in living organisms?

Hill: No, and there have been no studies on apodismutase either; usually only the enzymic activity is examined; the concentrations of holoprotein or apoprotein are not measured.

Hassan: Surely the activity of SOD is not impaired by removal of the zinc from the enzyme? (Forman & Fridovich 1973.)

Hill: Yes, it is. But if the zinc is removed copper can bind to the zinc site.

Mills: With what consequences?

Hill: In terms of the copper content, the catalytic activity of the protein that contains two copper atoms per subunit is about 50% that of the native enzyme. I suspect that the copper bound to the 'zinc' site would be labile.

Dormandy: Am I correct in believing that the CuSOD that you mentioned is only one of a large family of superoxide dismutases, some of which contain metals other than copper?

Hill: Yes, indeed. There are (Hill 1978) iron dismutases (FeSOD) and manganese dismutases (MnSOD), which do not require other metals for their function. The resting states seem to be Mn(III) and Fe(III). It is believed that there is only one manganese atom per dimer, which is different from the CuSOD. We shall hear more about this from Dr Hassan's paper.

Frieden: Isn't it true that the Cu-ZnSOD in the cytoplasm is present in much greater concentrations than the other dismutases, and that it therefore might be the most important one physiologically?

Hill: Even people working on dismutase might take issue with that view because assault on a eukaryote system by oxygen, or by entities derived from oxygen, seems to affect the manganese form of the enzyme more than the copper form (Stevens & Autor 1977).

Hurley: When we studied manganese-deficient chickens we found a reduced activity of MnSOD but an increased activity of Cu-ZnSOD (de Rosa et al 1980). This implies that there is some interaction between the two enzymes in their protective function for the animal. We are now trying to find out what happens in zinc-deficient animals.

Hassan: In the manganese-deficient chicks, was the increase in Cu-ZnSOD in the mitochondria or in the cytosol? You have implied that the Cu-ZnSOD will replace the MnSOD, which has a mitochondrial function. I don't see how this would be possible.

Hurley: There is some recent work that suggests that the functional distinction between the cytosolic and mitochondrial enzymes is not as clear-cut as was formerly believed (de Rosa et al 1980).

Hassan: I believe that there is evidence that supports the reverse possibility, i.e. if there is a copper deficiency, MnSOD may be present in the cytosol (Shatzman & Kosman 1978).

Bremner: Dr Hill, why is it that zinc is so often found as a secondary metal on metallothionein? When synthesis of metallothionein is induced by copper, cadmium or by other metals, zinc is also incorporated into the protein. Do you think that zinc may have some part to play in maintaining the structure of the protein?

Hill: No. Some people claim that metallothionein is a detoxifying protein. If a site for copper or zinc attachment is present, then cadmium will almost certainly bind there, because the site is not very selective. Dr Österberg was talking earlier about the selectivity between the metal ions and the amino acids, but it is likely that for the sulphur-containing amino acids, at any rate, there might be a considerable loss of selectivity.

Österberg: There are two different forms of the sulphur-containing amino acids. In the blood plasma, the sulphur is present generally in the form of disulphides and so 'the sulphur' is more or less inactive. But within the cell sulphur is, to a certain extent, in the form of the thiol groups of glutathione and cysteine, and here Cu(II) may be reduced simultaneously with disulphide formation. Some years ago we found that Cu(I) binds very weakly to amino acids and peptides except to those which contain a thiol group (R. Österberg, unpublished results). When copper ions are bound to dismutase they are fairly inert although the dismutase binds the copper ions via nitrogen ligand atoms that in general bind Cu(I) more weakly than thiol groups do. Cu(II) binds to amino acids a million times more strongly than zinc(II) does, while zinc ions bind a thousand times more strongly than cadmium ions. On the other hand, cadmium, zinc and Cu(II) might exhibit the reverse stability in binding, for instance, to metallothionein, which contains thiol ligands (Sillén & Martell 1964, 1971); thus cadmium would be bound preferentially to zinc. But I am not sure about the order of preference between zinc and Cu(II), since Cu(II) may be reduced to Cu(I).

Mills: I want to enquire about the relationship between copper and changes in peroxide or hydroxyl free radicals in biological systems. The effects of copper deficiency on systems that prevent the accumulation of these species can be predicted by arguments similar to those used to explain the adverse effects of selenium deficiency on lipid peroxidation and membrane integrity. If these arguments are taken to their conclusion, then a selenium deficiency should aggravate copper deficiency and an adequate amount of selenium should partially protect against copper deficiency. Is there any evidence of mitochondrial or pathological changes to support this idea? (I believe that these two deficiencies have some common effects on neutrophil function (Boyne & Arthur 1981), but as far as I know there are no other examples.)

Hill: Well, we know (Flohé 1976) that one of the roles of glutathione peroxidase is to protect against the noxious effects of reduced oxygen.

Mills: Surely α-tocopherol (vitamin E) should protect against some effects of both copper and selenium deficiencies and not just against selenium deficiency? Perhaps no one has ever examined this.

Dormandy: Antioxidant protection is complex and multifactorial (Dormandy 1978). One protective agent does not necessarily substitute for another; but there has been little study of what happens when a protective agent like α-tocopherol is given in large pharmacological doses rather than in the small doses given to correct specific vitamin deficiency. There is some evidence to suggest that α-tocopherol may protect against selenium deficiency if it is given in doses 1000-fold higher than those normally administered to supposedly vitamin E-deficient subjects. There is a possible analogy here with the small doses of adrenocortical hormones that are needed in replacement therapy and the much larger doses required when cortisol is used as a drug in arthritis.

Mills: Recent observations have shown that more than just a simultaneous deficiency of selenium and vitamin E is needed to provoke myopathy (Arthur 1979).

Hurley: Are you assuming that all of the effects of copper deficiency are caused by the lack of antioxidant activity?

Mills: No, obviously one could not claim that. I was simply trying to determine what metabolic similarities there might be between copper deficiency and selenium deficiency, particularly with respect to the appearance of potentially damaging free radicals generated by reactions involving oxygen.

References

Arthur JR 1979 Selenium deficiency in Friesian steers. Proc Nutr Soc 38:13A
Beem KM, Rich WE, Rajagopalan KV 1974 Total reconstitution of copper-zinc superoxide dismutase. J Biol Chem 249:7298-7305
Boyne R, Arthur JR 1981 Effects of selenium and copper deficiencies on neutrophil function in cattle. J Comp Pathol 91(2): in press
Cass AEG, Hill HAO, Bannister JV, Bannister WH 1979 Zinc(II) binding to apo-(bovine erythrocyte superoxide dismutase). Biochem J 177:477-486
Cass AEG, Hill HAO 1980 Anion binding to copper(I) superoxide dismutase: a high resolution ^{1}H nuclear magnetic resonance spectroscopic study. In: Bannister JV, Hill HAO (eds) Chemical and biochemical aspects of superoxide and superoxide dismutase. Elsevier/North-Holland, Amsterdam (Developments in biochemistry series, vol 11A) p 290-298
de Rosa G, Keen CL, Leach RM, Hurley LS 1980 Regulation of superoxide dismutase activity by dietary manganese. J Nutr 110:795-804
Dormandy TL 1978 Free radical oxidations and antioxidants. Lancet 1:247-250
Dormandy TL 1980 Free radical reactions in biological systems. Ann R Coll Surg Engl 62:188-194
Flohé L 1976 Proceedings of the symposium on selenium and tellurium in the environment. Industrial Health Foundation Inc, p 138-157
Forman HJ, Fridovich I 1973 On the stability of bovine superoxide dismutase. The effect of metals. J Biol Chem 248:2645-2649
Frieden E 1974 The evolution of metals as essential elements (with special reference to iron and copper). Adv Exp Med Biol 48:1-29
Hill HAO 1978 The superoxide ion and the toxicity of molecular oxygen. In: da Silva JRRF, Williams RJP (eds) New trends in bioinorganic chemistry. Academic Press, London, p 173-208
Shatzman AR, Kosman DJ 1978 The utilization of copper and its role in the biosynthesis of copper-containing proteins in the fungus, *Dactylium dendroides*. Biochim Biophys Acta 544:163-179

Sillén LG, Martell AE 1964 and 1971 Stability constants, 2nd edn and Suppl No 1. Stability constants of metal-ion complexes. The Chemical Society, London (Chem Soc Spec Publ No 17 and 25)

Sorenson JRJ 1976 Copper chelates as possible forms of antiarthritic agents. J Med Chem 19:135-148

Stevens JB, Autor AP 1977 Oxygen-induced synthesis of superoxide dismutase and catalase in pulmonary macrophages of neonatal rats. Lab Invest 37:470-478

Caeruloplasmin: a multi-functional metalloprotein of vertebrate plasma

EARL FRIEDEN

Department of Chemistry, Florida State University, Tallahassee, Florida 32306, USA

Abstract Caeruloplasmin is a blue copper protein found in the α_2-globulin fraction of vertebrate plasma. It is a single-chain glycoprotein of molecular weight 132 000. It contains six copper atoms per molecule, comprising three or possibly four different types of copper. Its many functions may be related to the heterogeneous nature of these six copper atoms and to the various catalytic activities which they provide.

Caeruloplasmin resembles albumin and transferrin in that all three serum proteins are regarded primarily as transport proteins. However, each has numerous other actions as important as this transport function.

Caeruloplasmin directly mobilizes iron into the serum and provides the major molecular link between copper and iron metabolism; it is the most prominent serum antioxidant, preventing deleterious oxidation of polyenoic acids and other substrates; it scavenges superoxide radicals; it serves as an acute-phase reactant (an endogenous modulator) of the inflammatory response; finally, caeruloplasmin may regulate the serum concentration of the biogenic amines, adrenaline (epinephrine) and serotonin (5-HT).

SOME EVOLUTIONARY IMPLICATIONS

A blue protein from the α_2-globulin fraction of human serum, and possessing oxidase activity, was first reported by Holmberg (1944). It was named caeruloplasmin, or blue substance from plasma, by Holmberg & Laurell (1948, 1951) in later papers which described its purification and its chemical properties. Since caeruloplasmin accounts for over 95% of the circulating copper in a normal mammal and since its concentration fluctuates extensively in numerous diseases and hormonal states, its study has excited the imagination of biomedical scientists who have generated several thousand papers on its chemistry and biology. Caeruloplasmin is also an attractive protein for study because, like serum albumin and transferrin, it is multi-functional.

Caeruloplasmin has numerous actions that compare in importance to its role as the principal copper transport protein. These versatile activities cannot be related to any subunit structures since the caeruloplasmin molecule is a single-chain protein, yielding molecular fragments only after the proteolysis of susceptible peptide bonds. Rather, its many functions appear to be related to the various catalytic activities of its six copper atoms. I shall summarize the functions of caeruloplasmin and the current information that has extended the physiological significance of this copper protein. Two more complete reviews describing much of the earlier work on caeruloplasmin have appeared in recent years (Frieden & Hsieh 1976, Frieden 1979).

Caeruloplasmin may represent a current end-point in the parallel development of copper and iron biochemistry in the natural selection of aerobic cells during the past three billion years (Fig. 1). That copper and iron have dominated not only the recent unfolding of human civilization, but also the lengthy evolution of essential metalloproteins and metalloenzymes, is a remarkable coincidence. In a review tracing the evolution of the essential metal ions, Frieden (1974) has pointed out the many close associations between copper and iron that have evolved in aerobic cells. Enzymes developed to protect cells from unavoidable toxic oxygen by-products – superoxide ion, singlet oxygen, and hydrogen peroxide – hence the ubiquitous occurrence of

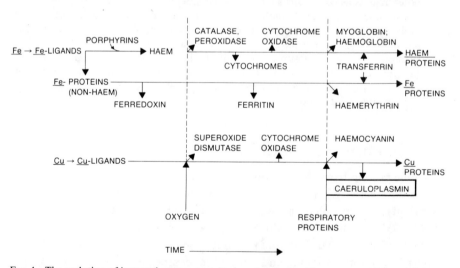

FIG. 1. The evolution of iron and copper proteins in animal cells. The sequence of events (left to right) starts with chelation of copper and iron in a reducing environment and ends with the complex metalloenzymes and metalloproteins necessary for storage, transport, and maintenance of respiration in an oxidizing environment. The horizontal axis indicates time; the advent of oxygen and respiratory proteins is marked.

the copper–zinc enzyme (superoxide dismutase, EC 1.15.1.1) and the haem enzymes (catalase, EC 1.11.1.6, and peroxidase, EC 1.11.1.7). The success of the aerobes was accompanied by the development of more sophisticated iron and copper enzymes, notably the cytochromes, cytochrome c oxidase (EC 1.9.3.1) and the numerous electron transferases in plants. As organisms increased in complexity, the cellular machinery utilizing iron and copper expanded to produce the oxygen-carrying proteins – haemoglobins, haemerythrins and haemocyanins. This adaptation required the elaboration of storage and transport proteins exclusively for copper and iron: caeruloplasmin, ferritin, and transferrin. During later stages of evolution crucial biosynthetic enzymes appeared in association with connective tissue and other more specific processes. A final example of the continuing close connection between iron and copper in the vertebrates is the ability of the copper protein of plasma, caeruloplasmin, to mobilize iron into transferrin for iron transport and distribution (Frieden 1971).

Dawson et al (1975) have discussed some interesting parallels that occur between caeruloplasmin and two other blue copper oxidases, ascorbate oxidase (EC 1.10.3.3) and laccase (EC 1.10.3.2), despite their different origins and biological roles. They note that all three of these multi-copper proteins have in common an oxidase activity towards ascorbate, although ascorbate oxidase has 10^3–10^4 times more enzymic activity than laccase or caeruloplasmin. Dawson et al (1975) also point out numerous similarities in various categories of amino acids (aromatic, hydrophobic, acidic, basic and bivalent), particularly between ascorbate oxidase and caeruloplasmin, and similarities in copper content (0.30%) as well. More information is needed to establish whether there is a close homology or common tertiary or quaternary structure between these three copper oxidases. On the basis of its more versatile substrate range, laccase is the primitive type of blue oxidase and perhaps it protected the early aerobes against excess oxygen. As the aerobes evolved the ability both to utilize molecular oxygen and to protect against toxic byproducts of oxygen, the need for this primal laccase disappeared. Later, modified forms of the primal laccase were used for other, more specialized functions – ascorbate oxidase as a terminal oxidase in plants, and caeruloplasmin in copper transport and iron mobilization.

An interesting observation has been made by Kingston et al (1977) on the amino acid sequence around a crucial invariant cysteine residue. This cysteine might be responsible for the blue colour of the caeruloplasmin type-1 Cu(II). Kingston et al (1977) noted a relatively good match between the cysteine, histidine, and methionine residues in sequences from caeruloplasmin, azurin (from *Pseudomonas aeruginosa*) and plastocyanin (from *Anabaena variabilis*). This match suggested a common site for Cu(II)–polypeptide interaction in these widely diverse copper proteins.

THE CHEMISTRY OF CAERULOPLASMIN

Molecular properties and the state of copper

Many fundamental chemical properties of caeruloplasmin are now well established (Fee 1975, Frieden & Hsieh 1976). Tables 1 and 2 summarize these

TABLE 1

Molecular properties of human caeruloplasmin

Property	Value	Property	Value
Molecular weight	132 000	Carbohydrate content	7–8%
Cu	0.30 ± 0.03%	Carbohydrate chains (no.)	9–10
Cu	6 atoms/molecule	Sialic acid	9/molecule
Type-1 Cu(II)	2 atoms/molecule	$E_{1cm}^{1\%}$ at 610 nm	0.69 ± 0.01
Type-2 Cu(II)	1 atom/molecule		
Type-3 Cu(II)	3 atoms/molecule	$E_{1cm}^{1\%}$ at 280 nm	15.0 ± 0.4
N-terminus	1 valine/molecule	Isoelectric point, pH	4.4
SH groups	4/molecule	Sedimentation constant	7.1
		Axial ratio	3.6

E, extinction.

TABLE 2

Proposed stoichiometry of the prosthetic copper of caeruloplasmin

No. of Cu atoms	Designation	E.p.r. signal	Other properties
1	Fast type-1 Cu^{2+}	Detectable	Blue, reoxidized fast
1	Slow type-1 Cu^{2+}	Detectable	Blue, reoxidized slowly
1	Permanent type-2 Cu^{2+}	Detectable	Non-blue, binds anions
2	Type-3 Cu	Non-detectable	Postulated to be spin-coupled pair of Cu^{2+} by analogy with other blue oxidases
1	Type-4 Cu	Non-detectable	Required by total Cu content

This summary is based on a molecular weight of 132 000 and a copper content of 6 atoms/molecule.
E.p.r., electron paramagnetic resonance.

key molecular properties. Uncertainties about the molecular weight of 134 000 and about the six copper atoms per molecule (Table 1) have been resolved by the work of Rydén & Björk (1976), Kingston et al (1977) and Huber & Frieden (1970). Amino acid and carbohydrate analyses suggest a large and typical serum glycoprotein. Data that I shall present later suggest that the 1065 residues are linked in a single chain with no subunits.

No real details of the copper-binding sites are yet available. Kingston et al (1977) have discussed a possible site for type-1 Cu(II), which is responsible for the intense blue colour of several copper proteins, including ascorbate oxidase, laccase, azurin, plastocyanin, stellacyanin, and umecyanin. The unique properties of type-1 Cu(II) suggest that it probably occurs in a similar environment in all these proteins. The SH groups from cysteine and the nitrogens of histidine are the logical ligand groups in the protein. Sequences in three of these proteins, azurin, caeruloplasmin and plastocyanin, show a reasonable match in the location of cysteine, histidine and methionine residues. This region, identified by Kingston et al (1979) in the 20 000 molecular-weight-fragment of caeruloplasmin, may be involved in binding one of the type-1 Cu(II) ions found in caeruloplasmin.

Table 2 summarizes a proposed stoichiometry of the prosthetic copper of caeruloplasmin (Rydén & Björk 1976). The designation of at least three different types of prosthetic copper agrees with Deinum & Vanngard (1973). Type-2 copper was found to be 33% of the total copper detectable by electron paramagnetic resonance (e.p.r.); furthermore, two different type-1 copper atoms could be distinguished. Therefore three e.p.r.-detectable copper atoms exist and the total copper content of six atoms requires that three e.p.r.-nondetectable copper ions be present. Two of these may form a spin-coupled pair of cupric ions, like that postulated for the laccases. The proposed stoichiometry is suggested also by the many fundamental similarities among all the blue copper-containing oxidases. In contrast, the third e.p.r.-nondetectable copper ion cannot be included in such a model, and Rydén & Björk (1976) propose a type-4 copper, either Cu(I) or Cu(II), to account for the sixth ion. It is generally agreed that only four of the six ions need to be part of an active site, with the slow type-1 copper and the type-4 copper being the least essential of the six for the catalytic activity of caeruloplasmin.

CAERULOPLASMIN AS A SINGLE POLYPEPTIDE CHAIN

The evidence that human caeruloplasmin is a protein consisting of a single polypeptide chain of a molecular weight of 132 000 is now overwhelming. The fact that caeruloplasmin has at least five peptide bonds especially sensitive to proteolytic fragmentation (Moshkov et al 1979) was responsible for the release of several large peptide fragments during typical isolation procedures with human serum. Rydén (1971) was the first to show that a negligible number of fragments was produced when caeruloplasmin was prepared from fresh blood supplemented with protease (EC 3.4.24.4) inhibitors (e.g. 20mM 6-aminohexanoate). His data have been amply confirmed by Kingston et al (1977) and by Moshkov et al (1979). Kingston et al (1979) have also reported the complete amino acid sequence of a histidine-rich proteolytic fragment of molecular weight 18 650 and a cysteine-containing sequence of molecular weight about 20 000, both of which are preferred side-chain ligands for Cu(II).

THE CATALYTIC ACTIVITY OF CAERULOPLASMIN

In 1948 Holmberg & Laurell explored the oxidase activity of caeruloplasmin on numerous reducing substances. Their preparations of caeruloplasmin increased the oxidation of aryldiamines, diphenols, and other reducing substances, including ascorbate, hydroxylamine, and thioglycolate. Later Curzon (1960) found that the oxidation of aryldiamines in the presence of caeruloplasmin could be activated or inhibited by ions of certain transition metals. Finally, Curzon (1961) reported that Fe(II) could reduce caeruloplasmin and he suggested a coupled iron–caeruloplasmin oxidation system.

Our interest at Florida State University was originally stimulated by the possibility that caeruloplasmin was a mammalian ascorbate oxidase, an enzyme that had been clearly identified in plants but had eluded detection in animal tissues. First, we showed that the ascorbate oxidase activity of caeruloplasmin was not due to traces of free Cu(II). However, we found that at low concentrations of ascorbate, oxidation was greatly stimulated by traces of iron ions present in most caeruloplasmin preparations unless special precautions were taken to eliminate the iron impurities. The most useful reagent for the removal of Fe(III) was apotransferrin. We proposed three major groups of substrates to describe the oxidase action of caeruloplasmin (McDermott et al 1968):

(1) Fe(II), the substrate with the highest V_{max} and the lowest K_m;

(2) an extensive group of bifunctional aromatic amines and phenols, which do not depend on traces of iron ions for their activity (this group includes the two classes of biogenic amines: the epinephrine (adrenaline) and 5-hydroxyindole series, and the phenothiazine series);

(3) numerous reducing agents that can rapidly reduce Fe(II) or partially ox-

idized (free radical) intermediates of class 2. We consider these compounds to be secondary substrates by way of an iron cycle or by the action of an aromatic diamine as a shuttle.

These three classes of substrates are shown in Fig. 2. In principle, any reductant can be a substrate if it can transfer an electron to oxidized caeruloplasmin without poisoning or blocking the autooxidizing capacity of reduced caeruloplasmin. For example, in our laboratory, D.J. McKee (unpublished results) has shown that 10mM $VOSO_4$ rapidly bleaches blue caeruloplasmin in its conversion to an ion that may thus represent an additional caeruloplasmin substrate that fits the first group along with Fe(II).

There has been uncertainty about whether certain organic compounds are true substrates (group 2) or pseudosubstrates (group 3). The issue is complicated by the fact that the cyclical iron-catalysed reactions are faster, in general, than direct electron transfer to caeruloplasmin. After the considerable early controversy and uncertainty, Curzon & Young (1972) reported that ascorbate was a true caeruloplasmin substrate with a rather high K_m (5.2mM) and a typical V_{max} (4.0 e^- Cu^{-1} min^{-1}). In their experiments, iron

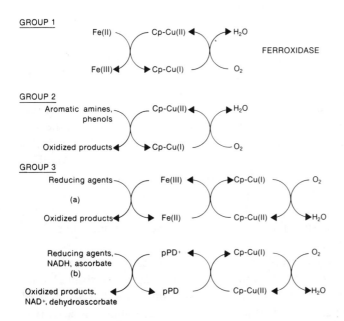

FIG. 2. The various substrate groups of caeruloplasmin and how they react. Groups 1 and 2 are true substrates, because they react directly with the oxidized form of caeruloplasmin (Cp). Groups 3a and 3b may be considered as pseudosubstrates since their reactions are mediated by a group 1 or 2 substrate. pPD, p-phenylenediamine.

impurities were assumed to be eliminated by the use of 100 μM EDTA (ethylenediaminetetraacetic acid). Young & Curzon (1972) also found that catechol was a caeruloplasmin substrate with a large K_m (282mM). Similarly, Løvstad (1972) reported that D- or L-dopa could also be catalytically oxidized, though very weakly, in the presence of the iron chelator, *Desferal* (desferrioxamine mesylate). A summary of K_m and V_{max} data on the various substrate groups of caeruloplasmin is presented in Table 3. A more detailed discussion is given by Løvstad (1975) and by Frieden & Hsieh (1976).

TABLE 3

Substrate groups of caeruloplasmin

Group	K_m μM	V_{max} $e^- Cu^{-1} min^{-1}$
Fe(II)	0.6, 50	22
p-Phenylenediamine derivatives	21–3000	1–7
Aminophenols	> 180	4–15
Catecholamines	> 2550	2–11
5-Hydroxyindoles	> 900	1.5–6
Phenothiazines	> 900	1–10

The reactions of group 3 pseudosubstrates have been used extensively in the study of the kinetics of caeruloplasmin. Ascorbate at concentrations well below its effective substrate range (100 μM) was used by Huber & Frieden (1970) to study Fe(II) oxidation, as in the reaction sequence in Fig. 2. Young & Curzon (1972) also used ascorbate (50 μM) as the reducing agent in the same reaction sequence for study of the oxidation of *N,N*-dimethyl-*p*-phenylenediamine. Walaas & Walaas (1961) used NADH and NADPH to provide the electrons necessary to reduce partially oxidized or free radical intermediates resulting from the action of caeruloplasmin on aromatic diamines, phenols or other oxidizable substrates. Since NADH does not react directly with caeruloplasmin it has been used as an electron donor for study of the aromatic substrates.

While the role of caeruloplasmin in iron mobilization is now widely documented, its catalytic activity towards any other class of substrates has not been related as directly to its biological function. We have suggested, therefore, that the name ferroxidase be used when describing the activity of caeruloplasmin as an enzyme (Osaki et al 1966). We further proposed that the enzyme be designated as a ferro-O_2-oxidoreductase. It now carries the International Union of Biochemists number EC 1.16.3.1. We realized, however, that in reference to the copper transport protein of the plasma, the name caeruloplasmin might be retained because of its historical significance and widespread familiarity.

Finally, *p*-phenylenediamine oxidase activity, presumably due to caeruloplasmin, has been reported in a wide variety of vertebrate plasma.

THE FERROXIDASE ACTIVITY OF CAERULOPLASMIN

The ferroxidase activity of human caeruloplasmin was first reported by Curzon & O'Reilly (1960). Its significance in iron metabolism and its substrate characteristics were explored extensively at Florida State University. Among other properties, the effect of oxygen and Fe(II) concentration on the caeruloplasmin and on the non-enzymic reaction under normal serum conditions were compared. The non-enzymic rate of Fe(II) oxidation was first-order with respect to both Fe(II) and oxygen concentrations. In contrast, Fe(II) oxidation catalysed by caeruloplasmin showed typical saturation kinetics, reaching zero-order at $> 10\mu M$ O_2 and $> 50\mu M$ Fe(II). From estimates of normal serum oxygen and Fe(II) concentrations, it was calculated that the caeruloplasmin-catalysed oxidation of Fe(II) was 10–100 times as fast as the non-enzymic oxidation. This estimate does not include any correction for the presence of reducing metabolites, such as ascorbate at $40\mu M$, which did not affect the caeruloplasmin-catalysed oxidation of Fe(II) but significantly reduced the net rate of the non-enzymic oxidation of Fe(II). Further studies of the kinetics of ferroxidase revealed biphasic curves in v versus $v/$Fe(II) plots, with two K_m values ($0.6\mu M$ and $50\mu M$) which differed by almost two orders of magnitude. While these data were originally interpreted in terms of two binding sites, Huber & Frieden (1970) reported a good fit between experimental points and calculated values based on a rate-determining substrate activation mechanism, described in detail in their paper.

ALTERNATIVE FERROXIDASE ACTIVITIES

Since the recognition of the possible importance of the ferroxidase activity of caeruloplasmin in relation to its function in iron mobilization (Osaki et al 1966), other ferroxidase activities have been proposed. These proposals fit into three categories:

(1) proteins, other than caeruloplasmin, that may have true ferroxidase activity (attempts to distinguish between these alternative ferroxidases usually require determination of the effect of 1mM azide, which inhibits caeruloplasmin by 98%);

(2) substances, including proteins, that have strong electron-accepting groups, thereby transforming Fe(II) to Fe(III);

(3) compounds that strongly and preferentially chelate Fe(III), thereby enhancing the Fe(II) to Fe(III) reaction. We designate these activities as pseudoferroxidase activity.

The two latter categories can be distinguished from true ferroxidases because they are consumed in the reaction and they exhibit a stoichiometric relationship between the oxidized Fe(II) and the chelated Fe(III). Therefore they do not satisfy the basic prerequisite for catalysis, since they end up in a different chemical state after each reaction cycle. The true ferroxidases also show strong binding for Fe(II) and typical Michaelis–Menten saturation kinetics for both Fe(II) and O_2.

PSEUDOFERROXIDASE ACTIVITY

At least four biologically important substances have been reported to have pseudoferroxidase activity – apoferritin, transferrin, phosvitin and citrate. The pseudoferroxidase activity of a dialysable, heat-stable component of human serum, identified as citrate, was first pointed out by Lee et al (1969). They proposed citrate as an alternative source of ferrous ion oxidizing activity in serum that had low caeruloplasmin levels. Phosvitin, a highly phosphorylated protein in hens, was also shown to possess significant Fe(II) oxidizing activity.

The uptake of iron by apoferritin, the iron storage protein, requires iron in the ferrous form (Macara et al 1972). Two groups (Macara et al 1972, Bryce & Crichton 1973) showed that apoferritin appeared to have a catalytic effect on the oxidation of Fe(II), which aided in the formation of Fe(III)-ferritin. Bates & Schlaback (1973) showed that Fe(II) oxidation was increased by apotransferrin, the iron transport protein. They proposed that Fe(II) forms a weak complex with apotransferrin, and that this complex results in a faster oxidation of Fe(II) to Fe(III) with subsequent formation of Fe(III)$_2$-transferrin.

The facilitation of Fe(II) autooxidation by Fe(III)-complexing agents was studied by Harris & Aisen (1973). They demonstrated that the rate of oxidation of Fe(II) by atmospheric oxygen at pH 7 was significantly enhanced by low-molecular-weight Fe(III)-complexing agents. The order of activity was EDTA ≃ nitrilotriacetate > citrate > phosphate > oxalate. Under the conditions used by Harris & Aisen (1973), Fe(II) had a $t_{1/2}$ of 2700 s; with EDTA (4×10^{-4}M), it had a $t_{1/2}$ of about 10 s. The authors pointed out that there was nothing unique about the ability of apotransferrin to stimulate Fe(II) oxidation. Thus the effect of preferential Fe(III) binding accounts for the 'ferroxidase' activity shown by apotransferrin, apoferritin and phosvitin and may be termed pseudoferroxidase activity.

Frieden & Osaki (1974) proposed a kinetic scheme, based on relatively simple assumptions, that accounted for pseudoferroxidase activity. In the oxidation of Fe(II) and Fe(III) any effect on the reaction system that tends to reduce the free Fe(III) concentration increases the rate of Fe(II) disappearance or the rate of formation of Fe(III) complexes.

F. Carver et al (personal communication 1980) analysed the role of several serum constituents which have been proposed to affect the following reactions *in situ:*

$$Fe(II) \xrightarrow{(A)} Fe(III) + Tf \xrightarrow{(B)} Fe(III) - Tf$$

These reactions were monitored by measuring the rate of Fe(II) oxidation (reaction A) and the rate of Fe(III)-transferrin (Fe(III)-Tf) formation (reaction B) at 465 nm. Reactions A and B were found to be kinetically equivalent. The results show that singly or in combination, bicarbonate, orthophosphate, citrate, apotransferrin and albumin have less than one-tenth of the ability to oxidize Fe(II) compared to the ability of the serum enzyme, caeruloplasmin. It was also found that the rate of Fe(II) oxidation by serum Fe-ligands was influenced by the efficiency of oxygen utilization. Whereas caeruloplasmin produces a 4:1 ratio of Fe(II) oxidized to oxygen utilized, the non-enzymic components yield a 2:1 to 3.09:1 ratio. These data also support the role of caeruloplasmin as an antioxidant that prevents the formation of active oxygen species by Fe(II) autooxidation.

A hitherto unrecognized factor in the control of non-enzymic oxidation of Fe(II) was the concentration of albumin. This protein at $> 25\mu M$ concentration was found to dampen sharply the rate of Fe(II) oxidation in the presence of a physiological concentration of bicarbonate, citrate and apotransferrin. Albumin did not appear to affect the caeruloplasmin-catalysed oxidation of Fe(II). The addition of caeruloplasmin effected a 13–44-fold increase in the rate of Fe(II) oxidation and in the rate of Fe(III)-transferrin formation, even in the presence of 0.6–1.2mM albumin. The data suggest that caeruloplasmin is the only effective ferroxidase in human serum.

OTHER OXIDASE ACTIVITIES OF CAERULOPLASMIN

The relationship between ferroxidase activity and the iron mobilization properties of caeruloplasmin has focused much recent attention on Fe(II) as a substrate. However, numerous investigators, stimulated by the fact that group 2 substrates include two important types of biogenic amines, the adrenaline and 5-hydroxyindole series, and the phenothiazine-type tranquillizers, have studied in detail the caeruloplasmin-catalysed oxidation of arylamines and phenols (e.g. Barrass et al 1974). Much of this work has been summarized by Frieden & Hsieh (1976) and by Løvstad (1978).

The possibility that there may be other classes of substrate for caeruloplasmin has been suggested from the work of Albergoni & Cassini (1975). They reported that bovine caeruloplasmin has a cysteine oxidase activity several-fold greater than the maximal activity reported for Fe(II) oxidation. A significant feature of this activity was a pH optimum of 7.4, which suggests that cysteine may be a significant substrate for caeruloplasmin in serum.

However, we have reported numerous differences between cysteine oxidation and the oxidation of other caeruloplasmin substrates, e.g. lack of sensitivity to 1.5mM azide. Many features of the cysteine oxidation are similar to those observed with free Cu(II). It seems likely that this oxidation is due to Cu(II) that is modified or liberated during reduction by cysteine of one or more Cu(II) ions in caeruloplasmin.

CAERULOPLASMIN AS A COPPER TRANSPORT PROTEIN

The role of caeruloplasmin as a copper transport protein was emphasized by Broman (1967). The abundance of caeruloplasmin – it contains over 90–95% of the copper in normal human sera – is a convincing argument. The remaining 5–10%, which exists primarily as copper complexes of albumin and histidine, could provide a necessary auxiliary transport mechanism. While there is good evidence (Frieden 1978, Frieden 1979) that caeruloplasmin is the major vehicle for copper transport, it cannot be the exclusive one. Wherever free copper ions occur, as in absorption of dietary copper from the gut or after the release of copper ions from caeruloplasmin or from tissue copper proteins (e.g. cytoplasmic superoxide dismutase, cytochrome c oxidase), there must be an alternative mechanism to transport copper ions to the liver, where they can be incorporated into caeruloplasmin. The copper chelates of albumin and histidine are assumed to serve this function.

Nevertheless, because of the short half-life of the only available copper isotopes, evidence for the transport–donator function of caeruloplasmin is still indirect. Marceau & Aspin (1973) reported that ^{67}Cu from partially purified caeruloplasmin was incorporated into cytochrome c oxidase. Hsieh

& Frieden (1975) fed to rats diets that were deficient in copper to reduce greatly the cytochrome c oxidase activity in most tissues. Restoration of the activity of this crucial enzyme occurred most rapidly in the animals injected with purified caeruloplasmin. Copper salts alone, or in complexes with albumin or histidine, produced a much slower return of cytochrome c oxidase activity.

The transport of copper by caeruloplasmin probably requires specific receptor mechanisms in the various target tissues. The greater lability of Cu(I) strongly suggests the presence of a reductive step in copper release. Ample reductive mechanisms are available once caeruloplasmin is within the reactive sphere of the cell.

First, there are the numerous endogenous substrates described earlier. Secondly, caeruloplasmin is able to tap the electron transport machinery of the cell. Brown & White (1961) reported that in the presence of cytochrome c and typical oxidative substrates (succinate and NADH), particles of heart muscle can reduce caeruloplasmin. The reaction occurs under anaerobic conditions and is reversed by oxygen; it is sensitive to cyanide, carbon monoxide and antimycin A. Under aerobic conditions, caeruloplasmin inhibits the electron transport system, possibly by a reaction with essential SH groups.

We have proposed a simplified mechanism, based on these ideas, for the incorporation of the copper from caeruloplasmin into an intracellular copper enzyme or copper protein (Fig. 3). The first step is the reduction of the Cu(II) of caeruloplasmin by any of the substrates or reduction sequences described earlier. If this occurs at the cell membrane, the Cu(I) is likely to be transferred to an intracellular Cu(I) acceptor, X. If caeruloplasmin penetrates to the inside of the cell, X might not be necessary for intracellular transport. In its Cu(I) form, the copper is added to an apoenzyme where it is fixed into the holoenzyme in the Cu(II) state with the aid of oxygen. (Oxidation to Cu(II) and direct addition of Cu(II) is less likely.) This preferred mechanism takes into account the primary role of caeruloplasmin as the Cu transport and donor molecule, the high exchangeability of Cu(I)-ligands, and the greater stability of the copper ion in Cu(II)-proteins (Frieden 1979).

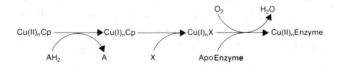

FIG. 3. Possible mechanism for the transfer of copper from caeruloplasmin to an intracellular copper enzyme. AH_2 is a reducing substrate and X is a hypothetical intracellular Cu(I) acceptor and/or a copper-ligand exchanger.

THE CONTROL OF IRON MOBILIZATION BY CAERULOPLASMIN

Over fifty years have passed since the recognition that the copper-deficient animal develops anaemia. Copper deficiency results in low plasma concentrations of caeruloplasmin and iron, reduced iron mobilization, and anaemia, despite adequate iron storage in the liver. On the basis of a study of the ferroxidase activity of caeruloplasmin, in 1966 (Osaki et al), we proposed that caeruloplasmin was the molecular link between iron and copper metabolism at the level of iron mobilization. The vital role of caeruloplasmin in controlling the release of hepatic iron to the plasma has been amply confirmed; the evidence has been summarized in detail in several reviews (Frieden 1971, Frieden 1979). Proposals that other serum components, particularly citrate, apotransferrin, bicarbonate ion, phosphate ion or albumin could play a significant role in the oxidation of Fe(II) to Fe(III) in serum have been refuted by F. Carver, D. Farb & E. Frieden (1980, unpublished work). Albumin (25μM) strongly inhibited the effect of citrate, bicarbonate and apotransferrin on the rate of Fe(II) oxidation. However, albumin (up to 1.20mM) had no effect on the caeruloplasmin-catalysed oxidation of Fe(II).

The crucial steps in the mobilization of plasma iron are shown in Fig. 4. Iron is released from ferritin by a specific ferritin reductase step involving $FMNH_2$ as the immediate reductant. Caeruloplasmin controls the rate at which the Fe(II) released is converted to circulating Fe(III)$_2$-transferrin. In this form, Fe is distributed to all cells, predominantly to the reticulocytes involved in haem biosynthesis which precedes the completion of the haemoglobin molecule.

REGULATION OF PLASMA OR TISSUE LEVELS OF BIOGENIC AMINES

Barrass & Coult (1972) have summarized the effects of drugs used in treatment of mental illness (tranquillizers and antidepressants) on the caeruloplasmin-catalysed oxidation of the biogenic amines – noradrenaline (norepinephrine) and 5-hydroxytryptamine (serotonin, 5-HT). It was suggested that caeruloplasmin, or an enzyme with similar properties, may affect the relative concentrations of noradrenaline and 5-HT in the serum and, eventually, in those areas of the brain where these compounds act as neurotransmitters. Linder & Moor (1977) found that caeruloplasmin is present in the brain and is taken up by brain cells. Thus, a caeruloplasmin-like enzyme, by its effect on the life time of biogenic amines, could function in the regulation of brain chemistry that is necessary for mental function; interference with this enzyme may lead to a chemical imbalance reflected in the appearance of abnormal mental states.

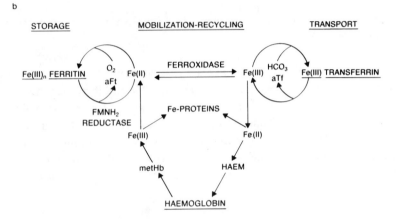

FIG. 4(a). Steps in the release of iron from liver storage cells to transferrin in the plasma and then to reticulocytes for haemoglobin synthesis.
(b). The central role of caeruloplasmin (ferroxidase) in iron mobilization and transport. The Fe(II) to Fe(III) cycles of iron are shown in relation to the metabolism of the major iron compound of the vertebrates, haemoglobin. aFt, apoferritin; aTf, apotransferrin.

A varied spectrum of drug effects is observed (Table 4). Hallucinogens such as lysergide (LSD) accelerated the caeruloplasmin-catalysed oxidation of noradrenaline but inhibited the enzymic oxidation of 5-HT. Tranquillizers of the phenothiazine type accelerated the enzymic oxidation of both substrates. Løvstad (1975) found that many phenothiazines were effective substrates for caeruloplasmin. Antidepressant drugs, e.g. imipramine, inhibited the enzymic oxidation of both substrates but only at relatively high concentrations (10^{-2}M). However, some phenylethylamines and anticholinergics with CNS activity had no effect on the enzymic oxidation of the two biogenic amines.

The mode of action of LSD was of particular interest since this drug increases 5-HT concentrations and reduces catecholamine concentrations in the brain.

TABLE 4

Effect of drugs on biogenic amine oxidation catalysed by caeruloplasmin

Drug	Noradrenaline %	5-HT %
Trifluoperazine (200µM)	200	200
LSD (200µM)	400	−50
Imipramine (10 mM)	−50	−50

When tested with caeruloplasmin, LSD at a concentration of one-tenth that of the substrate inhibited the oxidation of 5-HT by 50% and increased the oxidation of noradrenaline four-fold (Table 4). The K_m values for noradrenaline and 5-HT are similar − 3µM and 1µM, respectively. Caeruloplasmin may closely control the relative concentrations of these two compounds in key parts of the brain. Thus, LSD might produce its effects on the central nervous system through caeruloplasmin, by perturbing the balance between these two groups of biogenic amines.

Barrass & Coult (1972) also proposed a role for caeruloplasmin in Parkinson's disease. 3-Hydroxy-4-methoxyphenethylamine is one of the endogenous toxins that accumulate in patients with Parkinsonism and it produces tremors and hypokinesia. *In vitro,* this compound enhances dopamine oxidation catalysed by caeruloplasmin. Parkinson's disease can be associated with decreased catecholamine concentrations and with increased 5-HT and serum caeruloplasmin concentrations, and the latter has been suggested as a systemic basis for the aetiology of this disease.

Reduced caeruloplasmin concentrations in patients with Huntington's disease has been reported by Shokeir (1975). This had already been suspected, since many of the early symptoms of Huntington's chorea resemble those of another genetic disorder, Wilson's disease. The latter hepatolenticular disease is accompanied by copper accumulation and toxicity in the brain and liver and by decreased biosynthesis of caeruloplasmin with the result, almost invariably, of a low serum caeruloplasmin.

THE ROLE OF CAERULOPLASMIN AS A SERUM ANTIOXIDANT

Caeruloplasmin plays a dual role in the prevention of deleterious oxidations in serum. First of all, it prevents the formation of free radical intermediates that might be generated by non-enzymic (non-caeruloplasmin) oxidation of Fe(II) or Fe(II) complexes.

$$Fe(II) + O_2 \rightarrow Fe(III) + O_2^-$$

Even when Fe(II) is partially complexed by bicarbonate, citrate or phosphate ions, Fe(II) to O_2 ratios of between 1 and 3:1 are observed. In contrast, when Fe(II) oxidation is catalysed by caeruloplasmin, a ratio of Fe(II) to O_2 of 4:1 is observed, as predicted by the equation

$$4Fe(II) + O_2 + 4H^+ \rightarrow 4Fe(III) + 2H_2O$$

Thus, caeruloplasmin-catalysed oxidation of Fe(II) produces a rapid four-electron reduction of molecular oxygen to water, with no chance for formation of peroxides or superoxide anion free radicals.

Not only does caeruloplasmin affect the stoichiometry of the iron oxidation reaction, but it also affects the rate of Fe(II) oxidation and chelation in media that resemble serum. F. Carver, D. Farb and E. Frieden (1980, unpublished work) studied the following reaction

$$Fe(II) \xrightarrow[\text{oxidation}]{O_2} Fe(III) \xrightarrow[\text{chelation}]{Tf} Fe(III)\,Tf$$

By themselves, the serum constituents, bicarbonate, phosphate, citrate and apotransferrin (Tf) had less than one-tenth the ability to oxidize Fe(II) compared to the ability of normal levels of caeruloplasmin. When purified serum albumin was added at $25\mu M$, the rate of oxidation in the presence of bicarbonate, phosphate, citrate and apotransferrin was sharply reduced. The addition of $2\mu M$ caeruloplasmin effected a 13–44-fold increase in the rate of Fe(II) oxidation and in the rate of Fe(III)–Tf formation, even in the presence of 0.60–1.20mM serum albumin. Thus, the oxidation of Fe(II) in the complicated milieu of the serum appears to require caeruloplasmin for a rapid and complete conversion to Fe(III)–Tf with a minimum of deleterious free radical by-products.

Lipid peroxidation is significantly inhibited by the α_2-globulin fraction of human serum. The protein mainly responsible for serum antioxidant activity was shown to be caeruloplasmin (Stocks et al 1974). Dormandy (1978) observed a close correlation between serum antioxidant activity and the concentration of caeruloplasmin and copper in human serum. He measured antioxidant activity as % inhibition of the formation of malonylaldehyde from oxidized polyenoic acids, using the thiobarbituric acid reagent.

In addition, caeruloplasmin inhibited peroxidation in tissue homogenates and in systems containing Fe(II) and ascorbate, both of which are substrates of caeruloplasmin (Al-Timini & Dormandy 1977). The oxidation of Fe(II) and ascorbate by caeruloplasmin at less than 10^{-7}M inactivates an important free radical generating system in serum (Gutteridge 1978). Cu-ion-catalysed peroxidation required greater than 2×10^{-6}M caeruloplasmin for inhibition of the latter oxidation (Gutteridge 1978).

Caeruloplasmin can also serve as a scavenger for superoxide anion radicals (Goldstein et al 1979). Its mode of action is similar to but not identical with that of the enzyme superoxide dismutase and is much weaker. Caeruloplasmin inhibited, in a concentration-dependent fashion, two reactions that are dependent on the generation of oxygen free radicals by the aerobic action of xanthine oxidase (EC 1.2.3.2) on hypoxanthine – the reduction of cytochrome c and of blue tetrazolium. However, in contrast to superoxide dismutase, caeruloplasmin reduced the yield of H_2O_2 expected from the dismutase reaction by almost 50%. The destruction of oxygen free radicals by caeruloplasmin was only 1/3000 that by superoxide dismutase. Whether caeruloplasmin can be regarded as a significant extracellular disposer of oxygen free radicals is still an open question. Even at a serum concentration of 300μg/ml, caeruloplasmin would provide less scavenging activity than the level of superoxide dismutase (0.7μg/ml) that has been reported for human plasma (McCord 1974). Nevertheless, this activity represents another potential contribution of caeruloplasmin to the antioxidant activity of the serum.

CAERULOPLASMIN AS AN ACUTE-PHASE REACTANT

The role of caeruloplasmin as an acute-phase reactant may also be related to its antioxidant function. As an endogenous modulator of the inflammatory response, serum caeruloplasmin is increased two- to three-fold in response to inflammation. Oestrogens, which modulate inflammation, also increase caeruloplasmin two- to three-fold. Bonta (1978) has suggested that lipid peroxidation products in the serum may stimulate a greater output of acute-phase reactant.

One of the molecular participants in these inflammatory responses may be the leucocytic endogenous mediator (LEM) (Beisel 1976). LEM is a heat-labile protein released by the leucocytes and it has an extensive impact on cells, particularly hepatocytes. LEM stimulates the uptake of iron, zinc and amino acids by liver cells. It also stimulates the synthesis and release of acute-phase reactant proteins, including caeruloplasmin. Infection and inflamma-

TABLE 5

Multifunctions of caeruloplasmin

(1) *Transport of Cu:*
   See Fig. 3, p. 105 for details
(2) *Mobilization of iron into serum:*
   Ferroxidase activity
   Fe(II) receptor sites

(3) *Serum anti-oxidant:*
   Reduces peroxidation of polyenoates
   Scavenges superoxide ion

(4) *Endogenous modulator of inflammatory response:*
   Stimulated by LEM (leucocytic endogenous mediator)
   Acute-phase reactant

(5) *Regulation of serum biogenic amines:*
   Oxidation of catecholamines, hydroxyindoles, neuro-drugs

---

tion are believed to increase circulating LEM which in turn initiates the metabolic steps described above. Thus, increased serum concentrations of caeruloplasmin could increase its serum antioxidant function.

Table 5 provides a summary of the functions of caeruloplasmin.

ACKNOWLEDGEMENT

This work was supported by NIH grant AM 25451 and NSF grant PCM 7622171.

References

Albergoni V, Cassini A 1975 The oxidation of cysteine by ceruloplasmin. FEBS (Fed Eur Biochem Soc Lett) 55:261-264
Al-Timini DJ, Dormandy TL 1977 Inhibition of lipid autoxidation by human caeruloplasmin. Biochem J 168:283-288
Barrass BC, Coult DB 1972 Interaction of some centrally active drugs with caeruloplasmin. Biochem Pharmacol 21:677-685
Barrass BC, Coult DB, Rich P, Tutt KJ 1974 Substrate specificity of caeruloplasmin. Phenylalkylamine substrates. Biochem Pharmacol 23:47-56
Bates GW, Schlaback MR 1973 The reaction of ferric salts with transferrin. J Biol Chem 248:3228-3232
Beisel WR 1976 Trace elements in infectious processes. Med Clin N Am 60:831-849
Bonta IL 1978 Endogenous modulators of the inflammatory response. In: Vane JR, Ferreira SH (eds) Inflammation. Springer, Berlin (Handb Exp Pharmacol 50, part I) p 523-567
Broman L 1967 The function of caeruloplasmin – a moot question. In: Walaas O (ed) Molecular basis of some aspects of mental activity. Academic Press, New York, vol 2:131-146

Brown FC, White JB Jr 1961 The reduction of caeruloplasmin by the electron transport system. J Biol Chem 236:911-914

Bryce CFA, Crichton RR 1973 The catalytic activity of horse spleen apoferritin, preliminary kinetic studies and the effect of chemical modification. Biochem J 133:301-309

Curzon G 1960 The effects of some ions and chelating agents on the oxidase activity of caeruloplasmin. Biochem J 77:66-73

Curzon G 1961 Some properties of coupled iron-caeruloplasmin oxidation systems. Biochem J 79:656-663

Curzon G, O'Reilly S 1960 A coupled iron-caeruloplasmin oxidation system. Biochem Biophys Res Commun 2:284-286

Curzon F, Young SN 1972 The ascorbate activity of caeruloplasmin. Biochim Biophys Acta 268:41-48

Dawson CR, Strothkamp KG, Krul KG 1975 Ascorbate oxidase related copper proteins. Ann NY Acad Sci 258:209-220

Deinum J, Vanngard T 1973 The stoichiometry of the paramagnetic copper and the oxidation reduction potentials of type I copper in human caeruloplasmin. Biochim Biophys Acta 310:321-330

Dormandy TL 1978 Free-radical oxidation and antioxidants. Lancet 1:647-650

Fee JA 1975 Copper proteins – systems containing the 'blue' copper centre. Struct Bond 23:1-60

Frieden E 1971 Ceruloplasmin a link between copper and iron metabolism. Adv Chem Ser 100:292-331

Frieden E 1974 The evolution of metals as essential elements (with special reference to iron and copper). Adv Exp Med Biol 48:1-29

Frieden E 1978 Ceruloplasmin, a multifunctional protein. In: Kirchgessner M (ed) Trace element metabolism in man and animals. Arbeitskreis für Tierernährungsforschung, Freising-Weihenstephan, vol 3:36-37

Frieden E 1979 Ceruloplasmin: the serum copper transport protein with oxidase activity. In: Nriagu JO (ed) Copper in the environment, part II. Wiley, New York, p 241-284

Frieden E, Hsieh HS 1976 Ceruloplasmin: the copper transport protein with essential oxidase activity. Adv Enzymol 44:187-236

Frieden E, Osaki S 1974 Ferroxidases and ferrireductases: their role in iron metabolism. Adv Exp Med Biol 48:235-265

Goldstein IM, Kaplan HB, Edelson HS, Weissmann G 1979 Ceruloplasmin – a scavenger of superoxide anion radicals. J Biol Chem 254:4040-4045

Gutteridge JMC 1978 Caeruloplasmin: a plasma protein, enzyme and antioxidant. Ann Clin Biochem 15:293-296

Harris DC, Aisen P 1973 Facilitation of Fe(II) autoxidation by Fe(3) complexing agents. Biochim Biophys Acta 329:156-158

Holmberg CG 1944 On the presence of a laccase-like enzyme in serum and its relation to the copper in serum. Acta Physiol Scand 8:227-229

Holmberg CG, Laurell CB 1948 Investigations in serum copper: II. Acta Chem Scand 2:550-556

Holmberg CG, Laurell CB 1951 Investigations in serum copper: III. Acta Chem Scand 5:476-480

Hsieh HS, Frieden E 1975 Evidence for ceruloplasmin as a copper transport protein. Biochem Biophys Res Commun 67:1326-1331

Huber CT, Frieden E 1970 Substrate activation and the kinetics of ferroxidase. J Biol Chem 245:3973-3978

Kingston IB, Kingston BL, Putnam FW 1977 Chemical evidence that proteolytic cleavage causes the heterogeneity present in human ceruloplasmin preparations. Proc Natl Acad Sci USA 74:5377-5381

Kingston IB, Kingston BL, Putnam FW 1979 Complete amino acid sequence of a histidine-rich proteolytic fragment of human ceruloplasmin. Proc Natl Acad Sci USA 76:1668-1672

Lee GR, Nacht S, Christensen D, Hansen SP, Cartwright GE 1969 The contribution of

citrate to the ferroxidase activity of serum. Proc Soc Exp Biol Med 131:918-923
Linder MC, Moor JR 1977 Plasma ceruloplasmin. Evidence for its presence in and uptake by heart and other organs of the rat. Biochim Biophys Acta 499:329-336
Løvstad RA 1972 Interaction of ceruloplasmin with L- and D-dopa. Acta Chem Scand 26:2832-2836
Løvstad RA 1975 Kinetic studies on the ceruloplasmin-catalyzed oxidation of phenothiazine derivatives. Biochem Pharmacol 24:475-478
Løvstadt RA 1978 PhD thesis, Faculty of Science, University of Oslo (see p 5-56)
Macara TG, Hoy TG, Harrison PM 1972 The formation of ferritin from apoferritin. Kinetics and mechanism of iron uptake. Biochem J 126:151-159
Marceau N, Aspin N 1973 The association of the copper derived from ceruloplasmin with cytocuprein. Biochim Biophys Acta 328:351-358
McCord JM 1974 Free radicals and inflammation: protection of synovial fluid by superoxide dismutase. Science (Wash DC) 185:529-531
McDermott JA, Huber CT, Osaki S, Frieden E 1968 The role of iron in the activity of ceruloplasmin. Biochim Biophys Acta 151:541-544
Moshkov KA, Lakatos S, HajDu J, Zavodszky P, Neifakh SA 1979 Proteolysis of human ceruloplasmin — some peptide-bonds are particularly susceptible to proteolytic attack. Eur J Biochem 94:127-134
Osaki S, Johnson DA, Frieden E 1966 The possible significance of the ferrous oxidase activity of ceruloplasmin in normal human serum. J Biol Chem 241:2746-2751
Rydén L 1971 Human ceruloplasmin as a polymorphic glycoprotein. Tryptic glycopeptides from two forms of the protein. Int J Protein Res 3:131-138
Rydén L, Björk I 1976 Investigation of some physicochemical properties of human caeruloplasmin (ferroxidase). Biochemistry 15:3411-3417
Shokeir MHK 1975 Investigation on Huntington's disease. III: Biochemical observations, a possibly predictive test? Clin Genet 7:354-360
Stocks J, Gutteridge JMC, Sharp RJ, Dormandy TL 1974 Assay using brain homogenate for measuring the antioxidant activity of biological fluids. Clin Sci Mol Med 47:223-233
Walaas E, Walaas O 1961 Oxidation of reduced phosphopyridine nucleotides by p-phenylenediamines, catecholamines and serotonin in the presence of ceruloplasmin. Arch Biochem 95:151-162
Young SN, Curzon G 1972 A method for obtaining linear reciprocal plots with caeruloplasmin and its application in a study of the kinetic parameters of caeruloplasmin substrates. Biochem J 129:273-283

Discussion

Harris: Before we discuss the more specific properties of caeruloplasmin I'd like to raise a general question. Can we say that caeruloplasmin is a *universal* copper protein, i.e. in the serum of all animals, and is the protein always blue in all species that have been examined?

Those of us who do not work with human, pig or rat serum, for which the protein has been well characterized, find a different story emerging. In the normal chicken we find a low concentration of copper, a measurable but low level of *p*-phenylenediamine oxidase (PPD oxidase) activity, and absolutely no evidence of a blue copper protein.

Frieden: Caeruloplasmin has been detected in virtually all vertebrate sera. Early work indicated that there was no oxidase activity in frog serum but this has not been confirmed. We were able to isolate from frog serum a blue protein with the typical properties of caeruloplasmin (Inaba & Frieden 1967). In the rat, caeruloplasmin concentration in the serum is relatively low and the oxidase activity is also quite low; and yet Evans & Abraham (1973) have provided convincing evidence for caeruloplasmin in the rat. In terms of iron mobilization, the excess of caeruloplasmin determined both *in vivo* and *in vitro* is probably 100-fold; in terms of copper donation and antioxidant activity, the excess activity might be quite different. Therefore a relatively low level of caeruloplasmin is required to negotiate that particular part of the process, but I would also not exclude alternative mechanisms. Roeser et al (1970) reported extensive experiments *in vivo* and *in vitro* on copper-deficient pigs and they found that even at concentrations of caeruloplasmin lower than 10^{-7}M (as little as 1% of a normal level) there was still a significant iron-mobilizing effect. These facts may explain the low levels of caeruloplasmin in some animals.

Mangan: We have heard at this meeting from David Danks that one effect of copper deficiency is osteoporosis, and you have just mentioned, Professor Frieden, that oestrogens raise the plasma concentration of copper and caeruloplasmin. This has made me wonder whether oestrogen treatment for postmenopausal osteoporosis could in any way be effective by raising plasma caeruloplasmin levels. Do you know whether postmenopausal osteoporosis is associated with low plasma levels of caeruloplasmin?

Frieden: No; I'm not aware of that. We do not understand the mechanism or process by which oestrogens increase the level of caeruloplasmin. Some of these responses have probably arisen secondarily as a result of inflammatory responses produced by conditions of metabolic stress. If a strong antigen is administered to an animal, it produces an inflammatory response. In some animals so much caeruloplasmin is produced that the serum turns blue-green as a result of the increase in the blue protein.

Dormandy: You drew a distinction between oxygen utilization by a safe enzymic route that produces a stable molecule and that by a non-enzymic route that generates potentially dangerous free radicals. I believe that the non-enzymic route is essentially associated with damage to cells, i.e. with lesions that characteristically produce the inflammatory response. There has been a tendency to think that oxygen free radicals primarily *cause* structural damage. I think that the reverse is also true and equally important: structural damage from any cause will favour the generation of free radicals (Dormandy 1980).

My argument is teleological but perhaps it explains why caeruloplasmin is an acute-phase protein: caeruloplasmin concentrations may be increased when there is tissue destruction because then the generation of free radicals is greatly favoured and additional antioxidant protection is needed.

Frieden: Whenever cellular organization is disrupted there is a greater tendency for uncontrolled non-enzymic reactions to occur. That is true for lysosomal breakdown, for example, which causes intracellular destruction and perhaps the death of the cell itself. One of the effects of excess copper is the peroxidation and hence the destruction of lysosomal membranes, which then leads to the destruction of the cell.

Hill: As an example of that in plant biochemistry, one can interpret the action of paraquat (viologen) as one of molecular disruption that brings the oxygen part of the photosynthetic reaction into contact with the electron-transport apparatus, hence leading to the generation of hydroxyl radicals (Youngman et al 1980).

Mills: I would like to ask about the role of caeruloplasmin in copper transport. The entry of ^{64}Cu into the cytosol and into the mitochondrion is much more effective from de-sialated caeruloplasmin than from intact caeruloplasmin. Does this de-sialation reaction regulate copper accumulation in any way? Are there any physiological circumstances in which changes in neuraminidase activity (EC 3.2.1.18) could regulate copper influx into the mitochondrion?

Frieden: This is an attractive idea but it has not been related to any specific physiological process. It is desirable to keep these proteins in the circulation. De-sialation is peculiar to the liver, which is also the site of biosynthesis of caeruloplasmin. It would be counterproductive for caeruloplasmin to return to the same tissue somewhat destroyed, unless all that is required is for the copper to be recycled.

Mills: Do you envisage entry of caeruloplasmin in the form of the complete and intact protein or in its de-sialated form if it functions in copper transport to receptor sites?

Frieden: I can only speculate about this, but the intact form is certainly one way that these protein molecules, not only caeruloplasmin, can enter tissues.

Mills: Am I correct in saying that studies with de-sialated caeruloplasmin have revealed the entry not only of the copper but also of the protein (Gregoriadis et al 1970, Linder & Moore 1977)?

Frieden: Yes. There is a whole group of proteins which when de-sialated return to the liver.

Bremner: Linder & Moore (1977) injected rats with tritium-labelled caeruloplasmin and detected the tritium label in the cells of several tissues, including heart and brain, within a very short time. They argued that the protein was taken up in intact form from plasma by the tissues.

Frieden: But was it shown that de-sialation was the mechanism?

Bremner: No, not in this case.

Frieden: I accept the possibility that some caeruloplasmin may permeate a variety of tissues.

Bremner: Have you looked at the incorporation of copper from caeruloplasmin into any enzyme other than cytochrome *c* oxidase?

Frieden: No, we've not looked at other enzymic systems.

Mills: Marceau & Aspin (1973a, b) claimed that caeruloplasmin is also an effective copper donor for superoxide dismutase.

Frieden: Yes, but they did not test the restoration of enzymic activity by caeruloplasmin compared to other copper chelates.

Mills: Another question I'd like to ask relates to species differences in the way iron metabolism is affected by copper deficiency. The copper-deficient pig ultimately suffers a decline in iron concentrations in the plasma and the liver, and a decline in total body iron stores (Lee et al 1976). This is unique as far as I am aware. I do not know of any studies of changes in total body iron in the copper-deficient rat or ruminant but total hepatic iron certainly increases (Evans & Abraham 1973, Mills et al 1976) during copper deficiency. Do such species differences imply that the sites at which caeruloplasmin acts as a rate-limiting ferroxidase (EC 1.16.3.1) in iron metabolism differ between species (e.g. in the intestinal mucosa of the pig and in the liver, spleen and bone marrow of some other species), or is copper also involved in another and unidentified step that regulates iron absorption?

Frieden: Of the two possibilities you propose, a unique control mechanism for iron absorption in the pig appears to be the most logical.

Mills: An absorptive defect must be present in the pig because total mucosal iron rises and histochemically stainable iron granules appear in the mucosa of the copper-deficient pig (Lee et al 1976).

Frieden: The iron mobilization process in isolated liver systems certainly responds beautifully to caeruloplasmin as the key controlling factor, but I cannot immediately account for all the observations you have mentioned.

Owen: Cartwright's group (Lee et al 1968) directly measured iron absorption and found that it was sharply reduced in copper-deficient pigs, compared to that in other species, so it seems that perhaps the pig is unique. The first step in absorption is that the iron goes into the mucosal cell and the second step is passage of the iron through the cell and into the blood. These steps are similar to those observed for copper, the second being defective in Menkes' syndrome (Van den Hamer 1978).

Graham: You may know that two of our original cases of copper deficiency in babies were almost completely cured by pharmacological doses of vitamin B_{12}. They exhibited a reticulocyte response and a neutrophil response. Later we had a chance to examine that response again but we probably lost the opportunity to understand the mechanisms involved. We had an anaemic child with long-standing copper deficiency who had been receiving therapeutic doses of iron by mouth. We gave the child pharmacological doses of vitamin B_{12} which first produced a dramatic rise in serum iron, followed by the reticulocyte response. The copper concentration in the serum fell to zero. The mistake we made was that we continued to give him iron, so we don't know whether the effects were on iron absorption or on mobilization of iron from the liver. Another copper-deficient child who had not been preloaded with iron was also given vitamin B_{12} and showed no response. We suspect that in the first child the iron was in the liver and was mobilized by vitamin B_{12} but we have no idea why the vitamin should produce that effect.

Hill: You referred to caeruloplasmin as a superoxide scavenger, Professor Frieden, and yet you suggested that caeruloplasmin was not a dismutase. Could you elaborate on that?

Frieden: I was quoting some recent data from Goldstein et al (1979). Their evidence was convincing that caeruloplasmin could destroy superoxide, but their analyses for peroxide revealed that an appreciable fraction of the expected peroxide had disappeared in the process. They suggested that there could be a different mechanism by which caeruloplasmin deals with superoxide and with peroxide.

Hill: But do you think it would be possible to prepare caeruloplasmin that was completely free of dismutase?

Frieden: In view of the relatively low dismutase activity, Goldstein et al (1979) took some pains to check for dismutase in their preparations of caeruloplasmin and they believed that they did not have a contaminating amount of dismutase.

Hill: I am very interested in this because we have some preliminary results (H.A.O. Hill and M.J. Okolow-Zubkowska, unpublished) on detection of hydroxyl radical formation by neutrophils during the respiratory burst associated with phagocytosis. Dismutase is an extremely effective inhibitor of the formation of hydroxyl radicals, as detected by our methods. Caeruloplasmin also affects hydroxyl radical formation in our hands but the process is much slower and we have therefore been concerned to establish how free is our caeruloplasmin from superoxide dismutase (SOD). The effect is there, but we wouldn't yet like to attribute it directly to SOD.

Frieden: There are few substances that will inhibit SOD but the enzymic activities of caeruloplasmin can be inhibited completely by 10^{-3}M sodium azide. So that might help you to distinguish between the effects of caeruloplasmin alone and those that might be due to contamination of caeruloplasmin with SOD.

Hassan: Does cyanide inhibit caeruloplasmin?

Frieden: Yes. Cyanide is inhibitory at the relatively low concentration of 10^{-6}M.

Harris: Is it currently accepted that *one* protein performs all the functions that you have described or is there more than one caeruloplasmin that shares these functions?

Frieden: That is a good question in view of the confusion about the homogeneity of caeruloplasmin and whether it has subunits or not. Caeruloplasmin has about five highly sensitive peptide bonds that are easily cleaved, even by some of the plasma proteases that are encountered during isolation. Isolation should therefore be performed at 0°C and in the presence of 20mM ϵ-aminocaproic acid, which is a standard trypsin-like inhibitor. In this way it has been possible to isolate intact single-chain molecules. Some of the early conclusions that caeruloplasmin contained subunits were based upon the release of large polypeptide fragments as a result of proteolysis. We hope to establish whether all the enzyme activities will remain once we have an undegraded homogeneous preparation.

Harris: Do you therefore believe that the microheterogeneity that has been reported in purification of this protein is due to lack of care in isolating it?

Frieden: The so-called subunits range in molecular weight from 16 000 to 20 000, and they are all due to particular peptide bonds being cleaved, so these results cannot be attributed entirely to microheterogeneity.

Mills: Professor Frieden has mentioned the role of caeruloplasmin in catecholamine metabolism. Has any relationship that might support this role been observed between caeruloplasmin levels and noradrenaline levels in Menkes' syndrome in children or in mice?

Danks: We have not looked at that aspect in Menkes' children.

Hunt: And we have not measured serum levels of caeruloplasmin or noradrenaline in Menkes' mice.

Sourkes: I would question a significant role for caeruloplasmin in oxidation of amines in circulating blood. However, it may function in that way in shed blood. I have not made any measurement on this myself. Do you think that is possible, Professor Frieden?

Frieden: The case for a role of caeruloplasmin in biogenic amine oxidation is largely speculative. Shokeir (1975) claims that there is an association between caeruloplasmin levels and Huntington's disease. Our efforts to verify this relationship on a limited number of samples were negative.

Sourkes: There has, in fact, been very little recent work on a relationship between caeruloplasmin and Huntington's chorea.

Mills: One reason I am interested in this is that Dr J. Hesketh at Aberdeen (unpublished results) is detecting increases in noradrenaline concentrations in copper-deficient bovine blood, but how this effect arises is not clear.

Another point that Earl Frieden has mentioned is a role for caeruloplasmin in the oxidation of cysteine. Inadequately understood effects of copper deficiency are the metabolic origin of defects in – SH oxidation and the failure of disulphide bridge formation which lead to the appearance of kinky hair in the Menkes' child and of 'steely' wool in the copper-deficient sheep. Perhaps the proposed role for caeruloplasmin in cysteine oxidation may be related to this effect. Marston suggested, many years ago, on the basis of histochemical evidence, that the 'steely' wool defect is related to a loss of cytochrome *c* oxidase in the wool follicle.

Danks: Scientists at the C.S.I.R.O. Division of Wool Research in Melbourne have spent many years studying the structure of keratin (Fraser & Gillespie 1976). They still do not know what produces the disulphide bonds in keratin or how copper deficiency prevents their formation. Neither do they have any evidence that an enzyme is involved, and they are still unsure whether the process is enzymic or due to a direct effect of copper.

Harris: Could you clarify whether we are discussing cysteine or peptidyl cysteine because they are very different? I believe that Professor Frieden has been discussing cysteine itself.

Frieden: The rate of oxidation, catalysed by caeruloplasmin, is considerably greater for free cysteine than for cysteine in combination with peptide, e.g. glutathione.

Harris: If the K_m value for cysteine is in the mM range or higher, then the enzyme has an extremely weak affinity for this substrate. This might answer Dr Mills' question about whether the enzyme oxidizes cysteine *in vivo*.

Mills: So it seems that we remain without a biochemical explanation for kinky hair, one of the major defects in copper deficiency.

Danks: A large number of people in the Australian wool industry would be interested in an explanation, I'm sure!

Shaw: Might I ask about fetal caeruloplasmins? Milne & Matrone (1970) have described two types of caeruloplasmin in the developing piglet, which they thought differed in their copper content and enzymic activity; the neonatal type predominated at birth, but the postnatal rise was accounted for by an increase in the adult type. Is there any evidence for such fetal proteins in the human – perhaps proteins adapted for transporting the copper that crosses the placenta?

Frieden: I don't know of any recent work on identification of fetal caeruloplasmin or of its different properties.

Danks: Presumably there cannot be as large a quantity of fetal caeruloplasmin as is found postnatally and in the adult because the fetal concentrations of copper in the serum are very low.

Sourkes: Young & Curzon (1974) have demonstrated a neonatal form of caeruloplasmin that, when isolated, was not different from the adult form on kinetic grounds.

Dormandy: I think that the superoxide-dismutase-like activity of caeruloplasmin depends to some extent on how superoxide is generated and on how the dismutase-like activity is measured. With a standard superoxide-generating system, such as the xanthine-oxidase system, caeruloplasmin has a very weak dismutase-like action. But superoxide can also be generated by the iron-catalysed peroxidation of lipids; and the powerful inhibitory action of caeruloplasmin in such a system can be interpreted as a superoxide-dismutase-like effect.

Hill: Could superoxide possibly be a substrate for caeruloplasmin, as it is for indole 2,3-dioxygenase (EC 1.13.11.17) (Hirata et al 1977)?

Frieden: We examined that possibility years ago and found no catalytic activity of caeruloplasmin toward superoxide or peroxide.

Hill: Is one of the copper ions on caeruloplasmin bound very weakly?

Frieden: When caeruloplasmin is isolated there is one ion that is removed by Chelex resin. That copper ion is relatively weakly bound compared to the other copper ions.

Hill: I wondered if it was possible that you were observing reactions of the small amount of free copper ions.

Frieden: No; we believe we have excluded that possibility.

Mills: An unexplained recent observation by Kellerher & Mason (1979) is that the inhibitory effect of tetrathiomolybdate on caeruloplasmin appears to be much greater when the assay substrate is *o*-dianisidine than when *p*-phenylenediamine (PPD) is used.

Frieden: Are the pHs the same in both assays?

Mills: I believe that they are. There's quite a marked difference in the rates at which activity falls with increasing concentrations of tetrathiomolybdate; at 7.5×10^{-6}M, 85% of the activity is lost when *o*-dianisidine is the substrate compared with only 15% when PPD is the substrate.

I would like to ask Dr Graham how long before the appearance of neutropenia in copper-deficient children is the decline in caeruloplasmin activity apparent? In other words, how useful is caeruloplasmin as an indicator of developing copper deficiency in children?

Graham: The most common cause of a reduced concentration of caeruloplasmin in the serum of children is protein malnutrition, which can confuse the diagnosis, because the malnutrition often coexists with copper depletion.

Frieden: Does the malnutrition reflect liver malfunction?

Graham: It reflects failure to synthesize the apocaeruloplasmin in the liver. The neutrophils are probably a more sensitive index of copper deficiency, but as the child becomes deficient, both neutrophils and serum caeruloplasmin are reduced almost simultaneously. During recovery (after administration of copper), the neutrophil count increases rapidly, and before there is any measurable increase in caeruloplasmin.

Terlecki: Is there a direct relationship between the levels of caeruloplasmin and copper, so that when there is a high level of caeruloplasmin there is a high level of copper?

Frieden: That relationship always holds, but I don't know whether the converse is true. If copper is not present, then no caeruloplasmin is found, although small and possibly insignificant amounts of apocaeruloplasmin have been reported to exist in the plasma.

Terlecki: I asked the question because you mentioned that one of the functions of caeruloplasmin is to transport copper, and also because the concentration of caeruloplasmin increases in infectious disease. However, Patterson & Sweasey (1973) found that in the (viral) Border disease of sheep, the serum copper of clinically affected lambs was lower than in control animals; unfortunately no measurements of caeruloplasmin were performed. In this disease there is also an abnormal distribution of copper: the concentration is low in the brain and liver, but high in the spinal cord (Patterson et al 1975). Can you explain these observations?

Frieden: If a catalytic test is used to measure the levels of caeruloplasmin in the plasma there will be no catalytic activity in the absence of copper, although it is not clear whether all the copper is necessary for catalytic activity. I would be sceptical of results that reveal a high caeruloplasmin activity in the absence of copper although the reverse might be true, i.e. free copper can exist in the absence of caeruloplasmin, although those circumstances might be considered to be pathological or at least idiosyncratic.

Graham: In our studies of recovery from copper deficiency (Holtzman et al 1970) we found that the serum concentrations of copper recovered faster than those of caeruloplasmin (i.e. the percentage of copper in the form of caeruloplasmin was relatively low).

Another tantalizing question in relation to the viral disease that Dr Terlecki mentioned is whether copper deficiency increases susceptibility to neurotropic viruses. Two of the copper-deficient children that we have encountered developed type II poliomyelitis, the least pathogenic type (Cordano et al 1964, and unpublished observations). A third child who developed encephalitis after measles vaccination was later revealed to have been copper-deficient at the time of vaccination.

Owen: I would like to return to the relationship between serum copper and serum caeruloplasmin. Free copper can be present in greater amounts than the copper contained in the caeruloplasmin, typically in Wilson's disease. But there is one disease in which the reverse is sometimes true: in primary biliary cirrhosis, PPD-oxidase activity (and therefore, presumably, the amount of caeruloplasmin) is usually higher than can be accounted for by the serum copper (Dickson et al 1979). This phenomenon can be explained by two possibilities: either there is a caeruloplasmin molecule in the blood that has less than six atoms of copper, which seems unlikely, or the liver might be releasing other oxidases that attack the PPD; the conventional PPD-oxidase tests cannot discriminate between caeruloplasmin and other oxidases.

Frieden: If sodium azide were used it would discriminate between them because it would selectively inhibit caeruloplasmin. In addition, one could isolate the protein to determine whether it is the usual form of caeruloplasmin.

Owen: That could be determined immunologically, too.

Dormandy: We occasionally observe a discrepancy between caeruloplasmin as measured by immunodiffusion and caeruloplasmin as measured by its oxidase-like activity in diseases like primary biliary cirrhosis and Wilson's disease (Gollan et al 1977).

Frieden: Different caeruloplasmins from different organisms certainly have different activities, but presumably there is not a significant change in the prosthetic group.

McMurray: We have consistently observed a difference between the concentrations of caeruloplasmin and of copper in plasma and serum (McMurray 1980, this volume). This seems to result from a loss of copper and caeruloplasmin during the clotting process. Would you like to comment, Professor Frieden, on how this could occur?

Frieden: I have not heard about this difference and its relationship to clotting. Perhaps a deep proteolysis into the active site might decrease the enzymic activity of caeruloplasmin.

Sourkes: I'd like to raise a question about the formation of excess caeruloplasmin in response to oestrogens and during pregnancy. Has anybody studied the nature of the inductive mechanism responsible for the increased synthesis of caeruloplasmin under those conditions?

Frieden: Yes. The mechanism is hormonal, and slow — it takes a couple of weeks.

Sourkes: Does it involve an adenylate cyclase mechanism?

Frieden: No. I think it depends on general protein synthesis and release of the protein from the liver.

Harris: The effect also appears to operate by a stimulation or a stabilization of mRNA or of the protein in a selective manner (Sunderman et al 1971, Sunshine et al 1971). So the answer to Ted Sourkes' question is that the response to oestrogens is an inductive mechanism in the classical sense.

References

Cordano A, Baerti JM, Graham GG 1964 Copper deficiency in infancy. Pediatrics 34:324-336

Dickson ER, Fleming CR, Ludwig J 1979 Primary biliary cirrhosis. In: Popper H, Schaffner F (eds) Progress in liver disease. Grune & Stratton, New York, vol 6:487-502

Dormandy TL 1980 Free radical reactions in biological systems. Ann R Coll Surg Engl 62:188-194

Evans JL, Abraham PA 1973 Anaemia, iron storage and ceruloplasmin in copper nutrition in the growing rat. J Nutr 103:196-201

Fraser RDB, Gillespie JM 1976 Wool structure and biosynthesis. Nature (Lond) 261:650-654

Goldstein IM, Kaplan HB, Edelson HS, Weissmann G 1979 Ceruloplasmin – a scavenger of superoxide anion radicals. J Biol Chem 254:4040-4045

Gollan JL, Stocks J, Dormandy TL, Sherlock S 1977 Reduced oxidase activity in the ceruloplasmin of two families with Wilson's disease. J Clin Pathol (Lond) 30:81-83

Gregoriadis G, Morell AG, Sternlieb I, Scheinberg IH 1970 Catabolism of desialylated ceruloplasmin in the liver. J Biol Chem 245:5833-5837

Hirata F, Ohnishi T, Hayaishi O 1977 Indoleamine 2,3-dioxygenase: characterization and properties of enzyme O_2^- complex. J Biol Chem 252:4637-4642

Holtzman NA, Charache P, Cordano et al 1970 Distribution of serum copper in copper deficiency. Johns Hopkins Med J 126:34-42

Inaba T, Frieden E 1967 Changes in ceruloplasmin during anuran metamorphosis. J Biol Chem 242:4789-4795

Kellerher CA, Mason J 1979 The effect of tetrathiomolybdate upon sheep caeruloplasmin amine oxidase activity in vitro: the influence of substrate upon apparent sensitivity to inhibition. Res Vet Sci 26:124-125

Lee GR, Nacht S, Lukens JN, Cartwright GE 1968 Iron metabolism in copper-deficient swine. J Clin Invest 47:2058-2069

Lee GR, Williams DM, Cartwright GE 1976 Role of copper in iron metabolism and heme biosynthesis. In: Prasad AS, Oberleas D (eds) Trace elements in human health and disease. Vol 1: Zinc and copper. Academic Press, New York, p 373-390

Linder MC, Moore JR 1977 Plasma ceruloplasmin: evidence for its presence in and uptake by heart and other organs of the rat. Biochim Biophys Acta 499:329-336

Marceau N, Aspin N 1973a The intracellular distribution of the radiocopper derived from ceruloplasmin and from albumin. Biochim Biophys Acta 328:338-350

Marceau N, Aspin N 1973b The association of copper derived from ceruloplasmin with cytocuprein. Biochim Biophys Acta 328:351-358

Mills CF, Dalgarno AC, Wenham G 1976 Biochemical and pathological changes in tissues of Friesian cattle during the experimental induction of copper deficiency. Br J Nutr 38:309-331

Milne DB, Matrone G 1970 Forms of ceruloplasmin in developing piglets. Biochim Biophys Acta 212:43-49

Patterson DSP, Sweasey D 1973 Hypocupraemia in experimental Border disease. Vet Rec 93:484-485

Patterson DSP, Terlecki S, Foulkes JA, Sweasey D, Glancy EM 1975 Spinal cord lipids and myelin composition in Border disease (hypomyelinogenesis congenita) of lambs. J Neurochem 24:513-522

Roesen HP, Lee GR, Nacht S, Cartwright GE 1970 The role of ceruloplasmin in iron metabolism. J Clin Invest 49:2408-2417

Shokeir MHK 1975 Investigation on Huntington's disease. III: Biochemical observations, a possibly predictive test? Clin Genet 7:354-360

Sunderman FW Jr, Nomoto S, Gillies CG, Goldblatt PJ 1971 Effect of estrogen administration upon caeruloplasmin and copper concentration in rat serum. Toxicol Appl Pharmacol 20:588-589

Sunshine GH, Williams DJ, Rabin BR 1971 Role of steroid hormones in the interaction of ribosomes with the endoplasmic membranes of rat liver. Nat New Biology 230:133-139

Van den Hamer CJA 1978 Results of a ^{64}Cu-loading test applied to patients with an inherited defect in their Cu-metabolism (Menkes' disease). In: Kirchgessner M (ed) Trace element metabolism in man and animals. Arbeitskreits für Tierernährungsforschung, Freising-Weihenstephan, vol 3:394-396

Young SN, Curzon G 1974 Neonatal human caeruloplasmin. Biochim Biophys Acta 336:306-308

Youngman RJ, Pallett KE, Dodge AD 1980 Active oxygen species in herbicide action. In: Bannister JV, Hill HAO (eds) Chemical and biochemical aspects of superoxide and superoxide dismutase. Elsevier/North-Holland, Amsterdam (Developments in biochemistry series, vol 11A) p 402-411

Superoxide dismutases

HOSNI MOUSTAFA HASSAN

Department of Microbiology & Immunology, McGill University, Montreal, Quebec, H3A 2B4 Canada

Abstract   Superoxide dismutases (EC 1.15.1.1) are metalloenzymes that catalytically scavenge the superoxide radical. They are essential for the aerobic survival of all forms of life. There are three types of superoxide dismutase, containing manganese, iron, or copper and zinc. The copper–zinc type has generally been isolated from eukaryotic cells except for the enzyme from the symbiotic marine bacterium *Photobacterium leiognathi*. The copper–zinc type, from different sources, has a molecular weight of about 32 000, and is composed of two identical subunits, each containing one atom of copper and one atom of zinc. The copper participates in the catalytic activity of the enzyme, while the zinc plays only a structural role.   The enzyme has been resolved reversibly.   Superoxide dismutases provide protection against oxygen toxicity, against compounds that cause exacerbation of oxygen toxicity, against ionizing radiation, and also against the damaging sequelae of prolonged inflammation.

Superoxide dismutases (EC 1.15.1.1) are ubiquitous among aerobic and aerotolerant organisms, and are essential for defence against oxygen toxicity. Superoxide dismutases are metalloenzymes that catalyse the dismutation of the superoxide radical (O_2^-), which is normally produced during aerobic metabolism. The enzyme catalyses the reaction:

$$O_2^- + O_2^- + 2H^+ \rightarrow H_2O_2 + O_2$$

This reaction can proceed spontaneously at a relatively rapid rate, but the enzyme increases this rate by more than 10 000-fold.

Transition metal cations can also catalyse the above reaction but they are neither as effective nor as plentiful inside the living cells as are superoxide dismutases. There are three classes of superoxide dismutase: those containing manganese (MnSOD), iron (FeSOD), or both copper and zinc (Cu-ZnSOD). MnSOD and FeSOD are characteristic of prokaryotes and are closely related to each other (Steinman & Hill 1973). MnSOD is also found in the mitochondria of eukaryotes and is extensively homologous with prokaryotic MnSOD. Cu-ZnSOD is characteristic of eukaryotes (McCord & Fridovich 1969, Weisiger & Fridovich 1973, Bannister et al 1973, Beauchamp & Fridovich 1973), with the exception of the bacterium *Photobacterium leiognathi,* which contains Cu-ZnSOD and is usually found in symbiotic association with the pony fish (Puget & Michelson 1974). No sequence homology is found between Cu-ZnSOD and FeSOD or MnSOD, which suggests that Cu-ZnSOD evolved from a different origin but under the same selective pressure. This paper is devoted mainly to the physiology and chemistry of the copper–zinc-containing superoxide dismutases.

NATURE AND SOURCES OF O_2^-

The superoxide radical (O_2^-) is the conjugate base of a weak acid, the perhydroxyl radical (HO_2^-) whose pKa is 4.8 (Behar et al 1970). It can act as a reductant (E_0' for O_2/O_2^- is about $-0.33V$) as well as an oxidant (E_0' for H_2O_2/O_2^- is $+0.87V$). O_2^- is generated in many biological reactions that reduce molecular oxygen (for complete survey see Fridovich 1975, 1979). Attempts to measure the rates of O_2^- generated inside cells are usually hampered by the ubiquity of SOD. A recent study in which a specific inhibitory antibody was used to neutralize the activity of the enzyme has demonstrated that 17% of the total oxygen consumed by crude extracts of *Streptococcus faecalis* is channelled through a univalent pathway and generates O_2^- (Britton et al 1978). Although the quantitatively most significant sources of O_2^- within any given type of cell remain unknown, it is safe to conclude that the superoxide radical is produced in every living cell that is capable of reducing oxygen.

DELETERIOUS EFFECTS OF SUPEROXIDE

The discovery of the protective role of superoxide dismutases against the toxicity of oxygen have led to the proposal that superoxide radical is a major contributor to oxygen toxicity (McCord et al 1971). However, the chemistry of O_2^- in aqueous solutions indicates that it is much less reactive than the hydroxyl radical, OH· (Bielski & Richter 1977). Nevertheless, it has been demonstrated that enzymically, photochemically, or electrochemically generated fluxes of O_2^- are toxic and destructive to living cells. Fluxes of O_2^- kill bacteria (Babior et al 1975), inactivate viruses (Lavelle et al 1973), lyse erythrocytes (Kellogg & Fridovich 1977), destroy granulocytes (Salin & McCord 1975), damage myoblasts in culture (Michelson & Buckingham

1974), depolymerize hyaluronate (McCord 1974), inactivate enzymes (Lavelle et al 1973, Kellogg & Fridovich 1977), damage DNA (Van Hemmen & Meuling 1975), decompose methional to ethylene (Beauchamp & Fridovich 1970) and initiate lipid peroxidation (Kellogg & Fridovich 1975). Furthermore, paraquat, which is known to exacerbate the production of O_2^-, was found to be highly mutagenic in the *Salmonella typhimurium* test strains of Ames and this mutagenic effect was oxygen-dependent (H.M. Hassan & J.E. Fein, unpublished paper, abstract no. K3p, Can Soc Microbiol Annual Meeting, June 1980). In most cases, superoxide dismutase, catalase (EC 1.11.1.6) or compounds known to scavenge $OH\cdot$, were found to protect against the deleterious effects of O_2^-. These findings led to the proposal that O_2^- and H_2O_2 must have interacted to produce $OH\cdot$ (Beauchamp & Fridovich 1970), as originally suggested by Haber & Weiss (1934). Attempts to demonstrate the direct interaction between O_2^- and H_2O_2 showed that iron-chelates are required to catalyse the generation of $OH\cdot$ from $O_2^- + H_2O_2$ (McCord & Day 1978). Thus O_2^- reduces the ferric chelate to the ferrous state which can then reduce H_2O_2 to $OH\cdot$, as shown in the following scheme:

$$O_2^- + Fe^{3+}\text{-chelate} \rightarrow O_2 + Fe^{2+}\text{-chelate}$$
$$Fe^{2+}\text{-chelate} + H_2O_2 \rightarrow Fe^{3+}\text{-chelate} + OH^- + OH\cdot$$

$$O_2^- + H_2O_2 \rightarrow O_2 + OH^- + OH\cdot$$

This reaction is also known among some chemists as the superoxide-driven Fenton chemistry! Regardless of the name given to this reaction, O_2^- and H_2O_2 can interact to generate $OH\cdot$, which can wreck the delicate architecture of the cell. Since O_2^- itself generates H_2O_2, O_2^- production would lead to $OH\cdot$ formation and cellular damage. Superoxide dismutases protect the cells by keeping the steady-state concentration of O_2^- vanishingly low, thus minimizing the production of the deleterious agent, $OH\cdot$.

PHYSICOCHEMICAL PROPERTIES OF Cu-ZnSOD

The copper–zinc superoxide dismutases were originally known as cupreins (erythrocuprein, haemocuprein, cerebrocuprein, cytocuprein) (Mann & Keilin 1938); because their catalytic activity had not been demonstrated at that time, they were considered as copper storage proteins (Mohamed & Greenberg 1954). About twelve years ago, McCord & Fridovich (1969) reported the catalytic function of erythrocuprein. This led to the subsequent isolation, from different living forms, of other superoxide dismutases that contained copper, manganese or iron at the active site. Copper–zinc-superoxide dismutase has been isolated from a wide range of eukaryotes including *Neurospora* and yeasts. This enzyme is characteristically found in the cytosol of eukaryotic cells; however, the symbiotic marine bacterium *Photobacterium leiognathi* is an exception to this general rule (Puget & Michelson 1974). The possibility that the bacterium acquired this enzyme from its host, the pony fish, via gene transfer is currently under investigation (J. Martin & I. Fridovich, personal communication, 1979). The forms of Cu-ZnSOD from different human organs are immunologically identical (Hartz et al 1973, Shields et al 1961). Immunological cross-reactivity was also seen between the human and monkey cupreins, while no cross-reactivity was found between the human and the pig, chicken or cow cupreins (Shields et al 1961). The human Cu-ZnSOD gene has been assigned to chromosome 21 (Tan et al 1973).

The Cu-ZnSODs from vertebrates, fungi, and plants have a very similar amino acid composition, with a high degree of structural homology. The enzyme is a homodimer with a subunit weight of 16 000 daltons and contains one copper and one zinc atom per subunit. Analysis of the amino acid sequence has revealed that the enzyme is devoid of tryptophan, and that each subunit consists of 151 amino acid residues with an acetylated amino terminus (Abernethy et al 1974, Steinman et al 1974, Evans et al 1974). The basic structure and the arrangement around the active site have been elucidated at 3 Å resolution by X-ray crystallography (Richardson et al 1975a, b, 1976). The subunit resembles a cylinder resulting from eight extended chains of antiparallel β-structure consisting of 75 residues. The segments of the sequence in this β-barrel structure are residues 2–11, 13–23, 26–35, 38–47, 80–88, 91–100, 112–118, and 142–149. The subunit also possesses two loops of nonrepetitive structure which project from the β-barrel; together the two loops enclose and constitute the active site. One of these loops (residues 48–79) contains the disulphide loop (residues 55 and 144) which participates extensively in the subunit contact. This same loop (residues 61–79) contributes three ligands to the zinc. The contact area between the two subunits includes part of the outside surface of the β-barrel, the last few residues at the carboxyl terminus, and the disulphide loop. There is a free thiol residue on each subunit (Cys 6) which is unreactive in the native enzyme. The contact association between the subunits involves primarily hydrophobic interactions. There are two main-chain and three side-chain hydrogen bonds between the subunits and 12–14 hydrophobic side-chains from each subunit that are in Van der Waals interaction. Also, both valine 146 and isoleucine 111 interact with their counterparts across the local two-fold axis. At the closest approach between the main chains, glycines 49 and 112 of one subunit are opposite glycine 148 of the other subunit.

Within each subunit, the copper and zinc are 6 Å apart and they are bridged by the imidazolate ring of histidine 61. In addition to histidine 61, the copper is liganded to the imidazole rings of histidines 44, 46 and 118 which are arranged in a distorted square-planar structure, while the zinc is liganded by imidazole rings of histidines 69 and 78 and the carboxyl group of aspartate 81 which are arranged in tetrahedral structure (Richardson et al 1975b). The copper is relatively exposed to the solvent whereas the zinc is buried within the subunit structure. The two copper atoms on the opposite subunits are approximately 34 Å apart.

The enzyme is remarkably stable and remains active in the presence of 9.0 M urea or 4% sodium dodecylsulphate (Forman & Fridovich 1973). However, dialysis of the native enzyme at low pH against EDTA (ethylenediaminetetraacetate) caused loss of both the copper and the zinc, with a concomitant loss of activity. Reconstitution studies revealed that the addition of copper alone restored full enzymic activity to the apoenzyme (McCord & Fridovich 1969, Beem et al 1974, Fee & Briggs 1975). Attempts to replace the copper by other metals have always resulted in loss of activity. On the other hand the zinc may be replaced by a number of other metals (i.e. Co^{2+}, Hg^{2+}, Cd^{2+}) without loss of activity (Forman & Fridovich 1973). Thus, the copper plays a catalytic role whereas the zinc plays a structural role. The zinc contributes mainly to the stability of the enzyme; the replacement of zinc by mercury gives an active enzyme that is more stable than the native enzyme (Forman & Fridovich 1973).

The list of effective inhibitors of Cu-ZnSOD is limited, and so far only one compound has been useful for studies *in vivo*. *Cyanide* causes reversible inhibition of Cu-ZnSOD (Rotilio et al 1972) by the binding of its carbon portion to the copper (Haffner & Coleman 1973). Cyanide at about 50 μM and at pH 8.2 causes 50% inhibition of the human enzyme. *Azide* also binds reversibly onto the copper of Cu-ZnSOD (Beem et al 1977), but it is a much less effective inhibitor than cyanide; 32 mM of azide is required to cause 50% inhibition of the enzyme (Misra & Fridovich 1978). *Hydrogen peroxide* (H_2O_2) irreversibly inhibits the enzyme by its interaction with copper (Hodgson & Fridovich 1975). The concentration of H_2O_2 required for inhibiting the enzyme is several orders of magnitude higher than the physiological concentration of H_2O_2 *in vivo*. Hydrogen peroxide reduces Cu(II) to Cu(I) which slowly reduces oxygen to O_2^-. Cu(I) at the active centre of the enzyme can also react with a second H_2O_2 to generate OH·, which can attack and inactivate the enzyme (Hodgson & Fridovich 1975). Finally, the chelating agent *diethyldithiocarbamate* (DDC) can inactivate Cu-ZnSOD by removing the copper from the enzyme (Heikkila et al 1976, Misra 1979). DDC has been used to decrease levels of Cu-ZnSOD in the lungs of young rats. DDC-treated rats were more susceptible to the lethal effects of hyperoxia (Frank et al 1978).

CATALYTIC MECHANISM OF Cu-ZnSOD

The mechanism of the enzymic dismutation of superoxide has been elucidated primarily by pulse radiolysis. These studies revealed cyclical changes in the valence of the copper brought about by O_2^- (Klug et al 1972, Rotilio et al 1972, Klug-Roth et al 1973). Thus, the absorbance of the enzyme at 680 nm, which is due to Cu(II), could be bleached by a pulse of O_2^- without loss of activity. This bleaching of the enzyme was interpreted as being due to the reduction at the active centre of cupric ion to cuprous. Furthermore, when the enzyme (E) was first reduced to the Cu(I) state by hydrogen peroxide, a pulse of O_2^- partially restored the absorbance at 680 nm, presumably by the reoxidation of the Cu(I) to Cu(II). These results have indicated that the copper undergoes cycles of reduction and reoxidation during successive encounters with the superoxide radical, as follows:

$$E\text{-}Cu^{2+} + O_2^- \rightarrow E\text{-}Cu^{1+} + O_2$$
$$E\text{-}Cu^{1+} + O_2^- + 2H^+ \rightarrow E\text{-}Cu^{2+} + H_2O_2$$

Overall effect: $O_2^- + O_2^- + 2H^+ \rightarrow O_2 + H_2O_2$

This is believed to be an *inner sphere mechanism* with the superoxide radical directly binding to the copper during or before the electron transfer reaction. The rate constant for the reaction catalysed by the bovine erythrocyte enzyme is approximately 2×10^9 M^{-1} s^{-1} and is apparently diffusion-limited. This rate is not affected by changes of pH between 5.5 and 9.5; however there exists an inverse relationship between the viscosity of the solvent and the rate of the reaction.

The two copper atoms on the opposite subunits act independently of each other. Thus, it was found that the Cu-ZnSOD of swordfish liver, when dissociated into subunits in 8 M urea, was as active as the native enzyme (Bannister et al 1978). More recently, hybridization studies have shown that a native subunit will exhibit the same catalytic activity, whether associated with another native subunit or with a chemically modified and catalytically inactive subunit (Malinowski & Fridovich 1979a). These results ruled out the possibility of inhibitory interactions between subunits, which would have resulted in *half-of-the-sites reactivity* as proposed by Fielden et al (1974).

The essential role of arginine 141 in the catalytic activity of Cu-ZnSOD has recently been probed by the use of reagents ($\alpha\beta$-diketones) that specifically modify arginine residues (Malinowski & Fridovich 1979b). Amino acid analysis of the modified and virtually inactive enzyme showed a loss of one arginine residue per subunit of the enzyme. A positive correlation was seen between loss of activity and modification of arginine 141. Cu-ZnSOD from various sources behaved similarly, while FeSOD and MnSOD were insensitive to these reagents. X-ray crystallography has previously shown that arginine 141 lies in the region of the active site of Cu-ZnSOD. Therefore, this arginine moiety seems to play an essential part in the catalytic function of this class of enzyme. It is proposed that arginine 141 might serve either in electrostatic guidance of the substrate (O_2^-) to the active centre or in proton conduction (Malinowski & Fridovich 1979b).

EFFECTS OF COPPER ON BIOSYNTHESIS OF SOD

Feeding chicks a diet low in copper (< 1 mg/kg dry matter) resulted in a significant decrease of SOD activity in erythrocytes compared to that in copper-supplemented (10 mg/kg dry matter) controls (Bettger et al 1979). A high correlation between the levels of cytochrome *c* oxidase and erythrocyte SOD was observed. On the other hand, changing the dietary levels of zinc (5–100 mg/kg dry matter) had no effect on the levels of erythrocyte SOD (Bettger et al 1979). In the brains of male mice, the SOD activity decreased, between the ages of 50 and 900 days, by 32–36%. Meanwhile, the copper content of the brain increased by 45% up to 600 days of age and continued to increase slightly from 600 to 900 days of age. The concentrations of copper or SOD in the brains of either young or old mice did not change after copper supplementation (Massie et al 1979). The effect of copper on SOD biosynthesis was also studied in the fungus *Dactylium dendroides*, which contains both Cu-ZnSOD and MnSOD (Shatzman & Kosman 1978). It contained less Cu-ZnSOD, when grown in a copper-deficient medium, but the fungus made more of the MnSOD, so that the total SOD remained unaffected by the copper deficiency. This clearly indicates that the organism requires SOD for survival, and that one type of SOD can substitute for another.

THE BIOLOGICAL ROLE OF SUPEROXIDE DISMUTASES

All evidence available to date leads us to conclude that the dismutation of the superoxide radical is the true and only biological function of superoxide dismutases. The superoxide radical, the substrate for SOD, can be made only in the presence of oxygen. Studies *in vitro* and *in vivo* have demonstrated that O_2^- is cytotoxic, and that superoxide dismutases protect against such toxicity. Thus, only strictly anaerobic microorganisms, which are oxygen sensitive, and organisms that are indifferent to oxygen were found to lack the activity of superoxide dismutase (McCord et al 1971). Furthermore, a positive correlation was found between the concentrations of superoxide dismutase in the anaerobes that have it, and their tolerance to oxygen exposures (Tally et al 1977). There are several lines of evidence which lend support to the superoxide theory of oxygen toxicity and the protective roles of superoxide dismutases:

(1) Induction of the enzyme by its substrate

Exposure to increased concentrations of oxygen elicits increased synthesis of SOD both in prokaryotes (Gregory & Fridovich 1973, Gregory et al 1973, Hassan & Fridovich 1977a) and in eukaryotes (Gregory et al 1974, Asada et al 1976, Crapo & Tierney 1974, Boveris et al 1978, Rister & Baehner 1975). Studies with *E. coli* have demonstrated that molecular oxygen *per se* is not the inducer of SOD, and showed that a product of oxygen metabolism, O_2^-, is the true inducer (Hassan & Fridovich 1977b, c, d). The rate of SOD biosynthesis in *E. coli* was shown to be dependent on the level of O_2^- inside the cells, and this in turn was dependent on the growth rate, the type of metabolism, and on the presence of some redox-active compounds which could increase the level of O_2^- generated within the cells (Hassan & Fridovich 1977b, c, d, 1978, 1979a, 1980).

(2) Mutants

Mutants of *E. coli* that were selected on the basis of their temperature-sensitive intolerance for oxygen were found to be temperature-sensitive with respect to biosynthesis of superoxide dismutase (McCord et al 1973). More recently, mutants of *E. coli,* deficient in superoxide dismutase and catalase or deficient in catalase alone, were found to be exquisitely sensitive to oxygen toxicity (Hassan & Fridovich 1979b). The mutants were found to revert to the wild type phenotype (oxygen-tolerant) at a frequency of 1 in 10^5 cells to 1 in 10^4 cells, which is much higher than the normal spontaneous rate expected for point mutations. This pointed to the possibility that the unscavenged O_2^- and H_2O_2 might be mutagenic agents. This prediction has recently been confirmed (H.M. Hassan & J.E. Fein, unpublished paper, abstract no. K3p, Can Soc Microbiol Annual Meeting, June 1980). There is also an oxygen-resistant mutant of *Chlorella sorokiniana* which contained 3.5 times more SOD than the wild type, and which was also more resistant to streptonigrin (Pulich 1974).

(3) Protection against oxygen toxicity

Increased levels of MnSOD in *E. coli* as induced by a variety of conditions (i.e. oxygenation, increase in growth rates, growth in presence of redox-active compounds) imparted an increased resistance against oxygen toxicity (Hassan & Fridovich 1977b, c, d, 1978, 1979a, b, 1980). Similar findings were reported in eukaryotes. Exposure of rats to 85% oxygen caused an increase in the levels of SOD in the lung and the increase was positively correlated with the tolerance of the rats to 100% oxygen (Crapo & Tierney 1974). Superoxide dismutase also protects against radiation damage to DNA (van Hemmen & Meuling 1975), to viruses and mammalian cells in culture (Michelson & Buckingham 1974), to *E. coli* (Misra & Fridovich 1976) and to mice (Petkau et al 1976).

It is clear from the above discussions that oxygen free radicals have an important function in oxygen toxicity and that aerobic survival is only possible by virtue of the ubiquitous availability of superoxide dismutases.

ACKNOWLEDGEMENTS

I would like to express my sincere appreciation to my friend and teacher Prof. Irwin Fridovich for introducing me to the field of oxygen metabolism and superoxide dismutases. Thanks are also due to Ms. C. Zanfino for reading this manuscript.

References

Abernethy JL, Steinman HM, Hill RL 1974 Bovine erythrocyte superoxide dismutase. Subunit structure and sequence location of the intrasubunit disulfide bond. J Biol Chem 249:7339-7347
Asada K, Kanematsu S, Takehashi M, Kona Y 1976 Superoxide dismutases in photosynthetic organisms. Adv Exp Med Biol 74:551-564
Babior B, Curnutte JT, Kipnes RS 1975 Biological defense mechanisms: Evidence for the participation of superoxide in bacterial killing by xanthine oxidase. J Lab Clin Med 85:235-244
Bannister JV, Bannister WH, Bray RC, Fielden EM, Roberts PB, Rotilio G 1973 The superoxide dismutase activity of human erythrocuprein. FEBS (Fed Eur Biochem Soc) Lett 32:303-306
Bannister JV, Anastasi A, Bannister WH 1978 Active subunits from superoxide dismutase. Biochem Biophys Res Commun 81:469-472
Beauchamp CO, Fridovich I 1970 A mechanism for the production of ethylene from methional: The generation of the hydroxyl radical by xanthine oxidase. J Biol Chem 245:4641-4646
Beauchamp CO, Fridovich I 1973 Isozymes of superoxide dismutase from wheat germ. Biochim Biophys Acta 317:50-64

Beem KM, Rich WE, Rajagopalan KV 1974 Total reconstitution of copper-zinc superoxide dismutase. J Biol Chem 249:7298-7305

Beem KM, Richardson DC, Rajagopalan KV 1977 Metal sites of copper-zinc superoxide dismutase. Biochemistry 16:1930-1936

Behar D, Czapski G, Rabani J, Dorfman LM, Schwarz HA 1970 Acid dissociation-constant and decay kinetics of perhydroxyl radical. J Phys Chem 74:3208-3213

Bettger WJ, Savage JE, O'Dell BL 1979 Effects of dietary copper and zinc on erythrocyte superoxide dismutase activity in the chick. Nutr Rep Int 19:893-900

Bielski BHJ, Richter HW 1977 A study of the superoxide radical chemistry by stopped-flow radiolysis and radiation induced oxygen consumption. J Am Chem Soc 99:3019-3023

Boveris A, Sanchez RA, Beconi MT 1978 Antimycin- and cyanide-resistant respiration and superoxide anion production in fresh and aged potato tuber mitochondria. FEBS (Fed Eur Biochem Soc) Lett 92:333-338

Britton L, Malinowski DP, Fridovich I 1978 Superoxide dismutase and oxygen metabolism in *Streptococcus faecalis* and comparison with other organisms. J Bacteriol 134:229-236

Crapo JD, Tierney DL 1974 Superoxide dismutase and pulmonary oxygen toxicity. Am J Physiol 226:1401-1407

Evans HJ, Steinman HM, Hill RL 1974 Bovine erythrocyte superoxide dismutase. Isolation and characterization of tryptic, cyanogen bromide, and maleylated tryptic peptides. J Biol Chem 249:7315-7325

Fee JA, Briggs RG 1975 Studies on the reconstitution of bovine erythrocyte superoxide dismutase. V. Preparation and properties of derivatives in which both zinc and copper sites contain copper. Biochim Biophys Acta 400:439-450

Fielden EM, Roberts PB, Bray RC, Lowe DJ, Mautner GN, Rotilio G, Calabrese L 1974 The mechanism of action of superoxide dismutase from pulse radiolysis and electron paramagnetic resonance. Biochem J 139:49-60

Forman HJ, Fridovich I 1973 On the stability of bovine superoxide dismutase. The effect of metals. J Biol Chem 248:2645-2649

Frank L, Wood DL, Roberts RJ 1978 Effect of diethyldithiocarbamate on oxygen toxicity and lung enzyme activity in immature and adult rats. Biochem Pharmacol 27:251-254

Fridovich I 1975 Superoxide dismutases. Annu Rev Biochem 44:147-159

Fridovich I 1979 Superoxide and superoxide dismutases. In: Eichhorn GL, Marzilli DL (eds) Advances in inorganic biochemistry. Elsevier-North-Holland, New York, p 67-90

Gregory EM, Yost FJ Jr, Fridovich I 1973 Superoxide dismutases of *Escherichia coli:* Intracellular localization and functions. J Bacteriol 115:987-991

Gregory EM, Goscin SA, Fridovich I 1974 Superoxide dismutase and oxygen toxicity in a eukaryote. J Bacteriol 117:456-460

Gregory EM, Fridovich I 1973 Induction of superoxide dismutase by molecular oxygen. J Bacteriol 114:543-548

Haber F, Weiss J 1934 The catalytic decomposition of hydrogen peroxide by iron salts. Proc R Soc Lond A Math Phys Sci 147:332-351

Haffner PH, Coleman JE 1973 Cu(II)-carbon bonding in cyanide complexes of copper enzymes: ^{13}C splitting of the Cu(II) electron spin resonance. J Biol Chem 248:6626-6629

Hartz JW, Funakoshi S, Deutsch HF 1973 The levels of superoxide dismutase and catalase in human tissues as determined immunochemically. Clin Chim Acta 46:125-132

Hassan HM, Fridovich I 1977a Enzymatic defenses against the toxicity of oxygen and of streptonigrin in *Escherichia coli* K12. J Bacteriol 129:1574-1583

Hassan HM, Fridovich I 1977b Physiological function of superoxide dismutase in glucose-limited chemostat cultures of *Escherichia coli*. J Bacteriol 130:805-811

Hassan HM, Fridovich I 1977c Regulation of superoxide dismutase synthesis in *Escherichia coli:* glucose effect. J Bacteriol 132:505-510

Hassan HM, Fridovich I 1977d Regulation of the synthesis of superoxide dismutase in *Escherichia coli:* induction by methyl viologen. J Biol Chem 252:7667-7672

Hassan HM, Fridovich I 1978 Superoxide radical and the oxygen enhancement of the toxicity of paraquat in *Escherichia coli*. J Biol Chem 253:8143-8148

Hassan HM, Fridovich I 1979a Intracellular production of superoxide radical and of hydrogen peroxide by redox active compounds. Arch Biochem Biophys 196:385-395

Hassan HM, Fridovich I 1979b Superoxide, hydrogen peroxide, and oxygen tolerance of oxygen-sensitive mutants of *Escherichia coli*. Rev Infect Dis 1:357-367

Hassan HM, Fridovich I 1980 Exacerbation of oxygen toxicity by redox active compounds. In: King TE et al (eds) Oxidases and related redox systems. University Park Press, Baltimore, in press

Heikkila RE, Cabbat FS, Cohen G 1976 In vivo inhibition of superoxide dismutase in mice by diethyldithiocarbamate. J Biol Chem 251:2182-2185

Hodgson EK, Fridovich I 1975 The interaction of bovine erythrocyte superoxide dismutase with hydrogen peroxide: inactivation of the enzyme. Biochemistry 14:5294-5299

Kellogg EW 3rd, Fridovich I 1975 Superoxide, hydrogen peroxide, and singlet oxygen in lipid peroxidation by a xanthine oxidase system. J Biol Chem 250:8812-8817

Kellogg EW 3rd, Fridovich I 1977 Liposome oxidation and erythrocyte lysis by enzymically-generated superoxide and hydrogen peroxide. J Biol Chem 252:6721-6728

Klug D, Rabani J, Fridovich I 1972 A direct demonstration of the catalytic action of superoxide dismutase through the use of pulse radiolysis. J Biol Chem 247:4839-4842

Klug-Roth D, Fridovich I, Rabani J 1973 Pulse radiolytic investigation of superoxide catalyzed disproportionation. Mechanism for bovine superoxide dismutase. J Am Chem Soc 95:2786-2790

Lavelle F, Michelson AM, Dimitrejevic L 1973 Biological protection by superoxide dismutase. Biochem Biophys Res Commun 55:350-357

Malinowski DP, Fridovich I 1979a Bovine erythrocyte superoxide dismutase: diazo coupling, subunit interactions, and electrophoretic variants. Biochemistry 18:237-244

Malinowski DP, Fridovich I 1979b Chemical modification of arginine at the active site of the bovine erythrocyte superoxide dismutase. Biochemistry 18:5909-5917

Mann T, Keilin D 1938 Haemocuprein and hepatocuprein, copper-protein compounds of blood and liver in mammals. Proc R Soc London B Biol Sci 126:303-315

Massie HR, Aiello VR, Iodice AA 1979 Changes with age in copper and superoxide dismutase levels in brains of C57BL/6J mice. Mech Ageing Dev 10:93-99

McCord JM 1974 Free radicals and inflammation: protection of synovial fluid by superoxide dismutase. Science (Wash DC) 185:529-531

McCord JM, Fridovich I 1969 Superoxide dismutase: an enzymic function of erythrocuprein. J Biol Chem 244:6049-6055

McCord JM, Day ED Jr 1978 Superoxide-dependent production of hydroxyl radical catalyzed by iron-EDTA complex. FEBS (Fed Eur Biochem Soc) Lett 86:139-142

McCord JM, Keele BB Jr, Fridovich I 1971 An enzyme-based theory of obligate anaerobiosis: The physiological function of superoxide dismutase. Proc Natl Acad Sci USA 68:1024-1027

McCord JM, Beauchamp CO, Goscin S, Misra HP, Fridovich I 1973 Superoxide and superoxide dismutase. In: King TE et al (eds) Oxidases and related redox systems. University Park Press, Baltimore, p 51-76

Michelson AM, Buckingham ME 1974 Effects of superoxide radicals on myoblast growth and differentiation. Biochem Biophys Res Commun 58:1079-1086

Misra HP 1979 Reaction of copper-zinc superoxide dismutase with diethyldithiocarbamate. J Biol Chem 254:11623-11628

Misra HP, Fridovich I 1976 Superoxide dismutase and the oxygen enhancement of radiation lethality. Arch Biochem Biophys 176:577-581

Misra HP, Fridovich I 1978 Inhibition of superoxide dismutases by azide. Arch Biochem Biophys 189:317-322

Mohamed MS, Greenberg DM 1954 Isolation of purified copper protein from horse liver. J Gen Physiol 37:433-439

Petkau A, Chelack WS, Pleskach SD 1976 Protection of post-irradiated mice by superoxide dismutase. Int J Radiat Biol Relat Stud Phys Chem Med 29:297-299
Puget K, Michelson AM 1974 Isolation of a new copper-containing superoxide dismutase, bacteriocuprein. Biochem Biophys Res Commun 58:830-838
Pulich WM 1974 Resistance to high oxygen tension, streptonigrin, and ultraviolet irradiation in the green alga *Chlorella sorokiniana* strain ORS. J Cell Biol 62:904-907
Richardson JS, Thomas KA, Richardson DC 1975a Alpha-carbon coordinates for bovine Cu,Zn superoxide dismutase. Biochem Biophys Res Commun 63:986-992
Richardson JS, Thomas KA, Rubin BH, Richardson DC 1975b Crystal structure of bovine Cu,Zn superoxide dismutase at 3 Å resolution: Chain tracing and metal ligands. Proc Natl Acad Sci USA 72:1349-1353
Richardson JS, Richardson DC, Thomas KA, Silverton EW, Davies DR 1976 Similarity of three-dimensional structure between the immunoglobulin domain and the copper-zinc superoxide dismutase subunit. J Mol Biol 102:221-235
Rister M, Baehner RL 1975 Induction of superoxide dismutase activity *in vivo* by oxygen in polymorphonuclear leukocytes and alveolar macrophages. Blood 46:1016
Rotilio G, Bray RC, Fielden EM 1972 A pulse radiolysis study of superoxide dismutase. Biochim Biophys Acta 268:605-609
Salin ML, McCord JM 1975 Free radicals and inflammation. Protection of phagocytosing leukocytes by superoxide dismutase. J Clin Invest 56:1319-1323
Shatzman AR, Kosman DJ 1978 The utilization of copper and its role in the biosynthesis of copper-containing proteins in the fungus, *Dactylium dendroides*. Biochim Biophys Acta 544:163-179
Shields GS, Markowitz H, Klassen WH, Cartwright GE, Wintrobe MM 1961 Studies on copper metabolism XXXI. Erythrocyte copper. J Clin Invest 40:2007-2015
Steinman HM, Hill RL 1973 Sequence homologies among bacterial and mitochondrial superoxide dismutases. Proc Natl Acad Sci USA 70:3725-3729
Steinman HM, Naik VR, Abernathy JL, Hill RL 1974 Bovine erythrocyte superoxide dismutase. Complete amino acid sequence. J Biol Chem 249:7326-7338
Tally FP, Goldin HR, Jacobus NV, Gorbach SL 1977 Superoxide dismutase in anaerobic bacteria of clinical significance. Infect Immun 16:20-25
Tan YH, Tischfield J, Ruddle FH 1973 The linkage of genes for the human interferon-induced antiviral protein and indophenol oxidase-B traits to chromosome G-21. J Exp Med 137:317-330
Van Hemmen JJ, Meuling WJA 1975 Inactivation of biologically active DNA by γ-ray induced superoxide radicals and their dismutation products singlet molecular oxygen and hydrogen peroxide. Biochim Biophys Acta 402:133-141
Weisiger RA, Fridovich I 1973 Superoxide dismutase. Organelle specificity. J Biol Chem 248:3582-3592

Discussion

Frieden: If, in the presence of superoxide dismutase, you simultaneously give catalase or peroxidase to prevent the accumulation of hydrogen peroxide can you inhibit the Haber-Weiss reaction and the toxic action of OH·?

Hassan: Superoxide dismutase alone can prevent the generation of OH·. If enough catalase is added, it can also inhibit the generation of OH·, *in vitro*.

Frieden: In the work of Goldstein et al (1979) on the so-called scavenging action of caeruloplasmin they did not obtain stoichiometric yields of hydrogen peroxide. Can you suggest an alternative mechanism for removing superoxide, i.e. one that doesn't involve hydrogen peroxide production?

Hassan: I don't see how the dismutation of O_2^- can occur without generation of hydrogen peroxide unless the hydrogen peroxide is used in another reaction.

Hill: We should remember that water is a reduction product of superoxide, so if caeruloplasmin is reduced when it is added to the enzyme then water will be produced, provided that superoxide is a substrate for caeruloplasmin.

Dormandy: Catalase deficiency is known to be virtually harmless. Is there any hard evidence that superoxide dismutase deficiency in humans is associated with any functional disability? (I doubt it.)

Hassan: I would expect SOD mutations in humans to be lethal because we cannot survive anaerobically and that is why we don't see evidence of SOD mutations. We can detect mutations in bacteria because we can keep the bacteria alive by putting them in an anaerobic chamber. Some reports indicate that tumour cells lack the enzyme superoxide dismutase, especially the manganese (or mitochondrial) type (Oberley & Buettner 1979). But it is not yet known whether cancer is a cause or an effect of the MnSOD deficiency.

Hill: A Japanese study (Takahara & Ogata 1977) on a group of catalase-deficient people reported that they are healthy, as Dr Dormandy said, except that they suffer from severe gingival infections.

Dormandy: But this infection occurs only if they use gum paste that contains hydrogen peroxide.

Sourkes: I wonder whether the convulsions that occur under high oxygen pressures might be attributable to a reduction in superoxide dismutase. It is conceivable that this might be a condition that limits SOD concentrations.

Danks: I think it is misleading to assume that mutations affecting certain enzymes would be lethal in humans, Dr Hassan. One can suggest that a moderate reduction in the activity of one enzyme might be as serious as a nearly total loss of another, but mutations can produce either of these effects. One might predict that SOD mutations in humans will produce only partial loss of activity, but we cannot say that they will not occur.

Hassan: I was talking about deletion-type mutations, when the enzyme activity is nil.

Danks: I see. Can I change the subject and ask whether the subunits of copper–zinc-superoxide dismutase are identical?

Hassan: Yes, they are. Physicochemical studies showed that there are two subunits (Richardson et al 1975).

Frieden: When you removed all the metal ions from superoxide dismutase and then added copper back to it, enzymic activity was restored. Did the copper that you added enter the zinc sites as well as the copper sites, thus producing an all-copper enzyme?

Hassan: Yes, but the copper in the zinc sites does not act as a catalyst.

McMurray: Is there any biological regulation of the order in which the metal ions are bound, so that the zinc enters first? In enzyme synthesis *in vitro*, the copper–zinc enzyme always seems to be formed. Are there any control mechanisms known for ordered binding of metals to proteins?

Hill: That is not yet known for any metalloenzyme.

Frieden: There is a lot more zinc present which might explain why it gets to the binding sites first.

McMurray: Do you think that is the only limiting factor?

Frieden: Theoretically copper-transferrin could exist but I have not seen reports of its existence in any biological system.

Hill: It is possible (Cass et al 1979) to bind zinc to all the sites that normally bind the copper, and so to produce a protein with four zinc atoms.

McMurray: Do leucocytes or neutrophils that are deficient in SOD have their ability to kill bacteria altered? In the SOD-deficient cell is it the cell that is killed and not the bacteria?

Hassan: The leucocytes have an enzyme that makes superoxide radicals, which kill the bacteria, and a deficiency in this enzyme leads to a higher risk of infection (Babior 1978).

Hill: Has there been any work on killing by neutrophils in copper-deficient animals?

Mills: Yes, and evidence from work by Boyne & Arthur (1981) indicates that the ability of neutrophils to kill ingested *Candida* cells is reduced in copper deficiency as it is in selenium deficiency.

Hill: What happens to the leucocytes?

Mills: Their ability to ingest *Candida* is unimpaired but their ability to kill ingested cells is much reduced. In this situation, it appears that a loss of SOD through copper deficiency should produce the converse result if there is an accompanying increase in superoxide. Perhaps Dr Hill could enlarge on this in relation to the killing effect?

Hill: Well, the leucocyte has its own superoxide dismutase which is believed to be concerned only with the very small amount of oxygen that is normally required *in vivo* (Patriarca et al 1974). The superoxide that is produced in the process of phagocytosis is presumably directed against the target organism rather than the leucocytes themselves. Therefore I doubt that there is a relationship between the intrinsic dismutase and the use of oxygen to produce superoxide in the respiratory burst.

Hassan: However, if we assume that there is some leakage of superoxide during phagocytosis, then that leakage could destroy the phagocyte.

Mills: So it seems that both the sites of generation of superoxide and the location of superoxide dismutase activity are highly compartmentalized?

Hill: Yes but, as Dr Hassan implied, the compartmentalization may not be perfect and therefore the dismutase may have some protective role.

Harris: Superoxide dismutase has some intriguing biochemical properties. The classical theory indicates that an enzyme will work on a pre-existing level of substrate in the cell. But in the case of superoxide dismutases, the superoxide anions or any of the other components cannot be allowed to build up because of their potential for damage. There must therefore be a tight coupling between superoxide formation and dismutation, which suggests that the enzyme has an extremely high affinity for the superoxide anion. Has it been possible to quantify that affinity?

Hassan: I don't think it can be quantified.

Mills: In the biological Haber-Weiss-Fenton reaction, then, the important question is whether or not there is peroxide present to get rid of the superoxide. The fact that glutathione peroxidase often co-exists with SOD is presumably irrelevant.

Hill: But there will always be hydrogen peroxide if superoxide is present because there will always be the spontaneous disproportionation. I would emphasize (although this has not been characterized even chemically) that it may be possible to produce hydroxyl radicals without the direct intervention of hydrogen peroxide. This is because the hydroxyl radical is the two-electron reduction product of the superoxide. There are many two-electron reductants present in biological systems. It is therefore possible in certain circumstances for reduction of hydroxyl radicals to occur independently of the production of hydrogen peroxide. The hydroxyl radical is highly reactive (Anbar & Neta 1967) – its rate of reaction with most substrates is close to the diffusion rate – but it is not necessarily the most reactive radical that does the most critical damage; a longer-lived radical can diffuse to a more critical site and attack that site. It may be that indiscriminate damage to the cell is easier to deal with than potentially lethal damage to an important site. At a previous meeting at the Ciba Foundation I spoke about the direct reactions between hydrogen peroxide, superoxide and carbon dioxide (Esnouf et al 1979). These reactions produce the chemically curious peroxocarbonates which are reactive and which chemiluminesce as they decompose. It is possible that the reactions of superoxide, hydrogen peroxide and hydroxyl radicals produce other potentially noxious materials.

Frieden: We have been assuming that superoxide dismutase is strictly an intracellular enzyme so if it enters the circulation this indicates that cell destruction has occurred. There is no significant secretion of SOD. In contrast, caeruloplasmin is regarded as being extracellular. Has the amount of SOD detected in the circulation been used as an index of cell breakdown?

Hill: This is a puzzle. As Dr Hassan said, when neutrophils undergo phagocytosis there is leakage of hydroxyl radicals into the environment. (The precursors of OH· are the superoxide radicals.) Therefore one would expect to find a defence mechanism to protect the host against superoxide. I cannot envisage any specific defence against hydroxyl radicals, although many scavengers of hydroxyl radicals exist (Anbar & Neta 1967), such as mannitol and ethyl alcohol.

Frieden: When heart cells mutate we generally look for damage to a whole sequence of enzymes but I wonder why damage to superoxide dismutase has not been studied more as an index of cell destruction.

Dormandy: We were unable to detect measurable amounts of SOD in human serum.

McMurray: Similarly, we have examined bovine plasma and serum for SOD activity and found none (C.H. McMurray, unpublished results). This might not be the case however if there was active liver necrosis in which leakage of cellular contents occurred.

Hurley: There is, however, identifiable superoxide dismutase present in human milk, and by a sensitive new method (Lönnerdal et al 1979) we have detected the enzyme in plasma too (Keen et al 1980).

Riordan: I understand that there is a small amount of membrane-associated SOD. It is possible that this amount could contribute to the extracellular activity.

Hassan: Who showed membrane association?

Riordan: Some workers in Canada (at Atomic Energy in Manitoba) have described membrane-associated SOD. And I believe that a certain proportion of the SOD activity in red cells may be associated with the ghost.

Hassan: SOD can be incorporated into red cell vesicles (Lynch & Fridovich 1978); however, the enzyme does not seem to be closely bound to the cytoplasm.

Mills: Dr Hassan, in the study of the changes in SOD activity that occur with changes in substrate concentration, has the effect of a change in the purine load been examined? In many tissues, e.g. rat liver (Higgins et al 1956), there is a large excess of xanthine oxidase which apparently does nothing; it could however mediate the generation of substantially greater concentrations of superoxide and peroxide if tissue concentrations of xanthine or hypoxanthine were to rise. Has the effect of a high xanthine load on superoxide generation or on changes in SOD activity been examined?

Hassan: No.

Harris: Can we consider the possibility that superoxide dismutase and the various cupreins serve other functions besides enzymic ones? For instance, they might represent components in the pathways of copper metabolism. This would be consistent with the pH-dependent migration of copper between binding sites (Valentine et al 1979). Conceivably this could be one means by which the metal can be passed on from one component to another.

Hassan: Since the enzyme possesses only one tightly bound copper atom per subunit, I cannot believe that it functions as a copper-storage protein.

Frieden: I do not agree that your explanation excludes superoxide dismutase as a copper-storage protein. The ratio of copper atoms to weight of the protein subunit is the key factor.

Hill: Actually, the copper is not labile except (Fee & Phillips 1975) at pH 5 or below. Most simple copper proteins lose their copper at low pH values, Dr Hassan, and surely you would agree that some of those, e.g. plastocyanin and azurin, are electron-transport proteins, and that many zinc enzymes act as zinc transport systems, even though they lose their zinc at low pHs. Of course proteins *can* have more than one function and Nature may have been much more economical than we give her credit for! That SOD is a dismutase there is no doubt, and that it may have other functions is always possible. Obviously, the onus lies on those people who suggest other functions to find out what those functions are!

References

Anbar M, Neta P 1967 A comparison of specific biomolecular rate constants for the reactions of hydrated electrons, hydrogen atoms and hydroxyl radicals with inorganic and organic compounds in aqueous solution. Int J Appl Radiat Isot 18:493-523

Babior BM 1978 Oxygen-dependent microbial killing by phagocytes. I and II. N Engl J Med 298:659-668 and 721-725

Boyne R, Arthur JR 1981 Effects of selenium and copper deficiencies on neutrophil function in cattle. J Comp Pathol 91(2): in press

Cass AEG, Hill HAO, Bannister JV, Bannister WH 1979 Zinc(II) binding to apo-(bovine erythrocyte superoxide dismutase). Biochem J 177:477-486

Esnouf MP, Green MR, Hill HAO, Irvine GB, Walter SJ 1979 Dioxygen and the vitamin K-dependent synthesis of prothrombin. In: Oxygen free radicals and tissue damage. Excerpta Medica, Amsterdam (Ciba Found Symp 65), p 187-197

Fee JA, Phillips WD 1975 The behaviour of holo- and apo-forms of bovine superoxide dismutase at low pH. Biochim Biophys Acta 412:26-38

Goldstein IM, Kaplan HB, Edelson HS, Weissmann G 1979 Ceruloplasmin – a scavenger of superoxide anion radicals. J Biol Chem 254:4040-4045

Higgins ES, Richert DA, Westerfeld WW 1956 Molybdenum deficiency and tungstate inhibition studies. J Nutr 59:539-560

Keen CL, Lönnerdal B, Stein TS, Hurley LS 1980 Superoxide dismutase isoenzymes in bovine and human milk. Proc Soc Exp Biol Med, in press

Lönnerdal B, Keen CL, Hurley LS 1979 Isoelectric focusing of superoxide dismutase isoenzymes. FEBS (Fed Eur Biochem Soc) Lett 108:51-55

Lynch RE, Fridovich I 1978 Effects of superoxide on the erythrocyte membrane. J Biol Chem 253:1838-1845

Oberley LW, Buettner GR 1979 Role of superoxide dismutase in cancer. Cancer Res 39:1141-1149

Patriarca P, Dri P, Rossi F 1974 Superoxide dismutase in leukocytes. FEBS (Fed Eur Biochem Soc) Lett 43:247-251

Richardson JS, Thomas KA, Rubin BH, Richardson DC 1975 Crystal structure of bovine Cu,Zn superoxide dismutase at 3 Å resolution: Chain tracing and metal ligands. Proc Natl Acad Sci USA 72:1349-1353

Takahara S, Ogata M 1977 Metabolism in Japanese acatalasemia with special reference to superoxide dismutase and glutathione peroxidase. In: Hayaishi O, Asada K (eds) Biochemical and medical aspects of active oxygen. University Park Press, Baltimore, p 275-292

Valentine JS, Pantoliano MW, McDonnell PJ, Burger AR, Lippard SJ 1979 pH-dependent migration of copper(II) to the vacant zinc-binding site of zinc-free bovine erythrocyte superoxide dismutase. Proc Natl Acad Sci USA 76:4245-4249

Copper, biogenic amines, and amine oxidases

T.L. SOURKES*

Departments of Psychiatry and Biochemistry, Faculty of Medicine, McGill University, Montreal, Canada

Abstract Amine oxidases have been classified in the past on the basis of either (a) the structural requirements in the substrate or (b) the tissue (or species) of origin, or both. As knowledge about the chemistry of these enzymes grows, their classification on the basis of chemical structure is becoming possible. Currently, many amine oxidases can be categorized according to whether they contain riboflavin (e.g. the monoamine oxidases − EC 1.4.3.4) or copper (e.g. the amine oxidases of plasma and the diamine oxidases − EC 1.4.3.6 − found prominently in pig kidney cortex, placenta, and pea seedlings). The copper-linked oxidases are inhibited by cyanide and by semicarbazide. The nature of the carbonyl compound(s) in the various enzyme molecules is not yet known. Nutritional deficiencies of copper and treatment of animals with copper-chelating agents are reflected in reduced activity of one or more of these enzymes. The ultimate effects of copper deficiency and copper excess on amine metabolism *in vivo* are described.

AMINE OXIDASES

It is over fifty years since the discovery of the first amine oxidases: tyramine oxidase discovered by Hare (1928), and histaminase by Best (1929). Since then, many other such enzymes have been described. Usually they have been named on the basis of an especially rich source or for a particular substrate, and this has led to the accumulation of many trivial names. As a consequence there is a need for systematic classifications that will reflect important properties of the enzymes. The elimination of multiple names, such as tyramine oxidase, adrenaline oxidase and aliphatic amine oxidase for what is now known by the systematic number EC 1.4.3.4 and by the systematic name—amine: oxygen oxidoreductase (deaminating) (flavin-containing)—has been one of the first results of the more detailed study of these enzymes. Their fairly distinct specificity for substrates with a single amine function led Zeller (1951) to label

---

* Mailing address: 1033 Pine Avenue, West, Montreal, Quebec H3A 1A1, Canada

them monoamine oxidases (MAO) in contrast to diamine oxidases (DAO, EC 1.4.3.6), a name which was reserved for enzymes acting preferentially on compounds like putrescine (1,4-diaminobutane), and which eventually subsumed Best's histaminase. The systematic name for this group is amine:oxygen oxidoreductase (deaminating) (copper-containing).

The specificities suggested by the names MAO and DAO are by no means absolute. Thus, long-chain diamines are substrates for MAO, and some monoamines like mescaline are acted upon by DAO from various sources. Nevertheless, these categories have withstood the test of time and are in common use. The early papers of Zeller (1951, 1963) and Blaschko (1963) provide much information about the substrates of these enzymes as well as about other specific properties.

Ultimately, there should be no more precise a classification than one based on the structure of the various enzymes, particularly in the region of the active site. This classification is already being achieved for flavoenzymes, through characterization of short portions of the peptide attached covalently to the isoalloxazine ring. However, for the amine oxidases, categories based on cofactors are used at present. The first information on this subject has come from the study of plasma amine oxidases (benzylamine oxidase, spermine oxidase, histaminase), enzymes that contain copper as an essential constituent, as has since become known for the DAOs of kidney and pea seedlings. This feature distinguishes these enzymes from the well known MAOs.

Even before the copper function was recognized, enzymes in this group had been shown to be especially sensitive to the action of semicarbazide. Since the 1940's when pyridoxine coenzymes were discovered, biochemists have regarded such susceptibility to the action of carbonyl reagents as reason to suspect the presence of pyridoxal phosphate as coenzyme. In very many cases this compound has been identified as the coenzyme by the application of classical methods of enzymology: (a) resolution of the cofactor from the holoenzyme, with consequent loss of enzymic activity and (b) recovery of characteristic catalytic activity by the addition of pyridoxal phosphate to the apoenzyme preparation. However, these methods have not been successful in the case of DAO, and there is no definitive proof that the semicarbazide-sensitive group belongs to pyridoxal. Some investigators take the alternative view that pyridoxal phosphate is, indeed, present in the enzyme but covalently bound. If this be so, it would be exceedingly difficult to prepare a minimally altered apoenzyme, an omission that leaves the question of the cofactor open.

Some enzymes bear an essential carbonyl group in the form of covalently bound pyruvate, but thus far no DAO has been identified as a pyruvoyl enzyme.

MAOs are flavoproteins (Ciba Foundation 1976) usually with the 8α-methyl group of the isoalloxazine nucleus of FAD bound to the peptide chain. Iron has been implicated as playing a part in their structure or biosynthesis. Thus, mitochondrial MAO in certain tissues of rats and humans is sensitive to a nutritional deficiency of iron (Sourkes & Missala 1976). In fact, iron has been found in purified preparations of MAO under conditions in which minimal amounts of copper are present. However, we do not yet know if this enzyme requires the iron for its specific function. Haem iron has been identified in certain other bacterial amine oxidases, and non-haem iron in bacterial trimethylamine dehydrogenase (EC 1.5.99.7).

The more recent classifications of amine oxidases have been based simply on the presence of riboflavin or of copper (Blaschko 1974, Yasunobu et al 1976), these two cofactors apparently being mutually exclusive in the known amine oxidases. A thorough classification should leave room for other cofactors, and this is the basis of the system presented in Table 1. The descriptions 'riboflavin-containing' and 'copper-containing' correspond broadly to the monoamine oxidases and diamine oxidases, respectively.

A few partially characterized enzymes (those without a known coenzyme) have been placed in an open category.

TABLE 1

Classification of amine oxidases

Riboflavin-containing amine oxidases
    A.   Iron also present
           Examples: Amine oxidases of *Pseudomonas aminovorans, Serratia marcescens*
    B.   Presence of iron not in evidence or not established
           Examples: Various mitochondrial monoamine oxidases[a]; pyridoxamine phosphate oxidase; amine oxidases of *Micrococcus rubens* and *Sarcina lutea*

Copper-containing amine oxidases[b]
           Examples: Plasma amine oxidase; pig kidney diamine oxidase; placental diamine oxidase; plant diamine oxidase; connective tissue lysyl oxidase

Other amine oxidases without a cofactor characterized as yet
           Examples: Amine oxidases of fungal mycelia

[a]Some of these are influenced by iron nutrition
[b]These are susceptible to inhibition by semicarbazide and cyanide

COPPER-CONTAINING AMINE OXIDASES

Some properties of copper-containing amine oxidases, the cyanide-sensitive group, are set out in Table 2 (Blaschko 1974, Maśliński 1975, Yasunobu et al

TABLE 2

Properties of copper-containing amine oxidases[a]

Enzyme source	Molecular weight	Absorption maxima (nm)	Substrates	Copper (atoms per mole)	=CO groups (number per mole)
Beef plasma	170 000	280, 480	Spermine, spermidine	2	2
Pig plasma	195 000	280, 470	Benzylamine, histamine, mescaline	2.1	1
Sheep plasma[b]			Benzylamine, polyamines and many other aliphatic and aralkyl primary amines	+	+[c]
Human plasma[d]			Histamine, cadaverine, and many aliphatic and aralkyl primary amines		+[c]
Pig kidney cortex	185 000	280, 470	Histamine, putrescine, cadaverine, agmatine	2.17	2
Human placenta[e]	70 000		Histamine, benzylamine	1	
Aspergillus niger	252 000	280, 330, 410, 480	Benzylamine, aliphatic monoamines, diamines	1.9	2
Pea seedlings[f]	96 000	350, 437, 466[g]	Benzylamine, phenethylamine, diamines, polyamines, histamine, agmatine	1.2–1.4	+[c]

[a]This subject has been thoroughly reviewed by many authors. See Zeller (1963), Blaschko (1974), Yasunobu et al (1976) for many useful references.
[b]Mills et al (1966), Rucker & Goettlich-Riemann (1972)
[c]The enzyme is inhibited by carbonyl reagents
[d]McEwen (1965)
[e]Crabbe et al (1976). This enzyme also contains manganese, tightly bound
[f]Mann (1961), Hill (1967)
[g]These wavelengths apply to the enzyme–substrate complex. The de-coppered enzyme is orange-pink, with maximal absorption near 480 nm. Gaps in the table indicate incomplete characterization

1976). The metabolism of some of their important substrates is described in Fig. 1. Many of these enzymes have molecular weights around 200 000 daltons, this mass being made up of subunits. Generally there are two atoms of copper per mole. For the placental enzyme, a functional unit has been purified with a molecular weight of 70 000 containing only one atom of copper but also some tightly bound manganese (Crabbe et al 1976).

This group is characterized by its ready inhibition by semicarbazide; other carbonyl reagents, though also inhibitory, are not as specific for DAO. As I already mentioned, some investigators have offered evidence for the presence of pyridoxal or its phosphate in plasma amine oxidase and in DAO, but there are strong views to the contrary about this. Even for DAO from pig kidney cortex, the most actively investigated member of the group, the organic coenzyme has not been identified, although its copper function is now well established.

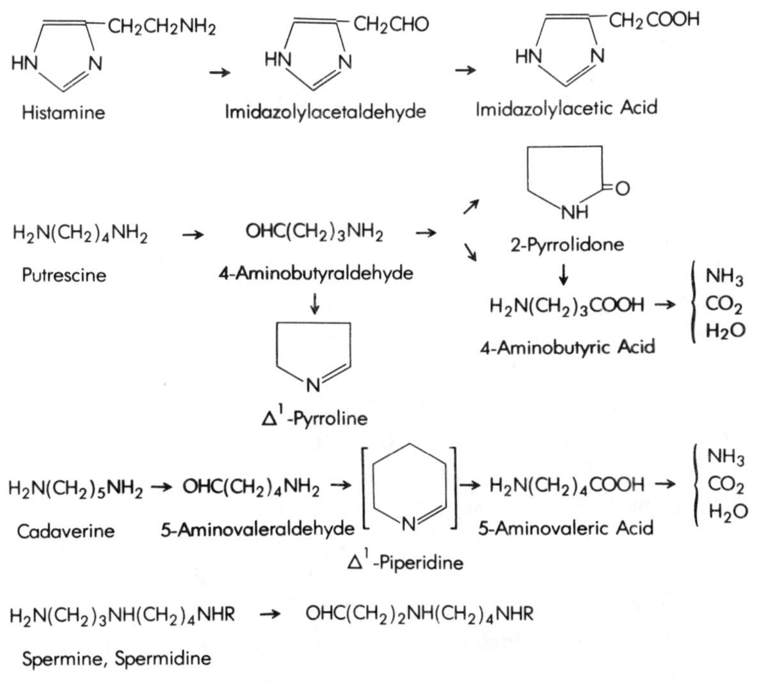

FIG. 1. Metabolism of some important substrates of copper-containing amine oxidases. In the structures shown for the two polyamines, spermine contains R = 3-aminopropyl; spermidine contains R = H. Plasma amine oxidase catalyses the oxidation of the primary amine function in these compounds, yielding a symmetrical di-aldehyde in the case of spermine.

THE USE OF PUTRESCINE IN THE STUDY OF DIAMINE OXIDASE ACTION *IN VIVO*

We have used ^{14}C-labelled putrescine (1,4-diaminobutane) as a tool for the physiological study of DAO action *in vivo*. Recently, we obtained evidence that this enzyme, i.e. its integrated action through various tissues of the rat, is rate-limiting in the conversion of the labelled diamine to radioactive CO_2 (Missala & Sourkes 1980), just as MAO is rate-limiting for the similar oxidation *in vivo* of pentylamine, a classical substrate for that enzyme. We have done these studies in both rat and guinea pig, but we prefer the former species because of the much higher rates of catabolism achieved. In nutritional investigations we have found that the rat can maintain normal rates of catabolism of putrescine *in vivo* only when adequate copper and pyridoxine are present in the diet (Sourkes & Missala 1976). Riboflavin and iron are also needed, but the effects of their deficiency are not as pronounced. These requirements are not necessarily to provide the cofactor; the nutrients might be required for synthesis of the enzyme or for other processes.

ALTERATION OF DIAMINE OXIDASE ACTIVITY OF TISSUES

Body fluids of mammals ordinarily contain very little DAO. However, after conception the female has a much increased plasma concentration of the enzyme, the source of which is the placenta. After parturition the concentration falls to normal. DAO activity increases sharply in the body fluids in anaphylactic shock. Heparin, one of the substances released from storage sites during anaphylaxis, itself causes the release of the enzyme (from storage sites in the villus cells of the intestinal tract of the rat) into the lymph and plasma (Maśliński 1975, Sourkes & Missala 1979). This release apparently decreases the putrescine-catabolizing ability of the rat and guinea pig, as studied by our method. It would seem that the redistribution of DAO caused by heparin should not directly affect the oxidation of putrescine to its primary oxidation products, and might even facilitate this process. However, many of these products would then be present in the body fluids and would have to await translocation to the liver or other organs for further degradation to radioactive CO_2, the product we detect. One might expect that any histamine released as a result of anaphylaxis would similarly undergo rapid oxidation in the plasma, but the resulting aldehyde would have to reach the liver for conversion by aldehyde dehydrogenase (EC 1.2.1.3) to imidazolylacetic acid (Fig. 1).

ENDOCRINE INFLUENCES ON DIAMINE OXIDATION

We have investigated the effects of endocrine glands on diamine oxidation *in vivo* by observing the putrescine-catabolizing ability of rats in some endocrine dyscrasias (Table 3). There is a decrease in the rate of catabolism of this amine in adrenalectomized rats, a result consistent with the previously demonstrated decrease of diamine oxidase in tissues of rat, guinea pig and cat, after removal of the adrenals. Adrenalectomy provokes an increase of enzymic activity in lymph and plasma (Maśliński 1975).

In contrast to the effect of adrenalectomy, excision of the thyroid gland in the rat does not significantly affect the catabolism of putrescine (Sourkes et al 1977). Moreover, rats made hyperthyroid by chronic administration of thyroxine did not have an altered rate of metabolism of putrescine. However, when cadaverine (1,5-diaminopentane) was tested, its normally slow rate of breakdown was accelerated by about 50% as a result of the hyperthyroidism.

TABLE 3

Endocrine influences on catabolism of putrescine in rats[a]

Treatment	Duration (days)	Number of rats	Percentage recovery of radioactivity as $^{14}CO_2$	Statistical significance
Thyroidectomy	26	5	10.8 ± 1.2[b]	
Controls[c]		5	12.2 ± 0.4	$P > 0.05$
Thyroidectomy	10–12	9	4.4 ± 0.2[d]	
Controls[c]		9	4.4 ± 0.3	$P > 0.05$
Thyroxine administration[e]	15–20	9	11.4 ± 0.8	
Controls		8	10.2 ± 1.0	$P > 0.05$
Thyroxine administration[e]	15–17	9	8.2 ± 0.05[d]	
Controls		9	5.4 ± 0.4	$P < 0.01$
Adrenalectomy	3–16	15	3.2 ± 0.2	
Controls[c]		15	5.4 ± 0.4	$P < 0.001$

[a]Groups of 3 rats (treated and controls) were studied in two or more replicated experiments for each treatment. The adrenalectomy data are based on unpublished results from my laboratory (1974). For thyroid data, see Sourkes et al (1977).
[b]Mean ± SEM for all results
[c]Sham-operated rats
[d]Substrate was cadaverine in this experiment
[e]Plasma thyroxine and triiodothyronine concentrations were significantly increased.

RATE OF SYNTHESIS OF DIAMINE OXIDASE

Aminoguanidine is a highly specific, potent and characteristic inhibitor of

DAO, as judged by many different types of experiment *in vivo* and *in vitro* (Sourkes & Missala 1977). In acute experiments as little as 0.01 mg/kg body weight, given intraperitoneally, inhibits the rate of catabolism of putrescine in the rat by 75%. The inhibition increases in proportion to the logarithm of the dose, until at 10 mg aminoguanidine/kg there is only 5% of the control activity (Missala & Sourkes 1980). The effect of the drug persists for periods that are proportional to the dose administered, but with 0.1 mg/kg the effect does not endure beyond 24 h. The recovery of putrescine metabolism from the inhibitory action of aminoguanidine can be used to estimate the rate of resynthesis of DAO in the rat. The time required to restore the rate of putrescine catabolism to 50% of the control value is 17.5 h (Table 4). In other experiments we have estimated this parameter in rats that were given heparin to reduce temporarily the rate of metabolism of labelled putrescine *in vivo*. In two such experiments the half-times were 15 and 18 h. These values compare well with those obtained by Shaff & Beaven (1976) for recovery of the DAO content of the small intestine in rats injected with heparin (Table 5).

Shaff & Beaven (1976) employed still another method for estimating the rate of resynthesis of DAO: they inhibited the protein synthesis acutely by administering cycloheximide to rats; this reduced the intestinal content of DAO sharply. They then observed the recovery of enzymic activity in the small intestine, and estimated a half-time for resynthesis of 16 h. Hence, there is good agreement in this measurement obtained by several independent methods.

TABLE 4

Rate of synthesis of enzymes concerned with diamine metabolism in the rat

Measurement	Conditions	Half-time for recovery (h)
Diamine oxidase activity of small intestine[a]	Depletion caused by inhibition of enzyme synthesis by cycloheximide	16
Diamine oxidase content of small intestine[a]	Depletion by heparin (4000 i.u./kg)	13, 16[b]
Putrescine-catabolizing ability in vivo[c]		
(a)	Depletion by heparin (4000 i.u./kg)	15
(b)	Depletion by heparin (15 000 i.u./kg)	18
(c)	Inhibition by aminoguanidine (0.1 mg/kg)	17.5

[a]Shaff & Beaven (1976)
[b]Two separate experiments
[c]Missala & Sourkes (1980)

TABLE 5

Effects of copper deficiencies on amine metabolism

Treatment	Species	Organ	Measurement	Effect
Experimental dietary deficiency[a]	Rat	Brain	Catecholamines	Decreased
			Tyrosine 3-monooxygenase	Decreased
			Monoamine oxidase	No change
	Rat	Heart	Dopamine β-monooxygenase (EC 1.14.17.1)	Decreased
Experimental dietary deficiency[b]	Chicken	Liver	Amine oxidase[c]	Decreased
	Chicken	Aorta	Amine oxidase	Does not appear after hatching
Experimental dietary deficiency[b]	Pig	Plasma	Amine oxidase	Decreased
Range deficiency[d]	Sheep	Plasma	Amine oxidase	Decreased
Disulfiram, 400 mg/kg i.p.[e]	Rat	Brain	Noradrenaline	Decreased
			Dopamine β-monooxygenase	Decreased
	Rat	Heart	Noradrenaline	Decreased
			Dopamine	Increased
			Noradrenaline uptake	No effect
Disulfiram, 50 mg/kg p.o., daily[f]	Rat	Adrenals	Dopamine	Increased
			Dopamine β-monooxygenase	Decreased
Fusaric acid[g], 25 mg/kg, i.p.	Rat	Brain	Noradrenaline	Decreased
			Serotonin	Increased
			Monoamine oxidase	No effect
α,α'-Dipyridyl, 5–140 mg/kg	Mouse	Brain	Noradrenaline	Decreased
α,α'-Dipyridyl, 30–75 mg/kg[h]	Rat	Brain	Noradrenaline	Decreased
			Dopamine β-monooxygenase	Decreased

[a]Sourkes (1979)
[b]Blaschko (1974)
[c]Benzylamine was the substrate in determinations of amine oxidase
[d]Mills et al (1966), O'Dell et al (1976)
[e]Musacchio et al (1964, 1966), Hashimoto et al (1965), Goldstein & Nakajima (1967)
[f]Drug was given for 4 or 6 months. There were no effects on catecholamines of heart or brain (Lippmann 1968)
[g]5-Butylpicolinic acid (Hidaka 1971)
[h]Mogilnicka et al (1975)

INHIBITION OF DIAMINE OXIDASE BY ANTIMALARIAL DRUGS

Most recently we have described the inhibition of DAO *in vitro* and *in vivo* by some antimalarial drugs. This work was stimulated by reports that amodiaquine is a potent inhibitor of histamine N-methyltransferase (EC 2.1.1.8) *in vivo* (Cohn 1965), yet it does not cause an increase of endogenous histamine (André & Schwartz 1973). We reasoned that this result might obtain if DAO were preventing an increase in the tissue content of histamine, and so we tested the effect of amodiaquine on diamine oxidation (i.e. on putrescine catabolism) in the intact rat. The result showed that the drug is also a DAO inhibitor, acting in a dose-dependent manner. In fact, chloroquine and mepacrine also inhibit the oxidation of putrescine *in vivo*, although more weakly. These effects are attributable to the action of the antimalarials on DAO itself (Ma & Sourkes 1980).

COPPER DEFICIENCY AND EXCESS

Copper deficiency generally produces a decrease of dopamine β-monooxygenase (EC 1.14.17.1) activity in the tissues where it has been measured so far, as well as a reduction in tissue content of noradrenaline (Table 5). This is true whether the deficiency results from experimental design, from range and fodder conditions in farm animals or from the administration of a copper-chelating agent. There are occasionally unexpected results, such as the decrease in activity of tyrosine 3-monooxygenase (EC 1.14.16.2) (Morgan & O'Dell 1977), an iron-linked enzyme (Table 5). This decrease would explain why, in the rat, the dopamine content of the brain also decreases.

Where MAO has been measured there is, as expected, no effect of the copper deficiency. However, the enzyme that catalyses the oxidation of benzylamine in sheep plasma and in chick tissues is diminished (Table 5). In chick connective tissues, this enzyme may be lysyl oxidase.

Chronic administration of copper to rats leads to accumulation of the metal in certain tissues. Aromatic L-amino-acid decarboxylase (EC 4.1.1.28) and MAO activities of the liver gradually decline over a four-week period; the brain does not show these changes even after six weeks. However, after six weeks of copper-loading the concentration of 5-hydroxyindoleacetic acid in the brain is decreased, a change that could signify a reduced rate of turnover of serotonin (5-HT).

In hepatolenticular degeneration there is increased output of monoamines and their metabolites in the urine, but this may simply result from pathological changes in the kidney. The few isolated cases of the disease in which measurements of monoamine metabolites in cerebrospinal fluid have been made have not shown abnormal values.

References

André JM, Schwartz JC 1973 Studies on the inhibition of histamine methylation in mice. Agents Actions 3:172-173
Best CH 1929 The disappearance of histamine from autolysing lung tissue. J Physiol (Lond) 67:256-263
Blaschko H 1963 Amine oxidase. In: Boyer PD, Lardy H, Myrbäck K (eds) The enzymes, 2nd edn. Academic Press, New York, vol 8:337-351
Blaschko H 1974 The natural history of amine oxidases. Rev Physiol Biochem Pharmacol 70:83-148
Ciba Foundation 1976 Monoamine oxidase and its inhibition. Excerpta Medica, Amsterdam (Ciba Found Symp 39)
Cohn VH 1965 Inhibition of histamine methylation by antimalarial drugs. Biochem Pharmacol 14:1686-1688
Crabbe MJC, Waight RD, Bardsley WG, Barker RW, Kelly ID, Knowles PF 1976 Human placental diamine oxidase. Improved purification of a copper- and manganese-containing amine oxidase with novel substrate specificity. Biochem J 155:679-687
Goldstein M, Nakajima K 1967 The effect of disulfiram on catecholamine levels in the brain. J Pharmacol Exp Ther 157:96-102
Hare MLC 1928 Tyramine oxidase. I: A new enzyme system in liver. Biochem J 22:968-979
Hashimoto Y, Ohi Y, Imaizumi R 1965 Inhibition of brain dopamine β-oxidase in vivo by disulfiram. Jpn J Pharmacol 15:445-446
Hidaka H 1971 Fusaric (5-butylpicolinic) acid, an inhibitor of dopamine β-hydroxylase, affects serotonin and noradrenaline. Nature (Lond) 231:54-55
Hill JM 1967 The inactivation of pea seedling diamine oxidase by peroxidase and 1,5-diaminopentane. Biochem J 104:1048-1055
Lippmann W 1968 Chronic oral treatment with disulfiram on dopamine β-hydroxylase and biogenic amines. Eur J Pharmacol 3:84-86
Ma K, Sourkes TL 1980 Inhibition of diamine oxidase by antimalarial drugs. Agents Actions, in press
Mann PJG 1961 Further purification and properties of the amine oxidase of pea seedlings. Biochem J 79:623-631
Maśliński C 1975 Histamine and its metabolism in mammals. Part II. Catabolism of histamine and histamine liberation. Agents Actions 5:183-225
McEwen CM Jr 1965 Human plasma monoamine oxidase. I. Purification and identification. J Biol Chem 240:2003-2010
Mills CF, Dalgarno AC, Williams RB 1966 Monoamine oxidase in ovine plasma of normal and low copper content. Biochem Biophys Res Commun 24:537-540
Missala K, Sourkes TL 1980 Putrescine catabolism in rats given heparin or aminoguanidine. Eur J Pharmacol 64:307-311
Mogilnicka E, Szmigielski A, Niewiadomska A 1975 The effect of α,α'-dipyridyl on noradrenaline, dopamine and 5-hydroxytryptamine levels and on dopamine-β-hydroxylase activity in brain. Pol J Pharmacol Pharm 27:619-624
Morgan RF, O'Dell BL 1977 Effect of copper deficiency on the concentrations of catecholamines and related enzyme activities in the rat brain. J Neurochem 28:207-213

Musacchio J, Kopin IJ, Snyder S 1964 Effects of disulfiram on tissue norepinephrine content and subcellular distribution of dopamine, tyramine and their β-hydroxylated metabolites. Life Sci 3:769-775

Musacchio J, Goldstein M, Anagnoste B, Poch G, Kopin IJ 1966 Inhibition of dopamine-β-hydroxylase by disulfiram in vivo. J Pharmacol Exp Ther 152:56-61

O'Dell BL, Smith RM, King RA 1976 Effect of copper status on brain neurotransmitter metabolism in the lamb. J Neurochem 26:451-455

Rucker RB, Goettlich-Riemann W 1972 Purification and properties of sheep plasma amine oxidase. Enzymologia 43:33-44

Shaff RE, Beaven MA 1976 Turnover and synthesis of diamine oxidase (DAO) in rat tissues. Studies with heparin and cycloheximide. Biochem Pharmacol 25:1057-1062

Sourkes TL 1979 Nutrients and cofactors required for monoamine synthesis in nervous tissue. In: Wurtman RJ, Wurtman JJ (eds) Nutrition and the brain. Raven Press, New York, vol 3:265-299

Sourkes TL, Missala K 1976 Nutritional requirements for amine metabolism in vivo. In: Monoamine oxidase and its inhibition. Excerpta Medica, Amsterdam (Ciba Found Symp 39), p 83-96

Sourkes TL, Missala K 1977 Action of inhibitors on monoamine and diamine metabolism in the rat. Can J Biochem 55:56-59

Sourkes TL, Missala K 1979 Effect of heparin on the metabolism of putrescine *in vivo*. Can J Biochem 57:959-961

Sourkes TL, Missala K, Bastomsky CH, Fang TY 1977 Metabolism of monoamines and diamines in hyperthyroid and hypothyroid rats. Can J Biochem 55:789-795

Yasunobu KT, Ishizaki H, Minamiura N 1976 The molecular, mechanistic and immunological properties of amine oxidases. Mol Cell Biochem 13:3-29

Zeller EA 1951 Diamine oxidase. In: Sumner JB, Myrbäck K (eds) The enzymes: chemistry and mechanism of action. Academic Press, New York, vol 2, part 1:544-558

Zeller EA 1963 Diamine oxidases. In: Boyer PD, Lardy H, Myrbäck K (eds) The enzymes. Academic Press, New York, vol 8:313-335

Discussion

Mills: Have there been any studies on the effects of copper deficiency on intestinal motility, on metabolic activity of mucosal cells or on activity of the intestinal musculature?

Sourkes: I am not aware of any work on the effects of copper deficiency on intestinal enzymes and intestinal function, but we know that a copper-dependent enzyme, diamine oxidase, is present in the intestinal mucosa.

Incidentally, I would like to mention the difficulties of finding a diet that will quickly produce copper deficiency in rats or mice. We have tried a great variety of diets and none of them is effective quickly. We found that a milk diet was good initially but it caused too many other problems to be of general use in this respect.

Harris: In your presentation you did not mention the existence of a flavin copper-protein that functions as an amine oxidase. Are we correct in assuming that such a protein does not exist?

Sourkes: Published work now indicates that riboflavin cofactors and copper are mutually exclusive: only one of them is present in any given enzyme. Over twelve years ago Yasunobu withdrew his claim that copper is present in monoamine oxidase (Yasunobu et al 1968), although some authors have continued to repeat the assertion since then.

Österberg: Does anybody know how copper is bound in these oxidase enzymes?

Sourkes: Mondovi & Calabrese (1973) obtained evidence for at least three forms of copper in the diamine oxidase present in pig kidney. The maximal content was estimated by atomic absorption spectroscopy, and about 2/3 of that amount was detected by e.p.r. spectroscopy. The e.p.r. spectrum could be resolved into two components by the use of mersalyl.

Frieden: In the reaction that you described on monoamine oxidation it seems that a mole of peroxide is produced. Are the corresponding aldehyde products stable in the presence of peroxide? I would expect aldehydes to be readily oxidized to the acid which would be the ultimate product of many of these reactions.

Sourkes: There is an aldehyde dehydrogenase that is more active than the enzyme responsible for peroxide oxidation, and there is also an alcohol dehydrogenase (or aldehyde reductase) which catalyses formation of the corresponding alcohols from some of the biogenic amines (Deitrich 1966, Tipton & Turner 1974).

Frieden: Is the sole function of these enzymes to produce aldehydes that are intermediates in other reactions?

Sourkes: This is a somewhat teleological argument, but what you said is possible. These aldehydes are intermediates in the formation of the isoquinolines which are ubiquitous in plants. Thus, dopamine reacts with the corresponding aldehyde (Holtz et al 1964) or with acetaldehyde, derived from ethanol during metabolism (Walsh et al 1970, Cohen 1976) to form biologically active isoquinolines that belong to the papaverine series. Other metabolites of L-dopa could also form such isoquinolines (Sourkes 1971). A number of people have perfected the preparation of these substances, including β-carbolines, from indoleamines (Rommelspacher et al 1978) by enzymic means.

Hunt: I would like to comment on catecholamine metabolism. In the animals that were given sodium diethyldithiocarbamate, the reduced activities of dopamine β-monooxygenase were presumably assayed *in vitro*. Have you any information about endogenous inhibitors of this enzyme?

Sourkes: There are endogenous inhibitors present but people who have attempted to study them have found great difficulty in identifying them. The reduction of dopamine β-monooxygenase activity that I discussed was not based on *in vitro* work alone. In our experiments we prepared rats that were known to be copper-deficient, on the basis of both the diet fed and, eventually, of copper analyses of the liver and heart. Administered radioactive dopamine was taken up into the heart. In non-deficient control rats about 50% of the catecholamine in the heart was present as labelled noradrenaline after about one hour. But in the copper-deficient animals there was much less conversion of dopamine to noradrenaline. I regard this reduction in dopamine conversion as due to reduced activity of dopamine β-monooxygenase, itself a result of deficiency of the copper cofactor. This is distinct from inhibition by exogenous compounds or by endogenous substances of unknown composition.

References

Cohen G 1976 Alkaloid products in the metabolism of alcohol and biogenic amines. Biochem Pharmacol 25:1123-1128
Deitrich RA 1966 Tissue and subcellular distribution of mammalian aldehyde-oxidizing capacity. Biochem Pharmacol 15:1911-1922
Holtz P, Stock K, Westermann E 1964 Pharmakologie des Tetrahydropapaverolins und seine Entstehung aus Dopamin. Naunyn-Schmiedebergs Arch Exp Pathol Pharmakol 248:387-405
Mondovi B, Calabrese L 1973 Copper heterogeneity in pig kidney histaminase. An EPR study on the effect of substrate, guanidine and mersalyl. Agents Actions 3:187
Rommelspacher H, Strauss SM, Rehse K 1978 β-Carbolines: a tool for investigating structure-activity relationships of the high-affinity uptake of serotonin, noradrenaline, dopamine, GABA and choline into a synaptosome-rich fraction of various regions from rat brain. J Neurochem 30:1573-1578
Sourkes TL 1971 Possible new metabolites mediating actions of L-dopa. Nature (Lond) 229:413-414
Tipton KF, Turner AJ 1974 Computer reconstruction of tyramine breakdown in brain. Biochem Pharmacol 23:1906-1910
Walsh MJ, Davis VE, Yamanaka Y 1970 Tetrahydropapaveroline: an alkaloid metabolite of dopamine *in vitro*. J Pharmacol Exp Ther 174:388-400
Yasunobu KT, Igaue I, Gomes B 1968 The purification and properties of beef liver mitochondrial monoamine oxidase. In: Costa E, Sandler M (eds) Advances in pharmacology. Academic Press, New York, 6A:43-59

General discussion

PATHOLOGY AND DIAGNOSIS OF COPPER DEFICIENCY

Mills: We have now considered in detail the function of some copper enzymes. In this general discussion we might consider how information about the function of copper in enzymes contributes to our understanding of the pathology and the clinical effects of copper deficiency. Sections of heart muscle from copper-deficient bovine animals that show no clinical signs of copper deficiency may contain grossly enlarged mitochondria and may show loss of cristae (Leigh 1975). Similar defects occur in cardiac mitochondria of copper-deficient rats (Dallman & Goodman 1970, D. Dinsdale, personal communication). Although a loss of cytochrome *c* oxidase activity occurs, osmotic effects due to changes in mitochondrial membrane permeability may also develop. If so, we know nothing of their origin and have no adequate parameter with which to monitor their development. Similarly, sections of *ligamentum nuchae* from animals that show no overt signs of copper deficiency often reveal a failure of normal elastin synthesis (Whiting et al 1974, Mills et al 1976) and some disorganization of collagen. We must ask whether changes in both these connective tissue proteins are caused by the same metabolic lesion – a 'deficiency' in oxidase – and also how we can improve our ability to detect such covert pathological changes in the living animal at the early stages of the syndrome.

Serum concentrations of copper or caeruloplasmin can be useful measurements but how are the sequential changes in physically accessible copper enzymes related to pathological changes? What effect does a change in amine oxidase activity have upon tissue concentrations of physiologically active amines that may influence blood pressure, intestinal motility and nerve conduction velocities?

Although there have been many investigations of the biochemical changes in the acute phases of the deficiency, investigation of the sequence of early changes is much less common. This may explain our limited ability to recognize deficiency at its early, but pathologically significant, stages.

Danks: I believe that the most logical way to identify a trace element deficiency is to examine the concentration of the metalloenzyme that is most sensitive to deficiency before and after a brief period of physiological replacement of the element in question. Every metalloenzyme is susceptible to variations in production of apoprotein, for reasons other than metal availability and, therefore, single measurements of metalloenzyme concentration can be misleading (e.g. oestrogen concentrations can give rise to considerable variations in caeruloplasmin concentrations). However, two measurements separated by a short period in which the appropriate metal is replaced in the diet may be informative, provided that the measurements do not coincide with acute change in the factors influencing production of apoproteins. (In practice this should not occur because we are discussing people with minimal rather than gross deficiency and they are not likely to be acutely ill.) Kirchgessner et al (1977) used this technique in detection of zinc deficiency. They found that alkaline phosphatase was the most sensitive index of zinc deficiency and that carbonic anhydrase in the red cells was the second most sensitive index. I am not aware that any of these enzymes can be superinduced by a moderate *excess* of metal, but the production of a number of them is restricted by a *deficiency* of the metal.

Mills: In cases of selenium excess, superinducibility has been reported for glutathione peroxidase in erythrocytes but not for the enzyme in the liver or kidney.

Hurley: I don't believe that the approach that you mentioned would be useful, Dr Danks. We have examined animals that were so zinc-deficient they were almost dead, yet many of their zinc-metalloenzymes were present in the same concentrations as those in normal animals. Carbonic anhydrase concentrations, for example, were completely normal. It is possible that the enzymes most crucial for life would be the last ones to be affected, although that principle may not always be a useful one! Alkaline phosphatase activity *was* influenced by zinc deficiency (Hurley 1980), but because it is affected by so many other variables too, I doubt its value as a diagnostic tool.

Frieden: In inflammation there is an increase in caeruloplasmin which might be accompanied by a decrease in the serum concentrations of zinc enzymes. This inflammatory response may therefore provide a disturbing influence on the measurements that Dr Danks proposed. Cytochrome *c* oxidase, however, may be a more reliable index of the nutritional state of copper.

Hill: I am not sure that this problem is really soluble. In, for example, a zinc-deficient animal, the polymerases (which seem to be zinc-dependent enzymes – see Galdes & Hill 1979) will be affected by the deficiency and this will therefore modify the synthesis of the very metalloenzymes that one might use as an index of the deficiency itself. This problem is exacerbated because most people measure activity as an indication of the concentration of the enzyme instead of measuring whether or not the apoenzyme is present. If an enzyme or a small molecule is responsible for insertion of the zinc into the metalloenzyme, the zinc deficiency may indirectly affect the synthesis of that enzyme or small molecule.

Danks: Surely one should survey a number of enzymes in order to find the best candidate to use as an indicator of deficiency? Kirchgessner et al (1977) found that in zinc-deficient rats the polymerase was only very slowly affected by the deficiency and was also slow to respond when the zinc was replaced. It was therefore relatively insensitive to zinc deficiency. The enzyme most sensitive to each deficiency may vary from species to species. It would be most useful for these techniques to be attempted with studies of copper deficiency.

Mills: There is evidence that some systems respond almost immediately to repletion after copper deficiency. The histological mitochondrial defects that I described earlier are also present in enterocytes and may well be associated with the diarrhoea that frequently occurs. Within 12 h of copper replacement therapy, normal intestinal function returns (although we have not yet shown that recovery depends upon a restoration of mitochondrial function). Correction of connective tissue problems would obviously take much longer; we understand so little of the biochemical basis for these events.

Frieden: Are the mitochondrial problems accompanied by alterations in cytochrome *c* oxidase activity?

Mills: Yes.

Bremner: There is a point we should bear in mind when we set up experiments to look at the patterns of change in the activity of various enzymes. As a rule our experiments are designed to induce a relatively severe copper deficiency within a short time. However, most copper deficiencies that arise naturally in humans or farm animals are not severe. Therefore we may obtain patterns of enzymic change that are different from those occurring naturally if we attempt to make our experimental animals too *severely* copper-deficient.

Mills: Another point worth mentioning is that although it may take five weeks before gross signs of copper deficiency appear in rats fed on diets low in copper, cardiac enlargement (by up to 30%) is evident after only 21 days (R.B. Williams and R. Dawson, personal communication). The biochemical origin of this enlargement is unknown.

Tanner: Is it useful to measure free copper in the serum (i.e. copper not bound to caeruloplasmin) as an index of the copper status of an animal?

Lewis: It's hard to say, because we do not know how *quickly* the concentrations of free copper would change. Most of the copper present in normal bovine and sheep plasma is bound to caeruloplasmin, although in experimental animals free copper may be present.

Frieden: It seems to me that only in pathological states are there significant differences between serum concentrations of free copper and caeruloplasmin-bound copper. In the normal animal, differences are not apparent.

Mills: It remains difficult to distinguish between an animal that is beginning to develop a pathological change from one that is not, when copper distribution in plasma is the only available measurement on which to base such a distinction. This is one reason why plasma copper is often regarded as a questionable index of the incidence of copper deficiency in farm animals.

Shaw: Is it possible to detect any changes in serum copper or caeruloplasmin in the animals that showed histological features of copper deficiency without any overt symptoms?

Mills: Yes – the concentrations are reduced to about one-third of the normal values by the time that the histological changes appear in rats or in cattle.

Shaw: Well, if I found a preterm infant to have a plasma caeruloplasmin of 10 mg/100 ml at three to six months of age, I would consider a diagnosis of copper deficiency to be virtually certain.

Mills: But do you meet the same problem with babies as we and others meet with other species: i.e. do you find low caeruloplasmin and yet fail to diagnose copper deficiency because there are no gross clinical symptoms?

Shaw: Sometimes; preterm infants are generally born with low levels of copper and caeruloplasmin in the plasma and though in some cases the levels rise after birth, as they do in full-term infants, this does not happen in all cases. For instance, I have observed an infant of 32 weeks gestation whose serum caeruloplasmin was 9–12 mg/100 ml and whose plasma copper was 42–52 μg/100 ml during the first 50 days of life, and yet she enjoyed apparently normal health. We do not know at present if the postnatal rise in caeruloplasmin is a birth-related or a gestation-related event. However, from studies of intestinal absorption and from measurements of the amount of copper in the body at birth, one can infer that preterm infants are likely to experience a period when body copper may be barely sufficient for their needs. In the published reports, severe copper deficiency occurred between three and six months after birth; the serum caeruloplasmin was generally 12 mg/100 ml and the plasma copper 2–10 μg/100 ml (al-Rashid & Spangler 1971, Karpel & Peden 1972, Seely et al 1972, Ashkenazi et al 1973, Heller et al 1978, Yuen et al 1979, Blumenthal et al 1980). Some premature babies die during this period, after they have been sent home, and they have a much higher mortality rate than full-term infants. The point we need to establish is whether this is a consequence of marginal copper deficiency and, if so, how we can detect it.

McMurray: I would like to point out that the timing of extraction of blood samples from neonates may be important in diagnosis of copper deficiency. In the newly born calf after a normal birth there is no change in the amount of caeruloplasmin in the serum during the first eight hours. It can take up to 96 hours for the amount of caeruloplasmin to increase and one could wrongly diagnose severe copper deficiency if samples were taken too early.

Shaw: Our measurements on premature babies were done serially, from birth to 60 days, and we therefore have avoided that problem.

References

al-Rashid RA, Spangler J 1971 Neonatal copper deficiency. N Engl J Med 285:841-843
Ashkenazi A, Levin S, Djaldetti M, Fisher E, Benvenisti D 1973 The syndrome of neonatal copper deficiency. Pediatrics 52:525-533
Blumenthal I, Lealman GT, Franklyn PP 1980 Fracture of the femur, fish odour, and copper deficiency in a preterm infant. Arch Dis Child 55:229-231
Dallman PR, Goodman JR 1970 Enlargement of mitochondrial compartment in iron and copper deficiency. Blood 35:496-505
Galdes A, Hill HAO 1979 Metalloenzymes. In: Hill HAO (Senior reporter) Inorganic Biochemistry. The Chemical Society, London, vol 1:317-355
Heller RM, Kirchner SG, O'Neill JA Jr, Hough AJ Jr, Howard L, Kramer SS, Green HL 1978 Skeletal changes of copper deficiency in infants receiving prolonged total parenteral nutrition. J Pediatr 92:947-949
Hurley LS 1980 Teratological aspects of manganese, zinc, and copper nutrition. Physiol Rev, in press
Karpel JT, Peden VH 1972 Copper deficiency in long term parenteral nutrition. J Pediatr 80:32-36
Kirchgessner M, Roth H-P, Spoerl R, Schnegg A, Kellner RJ, Weigand E 1977 A comparative view on trace elements and growth. Nutr Metab 21:119-143
Leigh LC 1975 Changes in the ultrastructure of cardiac muscle in steers deprived of copper. Res Vet Sci 18:282-287
Mills CF, Dalgarno AC, Wenham G 1976 Biochemical and pathological changes in tissues of Friesian cattle during the experimental induction of copper deficiency. Br J Nutr 38:309-331
Seely JR, Humphrey GB, Matler BJ 1972 Copper deficiency in a premature infant fed on iron fortified formula. N Engl J Med 286:109
Whiting AH, Sykes BH, Partridge SM 1974 Isolation of salt-soluble elastin from ligamentum nuchae of copper deficient calf. Biochem J 141:573-575
Yuen P, Lin HJ, Hutchison JH 1979 Copper deficiency in a low birth weight infant. Arch Dis Child 54:553-554

Copper and the synthesis of elastin and collagen

EDWARD D. HARRIS, JOHN K. RAYTON, JAMES E. BALTHROP, ROBERT A. DiSILVESTRO and MARGARET GARCIA-DE-QUEVEDO

Department of Biochemistry and Biophysics and the Texas Agricultural Experiment Station, Texas A & M University, College Station, Texas 77843, USA

Abstract Copper's role in connective tissue is linked to the enzyme lysyl oxidase. From a biochemical perspective, copper is a cofactor for the enzyme and a determinant of its activity in connective tissues. Lysyl oxidase catalyses a post-translational oxidation of certain lysine and hydroxylysine residues. The peptidyl aldehydes so formed become active centres for the formation of cross-links in collagen and elastin. Less well understood is how copper controls the steady-state activity of lysyl oxidase; the enzyme fails in copper deficiency. Giving copper to a deprived animal increases lysyl oxidase activity in aortic tissue. Such activation *in vivo* appears to require caeruloplasmin. Suspending aortic tissue in a copper-enriched growth medium also activates lysyl oxidase provided that tissue structure is kept intact. Activation *in vitro* occurs with the binding of copper to a large-molecular-weight component, presumably the enzyme. Binding will not occur if protein synthesis is blocked. These studies clearly show that the synthesis of mature elastin and collagen can be controlled by the availability of copper. They further suggest that transport of copper to aortic tissue and its engagement to lysyl oxidase are linked to the synthesis of lysyl oxidase, an extracellular carrier, or both.

The requirement for copper in the synthesis of collagen and elastin, the major proteins of connective tissue, was discovered after the metal had been recognized as essential for haemoglobin synthesis (Hart et al 1928), some 33 years later. The fibrous elastic lamina of major arteries of chicks and pigs was extensively fragmented when these animals were fed milk-protein diets deficient in copper (O'Dell et al 1961, Carnes et al 1961). Fig. 1 shows the typical structural defect that was observed at the time. The deficient aortic tissue was characterized as having a weakened internal structure with incomplete and fragmented elastic fibres. Such fibres were no longer able to provide flexible support to the blood vessels and the animals were prone to instant death through aortic rupture. Although no one doubted the symptoms, a molecular

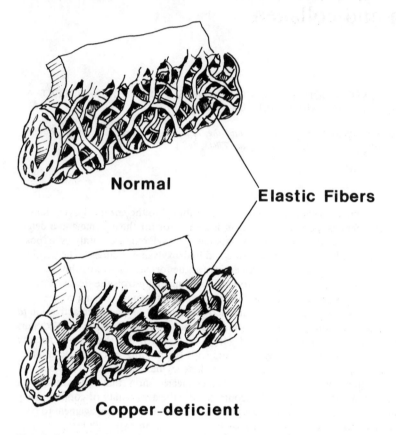

FIG. 1. Structural changes in the aorta associated with copper deficiency.

explanation of their cause could not be offered because of the then inchoate information on elastin structure and synthesis. Later, the formation of the elastic fibres was shown to require a post-translational modification of soluble tropoelastin, leading eventually to the appearance of cross-links between the peptide chains. One or more of the steps in the process appeared to require copper.

In 1968 Pinnell & Martin isolated from embryonic chick cartilage an enzyme which they called lysyl oxidase. The enzyme catalysed the oxidation of specific peptidyl lysine and hydroxylysine residues to aldehydes. Such aldehydes had been postulated earlier to be the active centres through which the intra- and intermolecular cross-links in collagen and elastin are formed.

The subsequent discovery that copper was required for the catalytic functioning of lysyl oxidase (Siegel et al 1970, Harris et al 1974) appeared to explain the function of the copper-sensitive site in the tissue. However, from a biochemical perspective, much more remained to be learnt about the function of copper in connective tissue.

In this paper we shall review more recent work on copper and lysyl oxidase. Recent studies have made it clear that copper ions are more than simple passive cofactors for lysyl oxidase. Rather, small amounts of copper appear to control the synthesis or degradation, or both, of the enzyme. These studies have opened the door to more intensive investigations of the control mechanism and should eventually yield more information on how copper ions are transported to and into the structure of metalloenzymes. More importantly, they intimate that there are important biological roles for copper yet to be discovered.

COPPER AND LYSYL OXIDASE

An understanding of the biological role of copper in connective tissue rests with understanding the function of lysyl oxidase. The enzyme, first observed in buffered saline extracts of embryonic cartilage (Pinnell & Martin 1968), has now been detected in a wide range of connective tissues from many different species (Table 1). Its extraction from these tissues requires a concentration of 4 M urea in the buffers. Lysyl oxidase activity has also been observed in various cell cultures both in the medium (which suggests a release from the cells) and in the cell layers. A recent discovery has revealed lysyl oxidase activity in the membrane-forming region of hen oviduct. The discovery suggests a requirement for copper and lysyl oxidase in the formation of the egg shell (Harris et al 1980). Such a widespread distribution of lysyl oxidase suggests an equally widespread need for copper.

The known biochemical function of lysyl oxidase is to catalyse the oxidative deamination of certain lysine and hydroxylysine (collagen only) residues in tropoelastin and tropocollagen before the formation of cross-links. The enzyme is classified technically as an amine oxidase. It requires oxygen as the recipient of electrons from the substrate. Products of the reaction are NH_3, H_2O_2 and the peptidyl derivative of the lysine or hydroxylysine residues (allysine and hydroxyallysine). The deamination of the lysine results in a loss of hydrogen from the epsilon carbon. The substitution of tritium for a hydrogen in this position has made it possible to assay the enzyme. The tritium is released (as tritiated water) during the course of the reaction. Spontaneous condensations between adjacent allysine residues, or allysine and

TABLE 1

Occurrence of lysyl oxidase in connective tissue and cultured cells

Site	Reference
Aortic tissue	
Fetal calf	Kagan et al 1974
Calf	Kagan et al 1979
Mature cattle	Rucker et al 1970
	Shieh et al 1975
Fetal dog	Kagan et al 1974
Mature chick	Harris et al 1974
Mature pig	E.D. Harris & L.B. Sandberg, unpublished
Mature rat	E.D. Harris & L.E. Yates, unpublished
Fetal human	E.D. Harris & J.E. Blount, unpublished
Cartilage	
Embryonic chick	Pinnell & Martin 1968
	Narayanan et al 1974
	Stassen 1976
Ligamentum nuchae	
Mature cattle	Harris & Garcia-de-Quevedo 1978
Lung	
Chick	E.D. Harris & L.E. Yates, unpublished
Hamster	Brody et al 1976
Mouse	Starcher et al 1977
Oviduct	
Hen	Harris et al 1980
Skin	
Mouse	Rowe et al 1977
	Royce et al 1978
Rat	Shoshan & Finkelstein 1976
Granulomatous tissue	
Human skin	Hayakawa et al 1976
Rat skin	Chvapil et al 1974
Cultured cells	
Human fibroblasts	Layman et al 1972
	DiFerrante et al 1975
Mouse lung fibroblasts	Starcher et al 1977
Mouse skin fibroblasts	Starcher et al 1977

lysine, give rise to aldol condensates and Schiff bases respectively, which form the major cross-links in collagen. More extensive condensation reactions lead to the formation of desmosine and isodesmosine, which form the cross-links unique to and prominent in elastin. Nutritional studies and direct chemical evidence (Murray et al 1977) have provided strong evidence that lysyl oxidase requires pyridoxal phosphate, in addition to copper, for its function. Siegel (1979) has recently reviewed other important properties of the enzyme.

Lysyl oxidase and copper deficiency

The critical dependence of lysyl oxidase on dietary copper is shown in Fig. 2. In the newly hatched chick the activity of lysyl oxidase in aortic tissue is substantial. However, when the chicks are put on diets containing less than

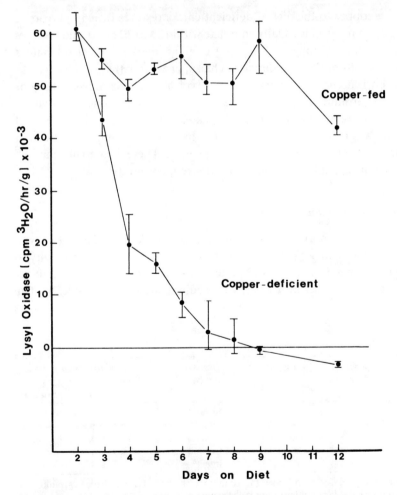

FIG. 2. Fall in aortic lysyl oxidase activity in response to copper-deficient diets. Each point represents the average of activity in three chick aortas.

1 mg/kg copper, lysyl oxidase activity falls to barely detectable levels in six to eight days. The chicks that receive enough copper do not experience a great loss in aortic lysyl oxidase activity. When the diet contains enough copper, lysyl oxidase has a half-life of 14–16 h (Harris et al 1977). Apparently the enzyme must be synthesized continuously in order to meet physiological requirements. Moreover, dietary copper could become a limiting factor in the expression and in the synthesis of the enzyme.

ACTIVATION OF LYSYL OXIDASE WITH COPPER

Returning copper to the diet of deficient chicks restores their lysyl oxidase activity. Such a reversal is usually complete within 24 to 30 hours and requires very small amounts of copper. When less than 0.5 μmol of copper sulphate is injected intraperitoneally into deficient chicks, enzyme activity is observed in aortic tissue within hours (Harris 1976). A typical response shows a delay of 1–2 h before substantial activity is detectable. The initial surge in activity is followed by a secondary response of greater intensity. Such a bimodal response to a single injection of $CuSO_4$ indicates a complex activation mechanism that requires a number of components. These data do not support the existence of a simple mechanism of passive incorporation.

Activation of lysyl oxidase in vitro

Suspending intact aortic tissue from deficient chicks in a defined growth medium enriched with copper salts (5 mg/kg) also restores enzyme activity to the tissue (Rayton & Harris 1979). The activation *in vitro* characteristically requires intact tissues, a serum-free growth medium and $CuSO_4$ (Table 2). The

TABLE 2

Conditions for lysyl oxidase activation *in vitro*[a]

Waymouth medium	Copper (5 μg/ml)	Buffer	Tissue status	Lysyl oxidase activity (values and errors $\times\ 10^{-3}$ absolute values)
+	–	–	Intact	4.2 ± 0.8
–	+	+	Intact	4.5 ± 0.8
+	+	–	Intact	17.5 ± 3.0
+	+	–	Ground	9.9 ± 1.9
+	+	–	Intact[b]	11.1 ± 1.3

[a]Aortas were incubated for 12 h under the conditions shown. Tissue labelled 'ground' was homogenized under sterile conditions until the structure of the aorta was destroyed.
[b]Incubated under 100% nitrogen.
Data ± SEM, $n = 3$.

response *in vitro* showed a lag period of 4 to 6 h before lysyl oxidase activity was detectable. Although serum proteins in their complete form had little effect on the activation of the enzyme, the system seemed to respond when a partially purified preparation of serum copper proteins provided copper at the physiological concentration of the metal (0.2 μg/ml). We have not determined whether any of these serum copper proteins perform a similar function *in vivo*.

Serum proteins in the activation of lysyl oxidase

We have also studied the requirement for serum copper proteins in the activation of lysyl oxidase. In copper deficiency there is a strong diminution in the level of serum copper. Serum from 10-day-old copper-deficient animals contains about half the amount of copper as that from copper-fed animals of the same age (Table 3). Both mature chicks and 10-day-old copper-fed chicks appear to have nearly identical copper protein profiles, as determined by gel filtration analysis (Fig. 3). Copper is present as two peaks: one near the void volume and a lesser peak of lower-molecular-weight material eluting at about the mid-volume of the column. Serum from deficient chicks shows diminished amounts of the void-volume copper component(s) and does not show lower-molecular-weight copper proteins. It remains to be determined whether the copper profiles are for a single protein that is present in different states of aggregation, or indeed whether they signify the presence of more than one copper protein in chick serum. Until this is resolved, our results suggest that a deficiency in copper is not associated with the loss of the metal from one serum protein.

TABLE 3

Changes in serum copper concentrations after various treatments

Treatment	n^a	μg *copper/ml serum*
Copper-fed	4	0.20 ± 0.07
Copper-deficient	6	0.095 ± 0.02
+ CuSO$_4$	6	0.19 ± 0.06
+ oestrogen	3	0.09 ± 0.01
+ CuSO$_4$, oestrogen	6	0.19 ± 0.07

[a]Each sample in the total no. (*n*) for each treatment contained pooled serum from at least six chicks. Serum was analysed for copper 16–20 h after treatment.
Data ± SD

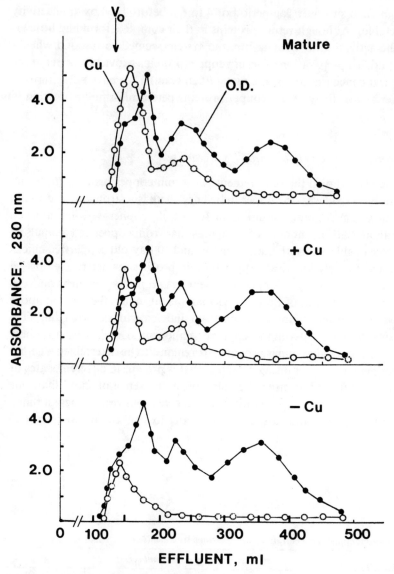

Fig. 3. Gel-filtration analysis (with Sephadex G-200) of proteins and copper in chick serum: response to copper deficiency.

Enhancement by oestrogen of lysyl oxidase activation

The changes in the serum copper profiles gave us inconclusive evidence about whether serum copper, or one or more specific serum copper proteins, mediates the activation of lysyl oxidase. Thus, additional experiments were needed to determine which of these factors is more influential. Oestrogen injections have been known to increase both serum copper levels and the p-phenylenediamine (PPD) oxidase activity of rat serum (Meyer et al 1958). Corticosteroids evoke a similar response in the chick (Starcher & Hill 1965). Since PPD oxidase activity is a measure of caeruloplasmin levels (Houchin 1958), the influence of depletion–repletion experiments on this serum copper protein can be evaluated directly. In Table 4 it can be seen that the general fall in serum copper levels noted earlier is accompanied by a loss in the PPD oxidase activity in the serum. This activity is normally quite low in the chick (Houchin 1958, Evans & Wiederanders 1967). When deficient chicks were given oestradiol-17β (1 mg/kg body weight, two injections, 24 h apart) there was no detectable change in the serum PPD oxidase activity or in the lysyl oxidase activity (Table 4). In contrast, injections of $CuSO_4$ raised the serum copper levels and, as expected, greatly increased the lysyl oxidase activity. However, when $CuSO_4$ at the same concentration (1 mg/kg body weight) was given to oestrogen-treated chicks, the lysyl oxidase activity increased much more than with $CuSO_4$ alone. The increase occurred when $CuSO_4$ was given with the second shot of oestrogen; enzyme activity was examined 15 to 20 h later. This increase in lysyl oxidase activity correlated directly with a further increase in PPD oxidase activity which also resulted from the combined $CuSO_4$ and oestrogen treatments. In contrast, the serum copper levels, although high, did not show a further increase, even with oestrogen. Thus the greater activation response correlated only with the PPD oxidase activity and not with serum copper levels in general. Details of these changes are brought out more clearly in a time-course study (Fig. 4). Here it can be seen that PPD oxidase activity increased in response to $CuSO_4$. However, the increase was stronger and of longer duration if the chicks had received oestrogen first.

TABLE 4

Correlation of lysyl oxidase activation with p-phenylenediamine (PPD) oxidase activity in serum

Treatment	n^a	PPD oxidase[b] (i.u.)	Serum copper (μg/ml)	Lysyl oxidase (^3H_2O released)
Copper-deficient	4	0.35 ± 0.01	0.09 ± 0.01	54 ± 15
+ oestrogen	4	0.37 ± 0.02	0.09 ± 0.01	53 ± 23
+ $CuSO_4$	4	0.92 ± 0.05	0.16 ± 0.01	1255 ± 146
+ $CuSO_4$, oestrogen	4	1.35 ± 0.05	0.14 ± 0.03	1773 ± 63

[a]Each sample contained pooled serum from six chicks.
[b]Pooled serum run in duplicate.
Data ± SD

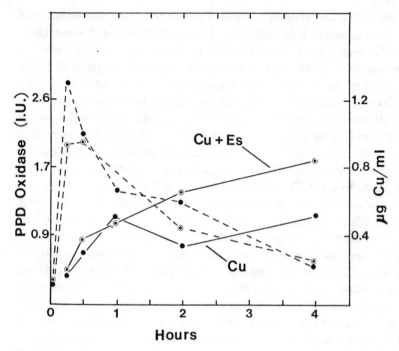

FIG. 4. Time-course of changes in serum copper and p-phenylenediamine (PPD) oxidase activity as a function of oestrogen treatment. At time zero, 10-day-old copper-deficient chicks were given intraperitoneal injections of $CuSO_4$ (1 mg/kg body wt) with and without oestradiol-17β (1 mg/kg) (Es), given subcutaneously at the same time. Those not receiving the oestradiol received an equal volume of 1,2-propanediol (propylene glycol), the solvent. Broken line is serum copper; solid line is PPD oxidase activity.

Fig. 4 also shows that the changes in the serum copper levels between oestrogen-treated and non-oestrogen-treated chicks were practically indistinguishable. Moreover, the further increase in PPD oxidase activity occurred within the first 2 to 4 h after injection, thus corresponding to the brief lag period before enzyme activity appeared.

DISCUSSION

With each new discovery of an enzyme system that requires copper, we reach a greater insight into the myriad functions of this metal. However, copper may be more than a simple cofactor for the biocatalysts. The metal or specific complexes of copper may also regulate important biochemical events — at least we suspect this to be true with connective tissue and the enzyme lysyl oxidase.

Precisely how copper regulates the activity of lysyl oxidase is uncertain at present. Certainly the changes in concentration of an enzyme or a protein in response to copper depletion are not unique to aortic tissue. In the liver, for example, specific copper-binding proteins in the cytosol decrease during copper deficiency and are strongly increased by copper loading (Bloomer & Sourkes 1973, Riordan & Gower 1975, Bremner & Young 1976). We need only recall the classic experiments of Granick (1946) and those of Fineberg & Greenberg (1955) to realize that a metal (in this case iron) is capable of regulating the synthesis of its storage protein (apoferritin). With copper we may be dealing with a similar phenomenon. The conclusion that the metal may be capable of inducing the synthesis of the respective protein seems justified, on the basis of observations (Premakumar et al 1975).

While the response is simple to describe, an explanation of what is happening biochemically when a metal ion controls a protein or the expression of enzyme activity is far from simple. In essence the question is reduced to how a system can detect the absence of copper. What biochemical events are set in motion in order to correct for the missing component? In the aorta the response to copper deficiency is a reduction in the amount of lysyl oxidase and perhaps of other proteins. In the liver a deficiency in copper diminishes the amount of cytochrome c oxidase (EC 1.9.3.1) (Gallagher et al 1956) and superoxide dismutase (EC 1.15.1.1) (Williams et al 1975), both copper-dependent enzymes. Extending the phenomenon further, we must explain the mechanism by which copper ions restore homeostasis of the enzyme levels. These are important biological questions whose answers will only come with time; but they must come if we are to know the biological role of this metal.

Delineation of the events through which copper finds its way back to the enzyme and restores enzyme activity will eventually define the biochemical pathways of the metal. In this respect the aortic system offers an advantage over other systems because the terminal event, the activation of an enzyme, can easily be monitored and its physiological significance to the organism cannot be questioned. Evidence for the existence of defined pathways for copper in metabolism have been greatly supported by the inborn diseases of copper metabolism, such as Wilson's disease and Menkes' syndrome. Surely specific proteins are required to handle transport of copper ions to the tissues and to control their uptake into the cells? We know very little about these processes other than that they are likely to be very complex, even in a tissue as simple as the aorta.

On the basis of the studies described in this report, we can summarize our current understanding of the events in the activation of lysyl oxidase (Fig. 5). The transport and delivery of copper to the aorta is likely to require a specific

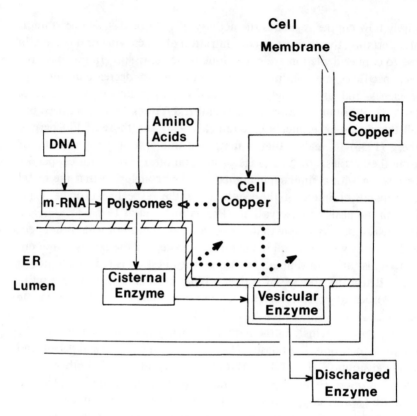

FIG. 5. Summary of molecular events in the activation of lysyl oxidase by copper. ER, endoplasmic reticulum.

serum copper protein. Our data suggest that this protein is caeruloplasmin. Thus, our studies concur with the postulated transport–delivery function of caeruloplasmin proposed by others (Owen 1965, Marceau & Aspin 1973, Hsieh & Frieden 1975), and extend these findings to the aortic system. The transport of copper into the tissue is unknown, but from studies of iron uptake from transferrin (Hemmaplardh & Morgan 1977) the process is likely to require specific receptors on the cell membrane, which bind the protein and internalize the protein–copper complex.

Fig. 5 brings out additional facets of the activation mechanism. Once inside the cell the copper must work its way to the site of enzyme synthesis or metal ion incorporation (the two may be identical or distinctly different). Copper salts returned to the tissue homogenates or to the urea extracts from the deficient aortic tissue fail to activate the enzyme. Thus the mere presence of copper is no guarantee that activity will be restored. Table 2 shows that disruption of the tissue seemed to block the activation or, at least, to retard the effectiveness of $CuSO_4$. Both these observations suggest that a preformed pool of apoenzyme does not exist as the recipient of the copper. Rayton & Harris (1979) reported that when protein synthesis was interrupted by cycloheximide, radioactive copper (^{64}Cu) failed to bind to aortic proteins that were identified by chromatography as lysyl oxidase. All these findings support a mechanism by which copper ions become incorporated into the protein structure during protein manufacture and not afterwards, i.e. co-translational, as opposed to post-translational, insertion of the metal occurs. Because lysyl oxidase has been found in the medium of cultured cells, the enzyme is generally believed to occur extracellularly, and to function within the fibres surrounding the cell (Layman et al 1972, Kagan et al 1974). Transport from the cell requires that the protein be synthesized on membrane-bound ribosomes and subsequently be secreted into the cisternal space, on its way to the Golgi apparatus, and eventually out of the cell (Blobel 1977). Since the completed protein will not bind copper, the metal must bind at some stage during the assembly of the amino acids into the protein structure. More important, however, is the possibility that once the protein is released into the internal cisternal space, it may no longer be capable of binding copper. The discharge mechanism is also time-dependent, which means that it can account for the lag period before the enzyme activity is detectable after exposure to copper.

Hence, the absence of copper may not prevent the synthesis of the enzyme, but it could result in a non-functioning enzyme being released from the cell. The identity of the intracellular site of copper incorporation may help to clear up the mystery of how copper regulates enzyme expression. Studies to meet that objective are currently in progress.

ACKNOWLEDGEMENTS

This research was funded in part by Future Leader Award Grant No. 532 from the Nutrition Foundation, Inc. and Grant No. PCM 77-25400 from the National Science Foundation. The authors acknowledge the excellent technical assistance of Mr Ben L. Pressly.

References

Blobel G 1977 Synthesis and segregation of secretory proteins: the signal hypothesis. In: Brinkley BR, Porter KR (eds) International cell biology 1976–1977. Rockefeller University Press, New York, p 318-325
Bloomer LC, Sourkes TL 1973 The effect of copper loading on the distribution of copper in rat liver cytosol. Biochem Med 8:78-91

Bremner I, Young BW 1976 Isolation of (copper, zinc)-thioneins from the livers of copper-injected rats. Biochem J 157:517-520

Brody JS, Kagan HM, Manalo AD, Hu CA, Franzblau C 1976 Lung lysyl oxidase and elastin synthesis during compensatory lung growth. Chest 69 (suppl):271-272

Carnes WH, Shields GS, Cartwright GE, Wintrobe MM 1961 Vascular lesions in copper-deficient swine. Fed Proc 20:118 (abstr)

Chvapil M, McCarthy DW, Misiorowski RL, Madden JW, Peacock EE Jr 1974 Activity and extractability of lysyl oxidase and collagen proteins in developing granuloma tissue. Proc Soc Exp Biol Med 146:688-693

DiFerrante N, Leachman RD, Angelini P, Donnelly PV, Francis G, Almazan A 1975 Lysyl oxidase deficiency in Ehlers-Danlos syndrome Type V. Connect Tissue Res 3:49-53

Evans GW, Wiederanders RE 1967 Blood copper variation among species. Am J Physiol 213:1183-1185

Fineberg RA, Greenberg DM 1955 Ferritin biosynthesis. III Apoferritin, the initial product. J Biol Chem 241:97-113

Gallagher CH, Judah JD, Rees KR 1956 The biochemistry of copper deficiency. I: Enzymological disturbances, blood chemistry and excretion of amino acids. Proc R Soc Lond B Biol Sci 145:134

Granick S 1946 Ferritin. IX. Increase of the protein apoferritin in the gastrointestinal mucosa as a direct response to iron feeding. The function of ferritin in the regulation of iron absorption. J Biol Chem 164:737-746

Harris ED 1976 Copper-induced activation of aortic lysyl oxidase in vivo. Proc Natl Acad Sci USA 73:371-374

Harris ED, Garcia-de-Quevedo M 1978 Reaction of lysyl oxidase with soluble protein substrates: effect of neutral salts. Arch Biochem Biophys 190:227-233

Harris ED, Gonnerman WA, Savage JE, O'Dell BL 1974 Connective tissue amine oxidase. II: Purification and partial characterization of lysyl oxidase from chick aorta. Biochim Biophys Acta 341:332-344

Harris ED, Rayton JK, DeGroot JE 1977 A critical role for copper in aortic elastin structure and synthesis. In: Sandberg LB, Gray WR, Franzblau C (eds) Elastin and elastic tissue. Plenum Press, New York, p 543-559

Harris ED, Blount JE, Leach RM Jr 1980 Localization of lysyl oxidase in hen oviduct: implications in egg shell membrane formation and composition. Science (Wash DC) 208:55-56

Hart EB, Steenbock H, Waddell J, Elevehjem CA 1928 Iron in nutrition. VII. Copper as a supplement to iron for hemoglobin building in the rat. J Biol Chem 77:797-812

Hayakawa T, Hino M, Fuyamada H, Nagatsu T, Aoyama H, Izawa Y 1976 Lysyl oxidase activity in human normal skins and postburn scars. Clin Chim Acta 71:245-250

Hemmaplardh D, Morgan EH 1977 The role of endocytosis in transferrin uptake by reticulocytes and bone marrow cells. Br J Haematol 36:85-96

Houchin OB 1958 A rapid colorimetric method for the quantitative determination of copper oxidase activity (ceruloplasmin). Clin Chem 4:519-523

Hsieh HS, Frieden E 1975 Evidence for ceruloplasmin as a copper transport protein. Biochem Biophys Res Commun 67:1326-1331

Kagan HM, Hewitt NA, Salcedo LL, Franzblau C 1974 Catalytic activity of aortic lysyl oxidase in an insoluble enzyme-substrate complex. Biochim Biophys Acta 365:223-234

Kagan HM, Sullivan KA, Olsson TA, Cronlund AL 1979 Purification and properties of four species of lysyl oxidase from bovine aorta. Biochem J 177:203-214

Layman DL, Narayanan AS, Martin GR 1972 The production of lysyl oxidase by human fibroblasts in culture. Arch Biochem Biophys 149:97-101

Marceau N, Aspin N 1973 The intracellular distribution of the radiocopper derived from ceruloplasmin and from albumin. Biochim Biophys Acta 293:338-350

Meyer J, Meyer AC, Horwitt MK 1958 Factors influencing serum copper and ceruloplasmin oxidative activity in the rat. Am J Physiol 194:581-584

Murray JC, Levene CI 1977 Evidence for the role of vitamin B-6 as a cofactor of lysyl oxidase. Biochem J 167:463-467

Narayanan AS, Siegel RC, Martin GR 1974 Stability and purification of lysyl oxidase. Arch Biochem Biophys 162:231-237

O'Dell BL, Hardwick BC, Reynolds G, Savage JE 1961 Connective tissue defect in chicks resulting from copper deficiency. Proc Soc Exp Biol Med 108:402-408

Owen CA Jr 1965 Metabolism of radiocopper (^{64}Cu) in the rat. Am J Physiol 209:900-904

Pinnell SR, Martin GR 1968 The cross-linking of collagen and elastin: enzymatic conversion of lysine in peptide linkage to α-amino-adipic-δ-semialdehyde (allysine) by an extract from bone. Proc Natl Acad Sci USA 61:708-714

Premakumar R, Winge DR, Wiley RD, Rajagopalan KV 1975 Copper-induced synthesis of copper chelatin in rat liver. Arch Biochem Biophys 170:267-277

Rayton JK, Harris ED 1979 Induction of lysyl oxidase with copper. Properties of an in vitro system. J Biol Chem 254:621-626

Riordan JR, Gower I 1975 Purification of low molecular weight copper proteins from copper loaded liver. Biochem Biophys Res Commun 66:678-686

Royce PM, Camakarix J, Mann JR 1978 The effect of copper therapy on lysyl oxidase activity in the mottled mouse mutants. Proc Aust Biochem Soc 11:31 (Abstr)

Rowe DW, McGoodwin EB, Martin GR, Grahn D 1977 Decreased lysyl oxidase activity in the aneurysm-prone, mottled mouse. J Biol Chem 252:939-942

Rucker RB, Roensch LF, Savage JE, O'Dell BL 1970 Oxidation of peptidyl lysine by an amine oxidase from bovine aorta. Biochem Biophys Res Commun 40:1391-1397

Shieh JJ, Tamaye R, Yasunobu KT 1975 A purification procedure for the isolation of homogeneous preparations of bovine aorta amine oxidase and a study of its lysyl oxidase activity. Biochim Biophys Acta 377:229-238

Shoshan S, Finkelstein S 1976 Lysyl oxidase: a pituitary hormone-dependent enzyme. Biochim Biophys Acta 439:358-362

Siegel RC 1979 Lysyl oxidase. In: Hall DA, Jackson DS (eds) International review of connective tissue research. Academic Press, New York, vol 8:73-118

Siegel RC, Pinnell SR, Martin GR 1970 Cross-linking of collagen and elastin. Properties of lysyl oxidase. Biochemistry 9:4486-4492

Starcher BC, Hill CH 1965 Hormonal induction of ceruloplasmin in chick serum. Comp Biochem Physiol 15:429-434

Starcher BC, Madaras JA, Tepper AS 1977 Lysyl oxidase deficiency in lung and fibroblasts from mice with hereditary emphysema. Biochem Biophys Res Commun 78:706-712

Stassen FLH 1976 Properties of highly purified lysyl oxidase from embryonic chick cartilage. Biochim Biophys Acta 438:49-60

Williams DM, Lynch RE, Lee GR, Cartwright GE 1975 Superoxide dismutase activity in copper-deficient swine. Proc Soc Exp Biol Med 149:534-536

Discussion

Danks: In our tissue cultures from patients with Menkes' syndrome or from mottled (brindled) mice we have obtained results similar to yours, Professor Harris (P.M. Royce, J. Camakaris & D.M. Danks, unpublished work, 1980). The only difference is that we can manipulate the copper level in our culture medium. We have difficulty in obtaining an abnormally low level but we can certainly raise it above the level normally present in standard media containing 10% fetal calf serum. We have shown that at all concentrations of copper, from that in 10% fetal calf serum to a considerably higher concentration obtained by the addition of copper, the copper concentration *within* the cell is increased 5–10 fold above normal. This is the established phenotypic effect of this mutation in cell culture (Danks 1977). This manipulation of copper levels does not alter the low lysyl oxidase activity in these cells. No matter what the level of copper is in the outside medium, lysyl oxidase activity stays at 10–20% of the level in normal fibroblasts treated in the same way. This fits in with your idea, Professor Harris, that the copper must actually be delivered to the site of synthesis of the enzyme. In the brindled mouse, delivery of copper is disrupted.

Dormandy: Professor Harris, you said that when you incubate the system in nitrogen (under anoxic conditions) there is only a slight diminution of the enzyme activity. This is more characteristic of a non-enzymic autoxidation. Oxygen is a substrate, i.e. a major component, of the reaction, so why does lack of oxygen have a more dramatic effect on enzyme activity?

Harris: I think this relatively small reduction in enzyme activity is due to our inability to achieve complete anaerobiosis. We try to drive the oxygen out of the environment of the cells. The tubes are gassed with nitrogen for about 15 seconds, and this is probably insufficient to remove all the oxygen, which is very difficult to do. Some work by Siegel et al (1970) has shown that an increase in the partial pressure of oxygen in an assay medium increases the activity of the enzyme during the course of the assay.

Dormandy: If you were to substitute pure oxygen for air would there be an appreciable difference in effect?

Harris: Yes; the enzyme activity, in a typical reaction with its substrate, would be affected. We must distinguish between activation of the enzyme, which requires intact tissue, and measurement of the enzyme activity; they are two different things. I was talking about the activation process.

Mills: What substrate do you use in this assay?

Harris: We use labelled aortic proteins which are a mixture of collagen and elastin. But the primary active component appears to be tropoelastin.

Mills: Have the assays done by other people contained substrates that had different ratios of elastin to collagen, and are there any differences in response between the two substrates? Waisman et al (1969), who examined aortic histology in copper-deficient pigs, found not only extensive defects in elastin but also what appeared to be a compensatory increase in collagen along the laminae.

Harris: As far as the enzyme specificity is concerned, no one has identified a species of lysyl oxidase that acts on elastin but not collagen. It has been shown that there are multiple forms of the enzyme in the aortic tissue of several species (Harris & O'Dell 1974, Kagan et al 1979, Vidal et al 1975), and highly purified lysyl oxidase works equally well on elastin substrates and on collagen substrates (Stassen 1976, Siegel & Fu 1976). In answer to your other point, I don't know why collagen would increase in copper-deficient pigs.

Hill: How well characterized is lysyl oxidase?

Harris: There are data on the characteristics of a homogeneous preparation of the enzyme. There is dispute about its molecular weight, because the pure enzyme has some unusual properties. For instance, it becomes more insoluble as it becomes more purified; even a urea solvent is ineffective in keeping it in solution. The amino acid composition of lysyl oxidase has been identified in a number of species (Stassen 1976, Kagan et al 1979) and there is good evidence to support the involvement of pyridoxyl phosphate (Murray & Levene 1977). Our own studies with inhibitors and with direct chemical analyses have demonstrated that copper is a cofactor (Harris et al 1974) but there is still a great deal to be discovered about this enzyme.

Hill: And are there any artificial substrates for the enzyme?

Harris: There are problems in measurement of the enzyme activity with artificial substrates, because the total number of substrate residues that are susceptible to oxidation is small. The tritium assay is useful because it is sensitive. Herb Kagan, in Boston, is now developing some artificial substrates for lysyl oxidase, and although a typical coupled assay for peroxide production does not reveal any lysyl oxidase activity, the fluorimetric procedure on the same assay increases the sensitivity by about 100-fold and shows that oxidation is taking place (Trackman & Kagan 1979). So there is now evidence that the enzyme *will* work on other substrates.

McBrien: When proteins that are going to be secreted are being synthesized, they start to cross the membrane as soon as the signal sequence of hydrophobic amino acids is attached to the protein (Freedman 1979). In most proteins, this signal sequence is right at the beginning of the molecule, so the protein starts to cross the membrane before the rest of the molecule is complete. Do you know the amino acid sequence of lysyl oxidase? Could there be a signal sequence in the *middle* of this protein? There was a report by Lingappa et al (1978) describing a protein in which the signal sequence was embedded in the middle of the molecule. (Generally the signal sequences are removed after the protein is completed, but if the sequence is in the middle of the molecule it cannot be removed and the protein does not cross the membrane until the synthesis is more complete.)

Harris: The amino acid composition of the different species of lysyl oxidase are not sufficiently well known for me to comment on that. However, it is possible that a precursor protein for lysyl oxidase exists.

Riordan: Is lysyl oxidase a glycoprotein?

Harris: That is not yet known.

Riordan: Have you analysed it for carbohydrate?

Harris: It has not been extensively analysed, but we did look for glucosamine in the protein and found very small amounts.

Riordan: Presumably glucosamine would be there if your postulated scheme is correct. Non-glycosylated proteins tend not to use the membrane-bound ribosome route.

Harris: That is correct. Incidentally, the elastin is comprised of two proteins – a microfibrillar component and an amorphous component. The microfibrillar component is supposedly a glycoprotein that forms a periphery around the fibre. We once believed (erroneously) that this protein itself was lysyl oxidase.

Riordan: In your repletion experiments, did you simply add large amounts of caeruloplasmin? Is caeruloplasmin the copper donor?

Harris: No; we did not add caeruloplasmin. In the experiments I described we injected small amounts of copper sulphate.

Hassan: You showed that cycloheximide inhibited the activation of lysyl oxidase. *In vitro*, is there a transcription of the DNA to mRNA? Did you try to inhibit transcription?

Harris: We have tried to inhibit transcription by using actinomycin D and we found that copper-induced activation of lysyl oxidase was relatively insensitive to actinomycin D, showing only about 15% inhibition (Harris 1976). Therefore I believe that if there is control of enzyme synthesis it occurs after the mRNA has been made; i.e. regulation occurs after translation and not during transcription.

McMurray: I understand that cross-linking of elastin and collagen takes place throughout life or over an extended period. Is lysyl oxidase important other than in young animals and can its activity be altered by copper deficiency in later life?

Harris: We do not yet know whether lysyl oxidase is essential to the changes that occur in the connective tissue in older animals. (The enzyme has certainly been isolated from tissues of older animals, so it is not restricted to neonates or to young animals.)

Mills: In view of your observations showing regional differences in lysyl oxidase activity along the chick oviduct, have you looked at the distribution of lysyl oxidase along the aorta?

Harris: No, because the chicken aorta is a tiny piece of tissue that is difficult to section, whereas the oviduct is more substantial.

Mills: I asked the question because recent work by Dr N.T. Davies at Aberdeen (unpublished results) on lysyl oxidase activity in copper-deficient cattle shows that whereas there was a large difference in activity of lysyl oxidase between *ligamentum nuchae* of control and copper-deficient animals, there was no detectable difference in the aortic enzyme activity in the two groups. No dissection of aortic tissue intima was made and perhaps this study may thus not have detected local differences in lysyl oxidase activity.

Harris: I can confirm that the amount of lysyl oxidase activity in *ligamentum nuchae* is comparable to that found in chick aorta.

Mills: One of the common features in the development of an aneurysm appears to be a change in collagen type as well as in its distribution. Does copper deficiency modify collagen type?

Harris: All I can say is that type III collagen appears in inflamed tissue and that this may be newly deposited collagen that is modified and eventually replaced by type I collagen. The whole sequence could simply be a general response of the cells to inflammation.

Danks: Sites of healing share this phenomenon, don't they?

Mills: I think so. In copper deficiency the phenomenon could represent an attempt to repair zones of weakness that are attributable, initially, to elastin defects.

Danks: We've looked at the activity of lysyl oxidase against both collagen and elastin substrates in mice (P.M. Royce, J. Camakaris & D.M. Danks, unpublished work). We found that it functions against both of them and that this activity is reduced in the mutant mice against both substrates. (This is consistent with the same enzyme being affected by copper, and with its acting on both substrates.)

Harris: Yes, that is reasonable, because the very nature of an oxidase is that it removes electrons from a substrate and momentarily houses them, before they are transferred to oxygen. If the metal ion cofactor is not present with the protein component, there would be damage to the protein.

Frieden: There are riboflavin oxidases, of course, that do not require a metal cofactor.

Mills: Since telopeptides of collagen and elastin are disulphide-bridged and, in copper deficiency, defects in disulphide bridge formation are so common, is it possible that copper deficiency may influence the initial alignment of telopeptides before other types of cross-linking are initiated?

Harris: Sulphydryl groups may be important in the initial reactions but mature collagen does not contain sulphydryl groups. It was once believed that the cross-linking was due to disulphide bonds but, of course, these bonds cannot occur without cysteine residues in the protein. Therefore other types of linkage must occur. The precursor proteins, which have extensions on their amino terminals, *do* form disulphide bonds, but I do not know whether these bonds are essential to formation of collagen fibres. However, the disulphide bonds may affect the alignment of the protein in the tissue; collagen has a precise 'quarter stagger' arrangement in the fibre. Lysyl oxidase is probably important after that stage, during the cross-linking and during the formation of the mature fibre.

Danks: Is copper essential for disulphide bonding in all proteins? I know of no evidence that insulin, another disulphide-bonded protein, is structurally damaged in copper-deficient animals. Copper is essential for the numerous disulphide bonds in keratin, but is that an exception or a rule?

Mills: We do not yet know the answer to that question!

References

Danks DM 1977 Copper transport and utilisation in Menkes' syndrome and in mottled mice. Inorg Perspect Biol Med 1:73-100
Freedman R 1979 Proteins have export problems too. New Sci 81:376-378
Harris ED 1976 Copper-induced activation of aortic lysyl oxidase *in vivo*. Proc Natl Acad Sci USA 73:371-374
Harris ED, O'Dell BL 1974 Copper and amine oxidases in connective tissue metabolism. In: Friedman M (ed) Protein metal interactions. Plenum Press, New York (Adv Exper Med Biol 48), p 267-284
Harris ED, Gonnerman WA, Savage JE, O'Dell BL 1974 Connective tissue amine oxidase. II: Purification and partial characterization of lysyl oxidase from chick aorta. Biochim Biophys Acta 341:332-344
Kagan HM, Sullivan KA, Olsson TA, Cronlund AL 1979 Purification and properties of four species of lysyl oxidase from bovine aorta. Biochem J 177:203-214
Lingappa VR, Shields D, Woo SLC, Blobel G 1978 Nascent chicken ovalbumin contains the functional equivalent of a signal sequence. J Cell Biol 79:567-572
Murray JC, Levene CI 1977 Evidence for the role of vitamin B-6 as a cofactor of lysyl oxidase. Biochem J 167:463-467
Siegel RC, Fu JCC 1976 Collagen cross-linking. Purification and substrate specificity of lysyl oxidase. J Biol Chem 251:5779-5785
Siegel RC, Pinnell SR, Martin GR 1970 Cross-linking of collagen and elastin. Properties of lysyl oxidase. Biochemistry 9:4486-4492
Stassen FLH 1976 Properties of highly purified lysyl oxidase from embryonic chick cartilage. Biochim Biophys Acta 438:49-60
Trackman PC, Kagan HM 1979 Nonpeptidyl amine inhibitors are substrates of lysyl oxidase. J Biol Chem 254:7831-7836
Vidal GP, Shieh JJ, Yasunobu KT 1975 Immunological studies of bovine aorta lysyl oxidase: evidence for two forms of the enzyme. Biochem Biophys Res Commun 64:989-995
Waisman J, Cancilla PA, Coulson WF 1969 Cardiovascular studies on copper deficient swine. XIII: Effect of chronic copper deficiency on the cardiovascular system of miniature pigs. Lab Invest 21:548-554

Copper deficiency in ruminants

CECIL H. McMURRAY

Veterinary Research Laboratories, Stoney Road, Stormont, Belfast BT4 3SD, UK

Abstract Copper deficiency can reduce the productivity of livestock. The effect of copper deficiency on a number of copper enzymes and copper-dependent systems is discussed, to highlight the areas where their role needs to be clarified. Special reference is made to cytochrome *c* oxidase, lysyl oxidase, superoxide dismutase and endoplasmic reticulum enzymes and to their role in the expression of disease. The modification of microbiological insult by a change in superoxide dismutase activity without any other direct metabolic consequences is discussed, to introduce the concept of an external challenge being necessary before any effect of an otherwise sub-clinical copper deficiency is observed. The changes in activity of the various copper enzymes are described in clinical and experimentally induced copper deficiency in sheep and cattle, two species in which copper deficiency can have economic consequences. The diagnostic value of various blood markers, such as copper, caeruloplasmin and erythrocyte superoxide dismutase is discussed. The measured degree of hypocupraemia is related to different types of sampling (e.g. plasma or serum), physiological status (e.g. in the pre- and postpartum cow), changes that occur in the neonate, and also to the effect of the acute-phase reaction. The use of erythrocyte superoxide dismutase as a marker for the copper status of sheep and cattle is compared with more conventional markers such as plasma concentration of copper. The use of blood markers to map the extent and location of hypocupraemia (due to reduced copper intake or availability) among suckler (beef) herds in Northern Ireland is also discussed.

Copper deficiency is the primary cause of specific disease states in a number of mammalian species and it could also play a secondary role in modulating the effects of other disease processes.

There are a number of diseases that owe their pathogenesis to copper deficiency and that have been attributed to molecular deficiency of key copper enzymes. These have recently been reviewed by Underwood (1977) and O'Dell (1976). Copper deficiency decreases (1) the activity of cytochrome *c* oxidase, necessary for normal oxidative phosphorylation (Erecińska & Wilson 1978),

(2) the activity of lysyl oxidase, essential for elastin and collagen maturation (see Harris et al 1980, this volume) and (3) the clearance of cholesterol from plasma (Lei 1978, Allen & Klevay 1978b). These effects may play an interactive or synergistic role in development of the final pathological state of aneurysm and subsequent heart rupture that has been reported in rats (Kelly et al 1974).

In this paper I shall consider the role of a few copper systems whose malfunction has been implicated in copper deficiency and under certain conditions of nutrition and development.

CYTOCHROME c OXIDASE

This enzyme, the terminal component of the mitochondrial electron-transfer chain, is located asymmetrically in the inner mitochondrial membrane. The enzyme couples electron transfer with the transmembrane proton pump required for the synthesis of ATP (Mitchell 1966, Erecińska & Wilson 1978).

There is a considerable body of published work that links copper deficiency to a reduction in activity of cytochrome c oxidase (EC 1.9.3.1) (Underwood 1977). While decreased enzymic activity may cause some of the symptoms of copper deficiency, equally important is the respiratory control ratio (Carroll & Racker 1977). This has not been examined definitively in preparations from copper-deficient animals. Alteration in mitochondrial structure is one of the features of copper-deficient animals (Gallagher et al 1973, Fell et al 1975, Leith 1975) and therefore the rate and control of respiration in these animals needs further examination in order to allow the relative importance of cytochrome c oxidase in copper deficiency to be quantified.

LYSYL OXIDASE

The first step in the biosynthesis of elastin cross-links is the oxidative deamination of lysyl residues. Reduction in activity of lysyl oxidase through copper deficiency (Harris 1976) results in the accumulation of soluble tropoelastin, the precursor of elastin. This accumulation has been demonstrated in the aorta in both the chick and the pig (Foster et al 1975, Rucker et al 1977).

The cross-linking mechanism of collagen is more complex than that of elastin, and some cross-links in collagen are dependent on oxidative deamination of lysine (see Fessler & Fessler 1978). Four different types of collagen exist and they have a differential distribution between and within tissues (Bailey et al 1979). Lysyl cross-links are apparently more important in type I and type III collagens than in the other forms. Normal development associated with collagen formation will be influenced by collagen defects due to copper deficiency and the consequences will depend on the age of the animal and on the tissue affected.

O'Dell (1976) distinguished three tissues, the skeleton, heart and vascular system, where reduction in the amount of cross-linking of elastin and collagen had important structural implications. Collagen integrity is important in bone because collagen accounts for 95% of the protein in the bone matrix (Aaron 1976). Skeletal defects have been reported in a number of species including chicks, pigs, sheep and cattle (Underwood 1977).

Similarly, a reduction in collagen cross-linking has been proposed to account for the thinning and deformation of egg shells produced by the copper-deficient hen (Baumgartner et al 1978). A similar defect in collagen cross-linking accounts for problems in tendon development (Chou et al 1969). However, Ganezer et al (1976) could find no change in the tensile strength of tendon obtained from the tails of copper-deficient pigs.

Cardiac defects due to copper deficiency can result in sudden death among the bovine species (Underwood 1977); the main lesion is usually myocardial atrophy. Cardiac hypertrophy, however, has been produced experimentally in copper-deficient rats (Kelly et al 1974) and in cattle (Mills et al 1976, Suttle & Angus 1976).

Both the quality and quantity of elastin and collagen are important in maintaining the integrity of the cardiovascular system (O'Dell 1976). Vascular aneurysm is a feature of copper deficiency in the rat and the pig (Waisman et al 1969). The structure of lung tissue is also altered in copper-deficient rats (O'Dell et al 1978) and chicks (Buckingham et al 1978), but O'Dell et al (1978) found that the lung lesions were not completely reversed by copper supplementation.

SUPEROXIDE DISMUTASE

Two isoenzymes of superoxide dismutase (SOD) are present in mammals (Fridovich 1975) and they have a differential cellular distribution. In the liver a cyanide-sensitive Cu-Zn-metalloenzyme is present in cytosol, whereas a cyanide-insensitive manganese form is present in the mitochondrion (Tyler 1975, Peeters-Joris et al 1975). A similar distribution of both metalloenzymes occurs in neutrophils (Rest & Spitznagel 1977) and in alveolar macrophages (Andrews et al 1980), but only the Cu-Zn form is present in the erythrocyte (McCord & Fridovich 1969).

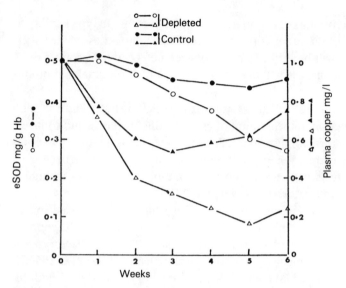

FIG. 1. Copper depletion in lambs: changes in plasma copper and erythrocyte superoxide dismutase (eSOD) in two groups of lambs; one group was fed a normal copper diet and the other the same diet except that thiomolybdate (Suttle & McMurray 1980) was added to reduce the copper availability. SOD is expressed as mg eSOD protein/g haemoglobin. 1 mg SOD protein is equivalent to 3000 units of SOD activity as measured by the cytochrome c assay (McCord & Fridovich 1969).

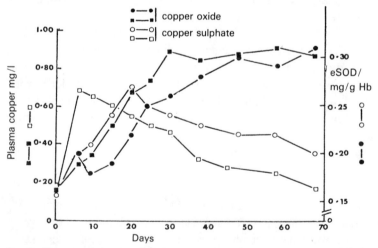

FIG. 2. Copper repletion in ewes: changes in plasma copper and erythrocyte superoxide dismutase as a result of single oral copper repletion by use of either copper sulphate or copper oxide needles in ewes that were being fed a copper-deficient diet (Cu 1.4 mg/kg DM, Mo 4 mg/kg DM and S: 3 g/kg DM (dry matter)). The same dose of copper (0.5 g) was used in both treatments (Suttle & McMurray 1980). SOD measurements explained in Fig. 1 legend (vertical axes reversed c.f. Fig. 1).

Copper nutrition affects the activity of SOD in the tissues and erythrocytes of pigs (Williams et al 1975) and sheep (Andrewartha & Caple 1980), and in erythrocytes (Bohnenkamp & Weser 1977), liver, heart-muscle (Paynter et al 1979) and brain (Morgan & O'Dell 1977) of rats.

The change in SOD activity in erythrocytes during copper depletion of normal lambs is shown in Fig. 1 and the change in erythrocyte SOD activity during repletion of copper-depleted ewes is shown in Fig. 2 (Suttle & McMurray 1980); similar measurements on copper-depleted calves are shown in Fig. 3 (Suttle & McMurray 1980). In both sheep and cattle the response to a changing copper status, as measured by SOD activity, is delayed with respect to the changes in plasma copper. This is probably due to an effect of haemopoiesis.

The presence of the manganese form of the enzyme may modify the animal's reaction to the fall in Cu-ZnSOD due to copper deficiency. However, the effect of changing the SOD status of the animal has to be considered. A reduction of Cu-ZnSOD activity in the cell may have clinical consequences for the animal either because of an alteration in the protective effect of SOD on the leucocytes or other cells against superoxide (O_2^-) produced during phagocytosis (Roos & Weening 1979) or, alternatively, because control over the inflammatory response is lost (McCord & Wong 1979). Further implications and consequences of superoxide production have recently been considered in detail (Chance et al 1979).

It is possible, therefore, that through an effective SOD mechanism an animal can be protected to some extent from microbiological and parasitological insult. As a result of the decreased phagocytosing efficiency, the SOD-deficient animal would be less resistant to infection, and the microbiological challenge required to initiate clinical infection might be lower. Alternatively, if SOD levels can be further increased by feeding animals high levels of copper, then this may explain, for instance, the increased growth observed when pigs are fed high levels of dietary copper (Braude 1975). This hypothesis may also explain why the response to copper therapy has been so variable in hypocupraemic cattle (see MacPherson et al 1979); i.e. an external challenge is needed in the absence of a direct metabolic effect before the benefits of copper supplementation can be demonstrated.

While most of this discussion is speculative it concurs with the role that SOD is currently thought to occupy in mammalian systems. If the self-limiting consequences of some diseases can be increased by the creation of maximum concentrations of SOD in tissues, there would be a strong argument for dietary supplementation of copper to copper-deficient animals, irrespective of the direct growth effects, which may or may not be obtained.

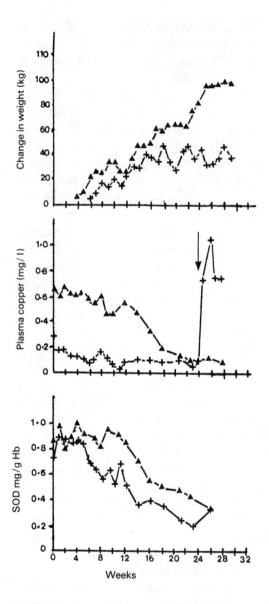

FIG. 3. Copper depletion in the calf: change in weight, in plasma copper and in erythrocyte superoxide dismutase (SOD) in two calves, A (+ + + +) and B (▲—▲—▲), which represent the extremes of a group of six animals fed a low copper diet (1.3 mg/kg dry matter). Point ↓ marks where the copper status of calf A was repleted (Suttle & McMurray 1980). SOD measurements explained in Fig. 1 legend.

MICROSOMAL ENZYMES AND OTHER SYSTEMS

The hepatic microsomal monooxygenase system, cytochrome P-450, functions in metabolism and in the detoxification of a range of substances foreign to the animal body. The activity of acyl-CoA desaturase (EC 1.14.99.5) is reduced by 40–50% in hepatic microsomes from the copper-deficient rat. Consequently a direct role for copper in the microsomal electron-transfer chain has been postulated (Wahle & Davies 1975).

The activity of the desaturase is dependent on a number of factors including the degree of unsaturation of dietary lipid (Holloway & Holloway 1975, Jeffcoat 1977). Therefore the right combination of contributory factors may not yet have been examined to demonstrate specific pathological symptoms associated with a change in the activity of the microsomal system. A fruitful area for future research may be the role of copper in the desaturases that are linked to the elaboration of essential unsaturated fatty acids, especially in ruminants whose unsaturated fatty acids in the diet are largely hydrogenated in the rumen before absorption. A further effect of copper deficiency on the rat microsomal enzyme system is to reduce both the activity of aniline hydroxylase and hexabarbitone metabolism; these changes are restored during copper repletion (Moffitt & Murphy 1973).

Lipid synthesis is depressed in the mitochondria of copper-deficient rats. Failure in the esterification of palmitoyl CoA and L-glycerophosphate to form phosphatidic acid occurs when there is an impairment of ATP synthesis as a direct consequence of the loss of cytochrome c oxidase activity. Microsomal lipid synthesis was unaffected by copper deficiency in the same animals (Gallagher & Reeve 1971). Hypercholesterolaemia is induced by altering the ratio of dietary zinc to copper (Klevay 1973) or by feeding a copper-deficient diet (Lei 1977). The effect is due not to an alteration in the rate of cholesterol synthesis, but to a shift in cholesterol from the liver to plasma (Lei 1977, Lei 1978, Allen & Klevay 1978a) by an unknown mechanism.

A reduction in catecholamine concentration in the copper-deficient rat brain can be explained (Morgan & O'Dell 1977) by loss of the copper-independent enzyme tyrosine 3-monooxygenase (EC 1.14.16.2) rather than by a change in concentration of the copper-dependent enzyme dopamine β-monooxygenase (EC 1.14.17.1), reported to be unreduced in copper deficiency (O'Dell 1976).

COPPER DEFICIENCY IN SHEEP

The classic copper-deficiency syndrome associated with sheep is swayback, or enzootic ataxia, which results from motor dysfunction caused by defective synthesis of myelin. Two forms are recognized — a congenital neonatal form with macroscopic cerebrospinal lesions and a delayed form in which the lesion is microscopic. The delayed form occurs between 1 and 12 weeks after birth. The two forms correlate with the distinct pre- and postnatal phases of myelin synthesis (Patterson et al 1971). The disease is associated with low levels of copper in the fetal brain and liver (Innes & Saunders 1962, Underwood 1977), and can be prevented by copper supplementation of the dam during late pregnancy (Allcroft et al 1959). The delayed condition has been reproduced experimentally in primary (Suttle et al 1970, Lewis et al 1974) and in secondary copper deficiency induced by molybdenum and sulphur (Mills & Fell 1960).

Howell et al (1964) observed a reduction in total lipid in the myelin of clinically copper-deficient lambs. The synthesis of both non-hydroxy cerebrosides and hydroxy cerebrosides is reduced in affected animals and alteration in the myelin protein also occurs (Patterson et al 1974). There is a reduction in cytochrome c oxidase activity in the central nervous system of lambs with swayback (Howell & Davidson 1959, Mills & Williams 1962, Barlow 1963), but Smith et al (1976) consider that the reduction in the oxidase concentration is not large enough to lead to respiratory constraint. O'Dell et al (1976) demonstrated that the concentration of catecholamines was depressed in the ataxic lamb, and that this correlated with a decrease in tyrosine 3-monooxygenase activity. It is possible that this condition is a multifunctional defect involving a number of copper-dependent enzymes.

Loss of crimp or steely wool has been a problem in Australia and it still occurs in areas of South Australia (Hannam & Reuter 1977). The condition has been observed in experimental primary copper deficiency (Suttle et al 1970) and in animals feeding on grass with a low copper content (Whitelaw et al 1979). While the lesion is due to the loss of disulphide bridges and to an increase in free sulphydryl groups (Marston 1952), I am not aware of any biochemical explanation for the lack of formation of the disulphide bridges. A growth response to copper during induced copper deficiency has been demonstrated in sheep in Australia (Hogan et al 1966) and in some experimentally fed lambs (Howell 1968, diet 0.3 mg Cu/kg dry matter) but not in others (Suttle et al 1970, diet 1.2 mg Cu/kg DM). Good evidence for reduced growth due to copper deficiency under grazing conditions (6.1 mg Cu/kg DM, 2.9 mg Mo/kg DM and 4.3 g S/kg DM) has been only recently shown in sheep in Great Britain (Whitelaw et al 1979). These are very similar to the conditions under which swayback has been found in Northern Ireland (Kavanagh et al 1972, 3.8 mg Cu/kg DM, 4.1 mg Mo/kg DM).

Osteoporotic lesions have been observed in experimental studies (Suttle et al

1970, Suttle et al 1972) and also in the field (Whitelaw et al 1979).

It is possible to distinguish between 'high copper' and 'low copper' breeds of sheep (Weiner et al 1976), and the susceptibility of an animal to either copper deficiency or copper toxicity is breed-dependent. In relation to swayback, the breed of the dam controls the amount of copper deposited in the liver of the fetus (Weiner et al 1978).

COPPER DEFICIENCY IN CATTLE

A number of clinical copper-deficiency symptoms have been observed both in the field and experimentally. One of the first changes seen in copper-deficient animals (Poole 1973) is achromotrichia (i.e. loss of hair colour) as a result of defective formation of melanin. Diarrhoea has been observed in experimental calves (Mills et al 1976, Suttle & Angus 1976). Fell et al (1975) consider the diarrhoea to be the result of changes in the cellular structure of the small intestine, mainly villus atrophy. Loss of cytochrome c oxidase in intestinal cells was also established, along with destruction of the enterocytic mitochondria. The presence of villus stunting must be considered as a qualitative rather than a quantitative change because direct measurements were not made of either villus length or crypt depth of intestinal mucosa in the treated and control animals. Neither has the quantity and quality of the diarrhoea been described adequately, and the microbiological effects which might be superimposed on the copper deficiency have not been properly examined.

One of the clinical signs of copper deficiency in animals grazing on pastures high in molybdenum is profuse diarrhoea. While molybdenum in conjunction with sulphur decreases copper availability (Dick et al 1975), the level of molybdenum in these pastures is much higher than the concentration used in the calf experiments (Lewis 1943). The diarrhoea produced by these high concentrations of molybdenum could be due either to a much more severe copper deficiency or to the presence of copper, sulphur and molybdenum complexes e.g. tetrathiomolybdates (Suttle 1974, Dick et al 1975) that exert direct toxic effects on the gut mucosa, as seen in the rat (Fell et al 1979).

The clinical symptoms associated with joint and bone disorders in copper-deficient cattle vary from stiffness of gait, swelling of joints, and lameness to an increase in the incidence of bone fractures (Underwood 1977). The most frequently described lesions have been attributed to reduced activity of the osteoblasts (Suttle & Angus 1978), which produces thin bone and overgrowth of cartilage similar to the lesions described in other species. In contrast the experiments of Mills et al (1976) produced little evidence of osteoporosis, although the tissue surrounding the metacarpophalangeal and carpometacarpal joints was swollen. Smith et al (1975), in presenting data from a field study (15.0 mg Cu/kg DM, 6.2 mg Mo/kg DM and 2.5 g S/kg DM), described lesions similar to those observed by Suttle & Angus (1978), but both the copper and molybdenum concentrations in the diet were much higher than those required for the experimental induction of the lesions.

Detailed explanations to account for the differences must be further investigated. In particular the role of copper, if any, in vitamin D metabolism (Suttle & Angus 1978) should be established and the effects of direct copper deficiency should be separated from those produced in the presence of thiomolybdates.

Cellular damage as a result of superoxide production could be amplified by two interacting effects: first, by a reduction in the activity of SOD, through decreased copper availability; and secondly, by an increase in the molybdenum enzyme xanthine oxidase (EC 1.2.3.2). Both effects could lead to an increase in the superoxide present in tissue.

Although progress is being made in understanding the nature of the interactions between copper, molybdenum and sulphur (Mills 1980), it may be some time before the details are completely understood.

Cardiac lesions similar to those observed in the rat, and attributable to failure of elastin and collagen cross-linking, have been described in copper-deficient cattle (Mills et al 1976, Leith 1975). This lesion may account for one of the original clinical conditions due to copper deficiency − 'falling disease', which occurs in Australia.

The rate of growth is reduced and the food conversion ratio (FCR) of young copper-deficient cattle is increased (Underwood 1977, Mills et al 1976, Suttle & Angus 1976). These studies have demonstrated that levels of plasma copper have to be maintained below 0.3 mg/l for variable lengths of time before growth rate or FCR is affected (Figs. 3 and 4).

Many claims implicating copper deficiency as the cause of reproductive disorders have been made. However, the only syndrome that has been consistently produced experimentally is fetal resorption in the rat (Underwood 1977). Spoerl & Kirchgessner (1976) reported an increase in the number of stillbirths produced by copper-depleted rats. The young from very deficient rats in that study died within three days of birth and had haemorrhages throughout the body. To my knowledge there are no reports that have been substantiated by carefully controlled trials of consistent beneficial effects of copper on reproductive efficiency in large animals.

Moderate to severe anaemia (haemoglobin < 5 g/100 ml) has been observed in severe copper deficiency in a number of studies on rats (Underwood 1977). However, only mild anaemia (Hb > 8 g/100 ml) has been reported in

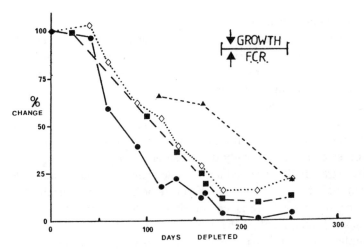

FIG. 4. Changes in copper markers in copper-depleted calves: caeruloplasmin (ferroxidase) (•——•); plasma amine oxidase (copper-containing) – EC 1.4.3.6 (■——■); whole blood copper (◊——◊) and hepatic cytochrome c oxidase (▲—▲). The horizontal bar marks the time period (and direction of change) over which growth and FCR (food conversion ratios) were altered in calves fed a copper-deficient diet (Cu 0.8 mg/kg dry matter). (Adapted from Mills et al 1976.)

experimental trials in cattle (Mills et al 1976, Suttle & Angus 1976). An exception to this could be the hypocupraemia and postparturient haemoglobinuria with accompanying Heinz body anaemia, reported from New Zealand (Smith & Coup 1973, Gardner et al 1976). Some haemolytic agents were eliminated in these studies but there is no record of the elimination of primary agents such as the blood parasites *Babesia* and *Eperythrozoon* which could cause similar problems (Kreiser 1977). Copper may modify the course of infection and account for the observed response to copper treatment.

Variable responses to copper have been obtained in comparisons of the growth rates of treated and untreated cattle in the field (see MacPherson et al 1979, Poole 1973). Responses range from zero in some studies to 60–70% in others. This variability is independent of how the low copper status is defined, e.g. plasma copper, blood copper, serum caeruloplasmin activity or hepatic cytochrome c oxidase activity.

In various trials in which responses have been tested on a number of farms the probability of achieving a significant growth response would appear to be approximately 0.25 (Rogers & Poole 1978, MacPherson et al 1979).

There is a fundamental difference between inducing a deficiency and treating an already imposed deficiency. With induction it is possible to monitor changes as a direct consequence of altering the intake of available copper. With treatment, however, the concern is about whether the changes are reversible. In copper deficiency there is good evidence that both reversible and irreversible changes occur. Diarrhoea is reversible owing to the continual turnover of cells in the gut mucosa; hair colour is reversible because of the continuous production of hair and presumably because the enzyme systems responsible change when copper status changes (Carrillo & Bingley 1976).

In contrast, swayback in lambs is irreversible, and copper appears to be necessary at key points in the formation of the myelin sheath. How far this applies to the other conditions already described is uncertain but it is of interest that O'Dell et al (1978) observed irreversible changes in elastin and collagen structure in the rat lung. There may therefore be a case for ensuring that hypocupraemic pregnant cows are given copper in order to protect the fetus and the subsequent development of the calf.

Another complication is that symptoms associated with copper deficiency can have other aetiological origins. For instance, changes in coat colour (one of the first indicators of copper deficiency) can also result from poor nutrition or from the presence of an enteroparasite such as *Coccidium,* and of course there are many causes of poor growth.

MARKERS OF COPPER STATUS

In order to provide a differential diagnosis of disease, markers are required which ideally should indicate the cause of the clinical condition, e.g. reduced copper status. Detailed investigations are necessary to rule out other agents which in themselves may be pathogenic. The rates of change of a number of markers, suggested as indicators of copper status, are shown in Fig. 4 (adapted from Mills et al 1976).

As in other species, the caeruloplasmin (ferroxidase, EC 1.16.3.1) level in the plasma of cattle rapidly falls on introduction of a diet that reduces copper intake or availability. The activity of plasma amine oxidase (copper-containing) (EC 1.4.3.6) falls more slowly and parallels the reduction in whole blood copper.

Qualitatively similar results have been obtained by other workers (Suttle & Angus 1976). The differences can be explained by the differences in the degree of deficiency imposed and, possibly, in the level of copper reserves in the liver at the start of the experiments.

A number of factors must be borne in mind in the use of markers for copper status. The steady-state concentration of an enzyme or protein in a tissue depends on both the rate of synthesis and the rate of degradation. For metalloenzymes and proteins another rate constant is important, i.e. the rate of entry of the metal into tissue (k_0), and so too are mechanisms, kinetic or otherwise, that influence the partition of the trace element between cells or between different compartments within the cell. This is illustrated below.

$$\xrightarrow{k_0} [M] \rightarrow \begin{cases} k_{s1} \rightarrow [E_1M] \rightarrow k_{d1} \rightarrow \\ k_{s2} \rightarrow [E_2M] \rightarrow k_{d2} \rightarrow \\ k_{s3} \rightarrow [P_1M] \rightarrow k_{d3} \rightarrow \\ k_{s4} \rightarrow [P_2M] \rightarrow k_{d4} \rightarrow \end{cases}$$

$$\text{or } \left(\xleftarrow{k_{s-4}} \right)$$

In this scheme, M is a metal, E_1M and E_2M are metalloenzymes while P_1M and P_2M are binding proteins. The rate constants of synthesis and degradation (without metal recycling) are k_s and k_d for each species. Some steps may also be reversible (k_{s-4}).

The real scheme will be more complex, because the rate of synthesis of protein may itself be controlled so that there is an increase (i.e. induction) or a decrease in the protein, dictated by cellular requirements, which may lead to alterations in the metal distribution between binding species.

The reduction in plasma copper or in caeruloplasmin activity after a reduction in the availability of dietary copper are models for this type of behaviour.

There is evidence that under certain circumstances, the k_s terms (i.e. synthesis), as well as responding to the rate of copper entry (k_0), can be altered, e.g. as in the acute-phase reaction (Koj 1974). The caeruloplasmin concentrations rise substantially in response to colibacillosis in the chick and *Salmonella* infection in calves (Piercy 1979).

A number of other factors also affect the degree of hypocupraemia found in animals in the field.

The degree of hypocupraemia of animals, as detected by plasma components, is not independent of the sample; substantial differences occur between serum and plasma (Fig. 5).

Our experiments show that there is a sequestering of caeruloplasmin during the early stages of blood clotting, which occurs within 30 min after the sample has been taken (C.H. McMurray & W.J. Blanchflower, unpublished results).

Another factor that affects the degree of hypocupraemia is the physiological status of the animal. We have observed reductions in caeruloplasmin and in plasma copper before birth, and rapid increases immediately after birth (Fig. 6) (McMurray et al 1980). The decrease in plasma copper before birth

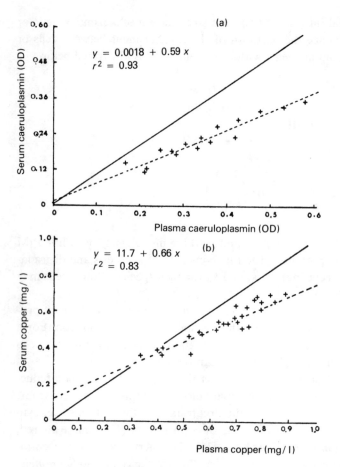

FIG. 5. Copper or caeruloplasmin in bovine plasma and serum: Interrelationships between plasma and serum content of caeruloplasmin (a) and copper (b). The dotted line is the fitted regression; the continuous line is the expected direct relationship. Each point represents the mean of measurements on not less than 10 animals (McMurray et al 1980). Caeruloplasmin activity (change in optical density (OD)at 530 nm/h) was measured at pH 6.5 and 37°C using p-phenylene diamine HCl as substrate (Sunderman & Nomoto 1970).

probably reflects an increasing requirement for copper by the dam, for deposition in the fetus. After birth the increase could reflect either an increase in the availability of copper for caeruloplasmin synthesis (i.e. one of the terms in the previous scheme has been removed) or a reaction of the animal to the stress of calving. Substantial changes in plasma copper also take place in the neonatal calf (Fig. 7) (McMurray et al 1978).

One factor that seems to be absent in the bovine species but present in other

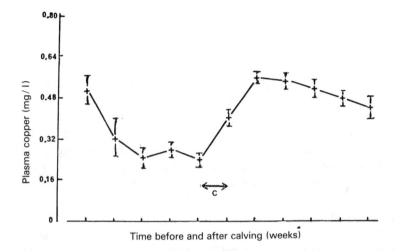

FIG. 6. The sequential changes in plasma copper (mean ± SEM) before and after calving in a group of 13 cows that had not been supplemented with copper. Calving occurred during the interval marked C. The copper levels in the plasma of supplemented animals were above 0.64 mg/l.

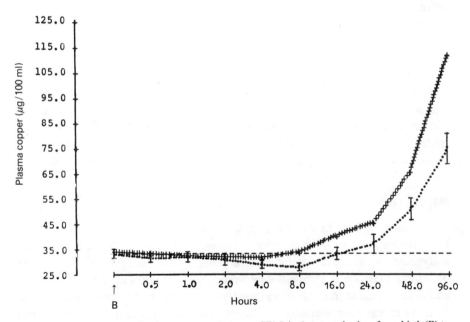

FIG. 7. Sequential changes in plasma copper (mean ± SEM) in 6 neonatal calves from birth (B) to 96 hours: (– – – –) concentrations of copper at birth; (....) concentrations at each period after birth; (+ + +) concentration corrected for plasma dilution (McMurray et al 1978).

species is the increase in caeruloplasmin during an oestrogenic stimulus (Underwood 1977). We have been unable to find any evidence of such an increase in cattle.

Erythrocyte SOD may be a better potential index of copper deficiency than plasma copper or caeruloplasmin activity (Suttle & McMurray 1980). There is a delay in the reduction of circulating SOD activity with respect to the decrease in plasma copper but it takes place approximately at the same time as the start of growth retardation (shown in Fig. 4). Measurement of SOD activity is not affected by the acute-phase reaction, and does not respond in the same way as plasma copper to physiological effects; it is basically an independent measurement. Unlike caeruloplasmin activity erythrocyte SOD activity is not directly correlated to plasma copper (Suttle & McMurray 1980). There may, however, be some advantage in measuring both plasma copper and erythrocyte SOD in order to estimate the degree and duration of copper deficiency. From these measurements a better reflection could be obtained of what happens in tissues when the rate of entry of metal into the system is reduced. Whether the SOD status of the animal has functional significance can be verified only by detailed clinical experience.

One potential disadvantage with measurement of erythrocyte SOD is that different concentrations of enzyme may be present in the red cells of animals of different ages. Therefore, the definition of a normal copper status for the blood will vary with age. Our evidence to date is that the erythrocyte SOD concentration in the neonatal calf is twice (1.2 mg SOD/g Hb) the activity found in the adult animal (0.6 mg/g Hb).

The copper content of hair has also been suggested as a marker of copper status (Kellaway et al 1978) but to obtain reliable results potential environmental contamination must be eliminated. Measurements of functional markers or of factors altered by functional changes must be obtained; broad markers such as hepatic content of copper have little application.

POPULATIONS

In Northern Ireland clinical copper deficiency is generally recognized by two clinical markers – coat colour and poor growth. Other abnormalities associated with copper deficiency are not often seen. Because of the widespread reports of copper-related syndromes, surveys have been performed of copper status in herds of suckler (beef) cattle (Thompson & Todd 1976). This study has been extended to the analysis of geographical and geological differences. Current information is summarized in Fig. 8 (a) and (b).

This evidence suggests that sub-regional problems of copper deficiency can

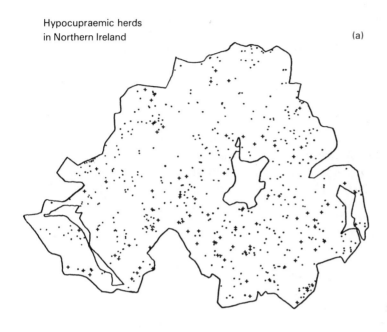
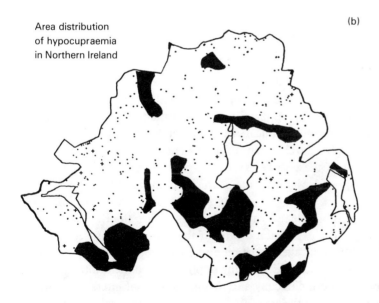

FIG. 8. Distribution of hypocupraemia in suckler herds in N. Ireland (plasma copper < 0.5 mg/l).
(a) Location of normocupraemic (•) and hypocupraemic herds (+) (672 herds sampled).
(b) Distribution of hypocupraemic areas, based on the location of the deficient herds in (a).

arise. The agreement between the regional geochemical map produced by Imperial College for Northern Ireland (Webb et al 1973) and the copper, molybdenum, zinc and iron concentrations in stream sediment is poor. We must continue to look for alternatives, such as herd management factors, that might restrict the intake of available copper.

The population mean for plasma copper of Northern Ireland suckler herds is 0.72 mg/l and the distribution is negatively skewed; 11.4% of the herds have plasma copper contents below 0.5 mg/l. Preliminary results from an investigation of the SOD levels in 180 herds (10 animals/herd) suggest that a significant 25% of the animals have erythrocyte SOD concentrations below 0.5 mg/g Hb, which is close to the activity that Suttle & McMurray (1980) considered important in relation to the onset of growth retardation in their experimental calves.

CONCLUSION

Clinical symptoms and pathology of trace-metal deficiencies result from defective functional or structural mechanisms that depend on trace metals. Three functions are worth distinguishing. (1) A direct deficiency syndrome can result from the absence of the key enzyme or transport protein which is rate limiting or becomes rate limiting in 'mainline' metabolic processes, e.g. in oxidative phosphorylation (cytochrome c oxidase) or in elastin and collagen maturation (lysyl oxidase). (2) A deficiency syndrome may become apparent only when a particular metabolic pathway is challenged and found to be insufficient to meet the degree of challenge (superoxide dismutase). (3) When the supply of nutrients other than the trace metal is limited, a trace metal deficiency could modify the efficient utilization of the nutrient. This modification would produce an exaggerated clinical syndrome relating to the primary deficiency or it could lead to development of a new clinical syndrome.

Deficiency in each of these three cases removes the integration of normal processes of metabolism. Metalloproteins may also have a direct structural function as distinct from a catalytic one and symptoms of deficiency may be caused therefore by subsequent structural defects rather than by direct enzymic insufficiency. The expression of any of these mechanisms depends on species, genetic disposition, age, nutrition and environment.

Copper deficiency has such wide-ranging implications to the health and performance of animals, both clinically and through potential interactions with other disease processes, that the cost—effectiveness of correcting copper status in copper-depleted animals must be evaluated. Further evidence is urgently needed to facilitate any such advice.

ACKNOWLEDGEMENT

I wish to thank D.A. Rice for helpful discussions and suggestions made during the preparation of this manuscript.

References

Aaron HE 1976 Histology and microanatomy of bone. In: Nordin BEC (ed) Calcium phosphate and magnesium metabolism; clinical physiology and diagnostic procedures. Churchill Livingstone, Edinburgh, p 298-356

Allcroft R, Clegg FG, Uvarov O 1959 Prevention of swayback in lambs. Vet Rec 71:884-889

Allen KGD, Klevay LM 1978a Copper deficiency and cholesterol metabolism in the rat. Atherosclerosis 31:259-271

Allen KGD, Klevay LM 1978b Cholesterolemia and cardiovascular abnormalities in rats caused by copper deficiency. Atherosclerosis 29:81-93

Andrewartha KA, Caple IW 1980 Effects of changes in nutritional copper in erythrocyte superoxide dismutase activities in sheep. Res Vet Sci 28:101-104

Andrews PW, Lowrie DB, Jackett PS, Peters TJ 1980 Analytical subcellular fractionation of rabbit alveolar macrophages with particular reference to the subcellular localisation of pyridine nucleotide dependent superoxide generating systems and superoxide dismutase. Biochim Biophys Acta 611:61-71

Bailey AJ, Restall DJ, Sims TJ, Duance VC 1979 Meat tenderness: Immunofluorescent localisation of the isomorphic forms of collagen in bovine muscles of varying texture. J Sci Food Agric 30:203-210

Barlow RM 1963 Further observations on swayback. II. Histochemical localisation of cytochrome oxidase activity in the central nervous system. J Comp Pathol 73:61-69

Baumgartner S, Brown DJ, Salevsky E, Leach RM 1978 Copper deficiency in the laying hen. J Nutr 108:804-811

Bohmenkamp W, Weser U 1977 2 Cu 2 Zn superoxide dismutase in copper depleted rats. In Michelson AM et al (eds) Superoxide and superoxide dismutases. Academic Press, London, p 387-394

Braude R 1976 Copper as a performance promoter in pigs. In: Copper in farming symposium. Copper Development Association, Potters Bar, Hertfordshire (Roy Zool Soc Symp), p 79-93

Buckingham K, Heng-Khoo CS, Lefevre M, Rucker RB, Julian LM 1978 Morphology of the copper deficient chick lung. Fed Proc 37:264

Carrillo BJ, Bingley JB 1976 Depigmentation of bovine hair in copper deficiency. Vet Med Rev 2:249-250

Carroll RC, Racker E 1977 Preparation and characterisation of cytochrome oxidase vesicles with high respiratory control. J Biol Chem 252:6981-6990

Chance B, Sies H, Boveris A 1979 Hydroperoxide metabolism in mammalian organs. Physiol Rev 59:527-605

Chou WS, Savage HE, O'Dell BLJ 1969 The role of copper in biosynthesis of intramolecular cross-links in chick tendon collagen. Biol Chem 244:5785-5789

Dick AJ, Dewey D, Gawthorne JM 1975 Thiomolybdates and copper molybdenum sulphur interaction in ruminant nutrition. J Agric Sci 85:567-568

Erecińska M, Wilson DF 1978 Cytochrome C oxidase; a synopsis. Arch Biochem Biophys 188:1-14

Fell BF, Dinsdale D, Mills CF 1975 Changes in enterocyte mitochondria associated with deficiency of copper in cattle. Res Vet Sci 18:274-281

Fell BF, Dinsdale D, El-Gallad TT 1979 Gut pathology of rats dosed with tetrathiomolybdate. J Comp Pathol 89:495-514

Fessler JH, Fessler LI 1978 Biosynthesis of procollagens. Annu Rev Biochem 47:129-162
Foster JA, Shapiro R, Voynow P, Crombie G, Faris B, Franzblau C 1975 Isolation of soluble elastin from lathyritic chicks. Comparison to tropoelastin from copper deficient pigs. Biochemistry 14:5343-5347
Fridovich I 1975 Superoxide dismutases. Annu Rev Biochem 44:147-159
Gallagher CH, Reeve VE 1971 Copper deficiency in the rat, effect on synthesis of phospholipids. Aust J Exp Biol Med Sci 49:21-31
Gallagher CH, Reeve VE, Wright R 1973 Copper deficiency in the rat: effect on the ultrastructure of hepatocytes. Aust J Exp Biol Med Sci 51:181-189
Ganezer KS, Hart ML, Carnes WH 1976 Tensile properties of tendons in copper deficient swine. Proc Soc Exp Biol Med 153:396-399
Gardner DE, Martinovich D, Woodhouse DA 1976 Haematological and biochemical findings in bovine post parturient haemoglobinuria and the accompanying Heinz body anaemia. N Z Vet J 24:117-122
Hannam RJ, Reuter DJ 1977 A note on the occurrence of steely wool in South Australia 1972–1975. Agric Rec 4:26-29
Harris ED 1976 Copper induced activation of aortic lysyl oxidase in vivo. Proc Natl Acad Sci USA 73:371-374
Harris ED, Rayton JK, Balthrop JE, DiSilvestro RA, Garcia-de-Quevedo M 1980 Copper and the synthesis of elastin and collagen. In: Biological roles of copper. Excerpta Medica, Amsterdam (Ciba Found Symp 79), p 163-177
Hogan KG, Ris DR, Hutchinson HA 1966 An attempt to produce copper deficiency in sheep by dosing molybdate and sulphate. NZ J Agric Res 9:691-698
Holloway CT, Holloway PN 1975 Stearyl Co A desaturase activity in mouse liver microsomes of varying lipid concentration. Arch Biochem Biophys 167:496-504
Howell JMcC 1968 The effect of experimental copper deficiency on growth, reproduction and haemopoiesis in sheep. Vet Rec 83:226-227
Howell JMcC, Davison AN 1959 The copper content and cytochrome oxidase activity of tissues from normal and swayback lambs. Biochem J 72:365-368
Howell JMcC, Davison AN, Oxberry J 1964 Biochemical and neuropathological changes in swayback. Res Vet Sci 5:376-384
Innes JRM, Saunders LZ 1962 Comparative neuropathology. Academic Press, New York
Jeffcoat R 1977 The physiological role and control of mammalian fatty acyl coenzyme A desaturases. Biochem Soc Trans 5:811-827
Kavanagh PJ, Purcell DA, Thompson RH 1972 Congenital and delayed swayback in lambs in Northern Ireland. Vet Rec 90:538-540
Kellaway RC, Sitorus P, Leibholz JML 1978 The use of copper levels in hair to diagnose hypocuprosis. Res Vet Sci 24:352-357
Kelly WA, Kesterson JW, Carlton WW 1974 Myocardial lesions in the offspring of female rats fed a copper deficient diet. Exp Mol Pathol 20:40-56
Klevay LM 1973 Hypercholesterolemia in rats produced by an increase in the ratio of zinc to copper ingested. Am J Clin Nutr 26:1060-1068
Koj A 1974 Acute-phase reactants. Their synthesis, turnover and biological significance. In: Allison AC (ed) Structure and function of plasma proteins, Vol 1, Plenum Press, London
Kreiser JP 1977 Parasitic protozoa Vol IV. Academic Press, New York
Lei KY 1977 Cholesterol metabolism in copper deficient rats. Nutr Rep Int 15:597-605
Lei KY 1978 Oxidation, excretion, and tissue distribution of [26–^{14}C] cholesterol in copper-deficient rats. J Nutr 108:232-237
Leith LC 1975 Changes in the ultrastructure of cardiac muscle in steers deprived of copper. Res Vet Sci 18:282-287
Lewis AH 1943 The teart pastures of Somerset. III: Reducing the teartness of pasture herbage. J Agric Sci Camb 33:58-63
Lewis G, Terlecki S, Parker BNJ 1974 Observations on the pathogenesis of delayed swayback. Vet Rec 95:313-316

MacPherson A, Voss RC, Dixon J 1979 The effect of copper treatment on the performance of hypocupraemic calves. Anim Prod 29:91-99

Marston HR 1952 Cobalt, copper and molybdenum in the nutrition of animals and plants. Physiol Rev 32:66-121

McCord JM, Fridovich I 1969 Superoxide dismutase; an enzymatic function for erythrocuprein (hemocuprein). J Biol Chem 244:6049-6055

McCord JM, Wong K 1979 Phagocyte-produced free radicals: role in cytotoxicity and inflammation. In: Oxygen free radicals and tissue damage. Excerpta Medica, Amsterdam (Ciba Found Symp 65), p 343-360

McMurray CH, McParland PJ, O'Neill DG 1980 Hypocupraemia in the bovine; pre and post parturient changes. Br Vet J, in press

McMurray CH, Logan EF, McParland PJ, McRory FJ, O'Neill DG 1978 Sequential changes in some blood components in the neonatal calf. Br Vet J 134:590-597

Mills CF 1980 Metabolic interactions of copper with other trace elements. In: Biological roles of copper. Excerpta Medica, Amsterdam (Ciba Found Symp 79), p 49-65

Mills CF, Fell BF 1960 Demyelination in lambs born of ewes maintained on high intakes of sulphate and molybdate. Nature (Lond) 185:20-22

Mills CF, Williams RB 1962 Copper concentration and cytochrome-oxidase and ribonuclease activities in the brains of copper deficient lambs. Biochem J 85:629-632

Mills CF, Dalgarno AC, Wenham G 1976 Biochemical and pathological changes in tissues of friesian cattle during experimental induction of copper deficiency. J Nutr 35:309-331

Mitchell P 1966 Chemiosmotic coupling of oxidative and photosynthetic phosphorylation. Glynn Research, Bodmin

Moffitt AE, Murphy SD 1973 Effect of excess and deficient copper intake on rat liver microsomal enzyme activity. Biochem Pharmacol 22:1463-1476

Morgan RF, O'Dell BL 1977 Effect of copper deficiency on the concentrations of catecholamines and related enzyme activities in the rat brain. J Neurochem 28:207-213

O'Dell BL 1976 Biochemistry and physiology of copper in vertebrates. In: Prasad AS, Oberlas D (eds) Trace elements in human health and disease vol 1, zinc and copper. Academic Press, New York, p 391-415

O'Dell BL, Smith RM, King RA 1976 Effect of copper status on brain neurotransmitter metabolism in the lamb. J Neurochem 26:451-455

O'Dell BL, Kilburn KH, McKenzie N, Thurston RJ 1978 The lung of the copper deficient rat, a model for developmental pulmonary emphysema. Am J Pathol 91:413-430

Patterson DSP, Sweasey D, Herbert CDJ 1971 Changes occurring in the chemical conformation of the central nervous system during foetal and post-natal development in sheep. J Neurochem 18:2027-2040

Patterson DSP, Foulkes JA, Sweasey D, Glaney FM, Terlecki S 1974 A neurochemical study of field cases of the delayed spinal form of swayback (enzootic ataxia) in lambs. J Neurochem 23:1245-1253

Paynter DI, Moir RJ, Underwood EJ 1979 Changes in activity of the Cu-Zn superoxide dismutase enzyme in tissue of the rat with changes in dietary copper. J Nutr 109:1570-1576

Peeters-Joris C, Van Devoorde AM, Baudhuin P 1975 Subcellular localization of superoxide dismutase in rat liver. Biochem J 150:31-39

Piercy DWT 1979 Acute phase responses to experimental salmonellosis in calves and colibacillosis in chickens. Serum iron and caeruloplasmin. J Comp Pathol 89:309-319

Poole DBR 1973 Studies in induced copper deficiency in cattle. PhD Thesis, Dublin University, Dublin

Rest RF, Spitznagel JK 1977 Subcellular distribution of superoxide dismutases in human neutrophils. Biochem J 166:145-153

Rogers PAM, Poole DBR 1978 The effects of Cu-EDTA injection on weight gain and copper status of calves and fattening cattle. In: Kirchgessner M (ed) Trace element metabolism in man and animals 3. Inst für Ernährungsphysiologie Technische Universität München, Freising-Weihenstephan, Germany, p 481-485

Roos D, Weening RS 1979 Defects in the oxidative killing of microorganisms by phagocytic leucocytes. In: Oxygen free radicals and tissue damage. Excerpta Medica, Amsterdam (Ciba Found Symp 65), p 225-262

Rucker RB, Murray JA, Riggins RS 1977 Nutritional copper deficiency and penicillamine administration. Some effects on bone collagen and arterial elastin crosslinking. Adv Exp Med Biol 86B:619-648

Smith B, Coup MR 1973 Hypocuprosis, a clinical investigation of dairy herds in Northland. N Z Vet J 21:252-258

Smith BP, Fischer GL, Poulos PW, Irwin MR 1975 Abnormal bone development and lameness associated with secondary copper deficiency in young cattle. J Am Vet Med Assoc 166:682-688

Smith RM, Osborne-White WS, O'Dell BL 1976 Cytochromes in brain mitochondria from lambs with enzootic ataxia. J Neurochem 26:1145-1148

Spoerl R, Kirchgessner M 1976 Cu-Mangelschaden bei reproduzierenden Ratten. Zentralbl Veterinärmed Reihe A 23:131-138

Sunderman FW, Nomoto S 1970 Measurement of human serum ceruloplasmin by its p-phenylenediamine oxidase activity. Clin Chem 16:903-910

Suttle NF 1974 Recent studies of the copper molybdenum antagonism. Proc Nutr Soc 33:299-305

Suttle NF, Angus KW 1978 Effects of experimental copper deficiency on the skeleton of the calf. J Comp Pathol 88:137-148

Suttle NF, Angus KW 1976 Experimental copper deficiency in the calf. J Comp Pathol 86:595-608

Suttle NF, Angus KW, Nisbet DI, Field AC 1972 Osteoporosis in copper-depleted lambs. J Comp Pathol 82:93-97

Suttle NF, Field AC, Barlow RM 1970 Experimental copper deficiency in sheep. J Comp Pathol 80:151-164

Suttle NF, McMurray CH 1980 The use of erythrocyte superoxide dismutase and hair or fleece Cu concentrations in the diagnosis of Cu-responsive conditions in ruminants. Br Vet J, in press

Thompson RH, Todd JR 1976 The copper status of beef cattle in N. Ireland. Br J Nutr 36:299-303

Tyler DD 1975 Polarographic assay and intracellular distribution of superoxide dismutases in rat liver. Biochem J 147:493-504

Underwood EJ 1977 Trace elements in human and animal nutrition, 4th edn. Academic Press, New York, p 56-108

Wahle KWJ, Davies NT 1975 Effect of dietary copper deficiency in the rat on fatty acid composition of adipose tissue and desaturase activity of liver microsomes. Br J Nutr 34:105-112

Waisman J, Cancilla PA, Coulson WF 1969 Cardiovascular studies on copper deficient swine. XIII. The effect of chronic copper deficiency on the cardiovascular system of miniature pigs. Lab Invest 21:548-554

Webb JS, Nichol I, Foster R, Lowenstein PL, Howarth RJ 1973 Provisional geochemical atlas of Northern Ireland. Technical Commun No 60. Applied Geochem Res Group, Imperial College, London

Weiner G, Herbert JG, Field AC 1976 Variation in liver copper and plasma copper concentrations of sheep in relation to breed and haemoglobin type. J Comp Pathol 86:101-109

Weiner G, Wilmut I, Field AC 1978 Maternal and lamb breed interactions in the concentrations of copper in tissues and plasma of sheep. In: Kirchgessner M (ed) Trace element metabolism in man and animals. Arbeitskreis für Ernährungsforschung, Freising-Weihenstephan, vol 3:469-472

Whitelaw A, Armstrong RH, Evans CC, Fawcett AR 1979 A study of the effects of copper deficiency in Scottish blackface lambs on improved hill pasture. Vet Rec 104:455-459

Williams DM, Lynch RE, Lei GR, Cartwright GE 1975 Superoxide dismutase activity in copper deficient swine. Proc Soc Exp Biol Med 149:534-536

Discussion

Hurley: I would like to comment on the irreversibility of the elastin lesion in copper-deficient animals. Presumably the irreversibility is due to the slow turnover of elastin, so that once the elastin is formed it lasts for the life of the animal and therefore there is no opportunity for the lesion to be reversed.

Harris: It has been generally believed that connective tissue proteins turn over very slowly. However, our experience with elastin has been in the growing chicken where we see a much more rapid turnover of protein. The time-course of turnover of mature collagen and mature elastin is considerably longer.

McMurray: Professor Harris, if cross-links are not formed, due to a reduction in lysyl oxidase activity, can these missing links be formed when the activity of lysyl oxidase is restored, after copper repletion?

Harris: Yes. If early nutrition is inadequate it is possible that the defects that are laid down as a result will be manifested later on in life.

Dormandy: It is not yet established that superoxide dismutase has any critical role in normal metabolism, although the enzyme is certainly present. No known disease is incontrovertibly associated with a deficiency or a defect of the enzyme. Some of my colleagues measured copper and superoxide dismutase activity in red cells from a wide spectrum of patients with various anaemias. We found no correlation between superoxide dismutase (whether measured as enzyme activity or as red cell copper) and any of the absolute indices (Scudder et al 1976). Perhaps there is no reason why there should have been a correlation, but the results led me to believe that either superoxide dismutase is not essential in red cells or that it is so essential that the enzyme is always present in great excess – i.e. it is never rate-limiting.

Hill: I don't understand, Dr Dormandy, what you mean by a *role* for superoxide dismutase.

Dormandy: Well, in the varied population that we examined there was a fairly wide range of copper and superoxide dismutase-like activity in the red cells but there was no correlation whatever with any functional abnormality.

McMurray: It depends on the cause of the abnormalities that were present. The copper in the red cells may have had no protective action against these abnormalities in any case. When a challenge occurs, the degree of challenge will have to be in excess of the amount of protection afforded. It may be possible to swamp any protective mechanism that exists.

Hill: For there to *be* a correlation there would have to be a relationship between, say, the role of superoxide dismutase in protection against the ravages of oxygen radicals and the defects that are observed.

Danks: Dr Dormandy, are you suggesting that there was a dissociation between red cell copper and superoxide dismutase activity?

Dormandy: No, I am not. The two parameters are virtually identical, but there is, of course, absolutely no relationship between plasma caeruloplasmin and red cell copper.

Hassan: Dr McMurray, you said that the superoxide dismutase is not subject to physiological reactions. What does this mean?

McMurray: Superoxide dismutase activity in erythrocytes should be insensitive to factors that rapidly change the apparent copper status, e.g. in the acute-phase reaction. In the *prepartum* changes that I mentioned in depleted cows, changes in SOD activity should occur on a different time-scale, i.e. they should be delayed with respect to changes in plasma copper or caeruloplasmin.

The fact that the SOD activity is in the erythrocytes means that there are constraints on its activity, which are different from those controlling copper concentrations in plasma.

Mills: Do you have proof of this?

McMurray: Yes; I have already mentioned the differing rates of change between plasma copper and erythrocyte SOD. The population of red cells contains cells of different ages, and as a result its sampling can be regarded as an integrated technique which would be less sensitive to short-term changes that occur before and after parturition.

Hassan: In your lesion experiment, the level of SOD started as 0.15 mg/g Hb and was doubled, to 0.30 mg/g Hb; it didn't go up to the normal value, which you claimed to be 0.6 mg/g Hb. Can you get inactivation of SOD?

McMurray: We know that in the bovine species there is an effect of age, i.e. older animals have about half the SOD per g haemoglobin that younger animals have. This explanation may account for the results because both lambs and adult ewes were used in these experiments. Sufficient copper was given to replete the animals fully. The depletion experiment was done with the lambs and the repletion experiment was done with the adult ewes.

Tanner: Most red cell enzymes have a higher activity in younger than in older red blood cells. Is that the case with SOD?

Dormandy: We have never found any difference in SOD activity between young and old red cells or between cells from patients with haemolytic disease and from normal people (Scudder et al 1976). In haemolytic disease, of course, one would expect a higher proportion of relatively young cells. Most red cell enzymes have a higher activity in a young than in an old population of cells.

McMurray: I think we are measuring two separate effects here, because we are talking about the young animal as distinct from the young red cell.

Frieden: Has anyone used the composition or enzymic activity of white cells as an index for the copper function in cytochrome c oxidase and superoxide dismutase?

Bremner: Some of my colleagues have looked at changes in cytochrome c oxidase and superoxide dismutase activities in white cells from animals of different copper status. Cytochrome c oxidase activity in the leucocytes did fall in animals deprived of copper but this was not a sensitive guide to the animals' copper status (Boyne 1978). Superoxide dismutase activity in the neutrophils of calves with mild copper deficiency was normal (J.R. Arthur & R. Boyne, personal communication).

References

Boyne R 1978 Changes in leucocyte cytochrome oxidase activity associated with deficiency of copper in laboratory and farm animals. Res Vet Sci 24:134-138

Scudder P, Stocks J, Dormandy TL 1976 The relationship between erythrocyte superoxide dismutase activity and erythrocyte copper levels in normal subjects and in patients with rheumatoid arthritis. Clin Chim Acta 69:397-403

Copper deficiency in humans

DAVID M. DANKS

Department of Paediatrics, University of Melbourne and Genetics Research Unit, Royal Children's Hospital, Melbourne, 3052, Australia

Abstract Copper deficiency of nutritional origin has been recognized as an important part of complex nutritional problems in Peru, as an occasional event in premature babies in Western countries, and as a real hazard of over-zealous zinc therapy or of prolonged parenteral alimentation in children or adults. Anaemia, osteoporosis and scurvy-like bone changes are recognized in the deficiency, and they respond to copper. Copper intake is falling in western countries as a result of prepackaging of foods, and low-grade chronic deficiency may become a problem. The features seen in Menkes' syndrome suggest that human beings may be rather susceptible to the vascular and neurological effects of copper deficiency; these effects may be encountered as a consequence of prolonged mild copper deficiency. Measurement of the serum concentrations of caeruloplasmin before and after moderate copper repletion is suggested as a method of detecting mild copper deficiency.

In addressing this subject I have decided to talk about four aspects:

(1) What are the known effects of copper deficiency in humans and what other effects may occur, especially in prolonged mild deficiency?

(2) In what circumstances is copper deficiency encountered or likely to be encountered?

(3) How can copper deficiency be diagnosed, especially when it is mild?

(4) What do we know about the pathogenesis of copper deficiency?

In documenting my statements I will refer only to evidence that is not accessible via Underwood (1977) or Danks (1977). Other general information can be found in Hambidge (1978).

There are four available sources of information about the effects of copper deficiency in humans: the documented effects that occur in other animal species provide a useful guide; consideration of the copper proteins known in the human body and of the likely effects of their malfunction provide a second method of predicting the range of effects that may be encountered; the features seen in Menkes' syndrome, a genetically determined defect in copper absorption and utilization, should stimulate the search for wider effects of nutritional copper deficiency; and, finally, apparent nutritional copper deficiency in humans provides at least a minimum list of observed features. These four sources of information are deliberately listed in this order to emphasize that the features currently attributed to copper deficiency may grossly underestimate the effects that may occur in humans.

The effects of copper deficiency vary widely from species to species (Underwood 1977) and it is pointless to guess which species humans might mimic most closely in this regard. However, it is worth listing all the features known to occur in any species and regarding them as likely to occur also in humans. People who have worked personally in animal nutrition may be able to expand the list in Table 1. However, humans may exhibit an effect that has not been seen in another species. Patients who suffer from copper deficiency should therefore be observed carefully for such additional effects.

It is relatively easy to predict the consequences of deficient activity of some of the copper enzymes in the human body — e.g. tyrosinase and lysyl oxidase. However, it is not very easy to predict the consequences of a relative deficiency of enzymes such as cytochrome c oxidase or superoxide dismutase. We should not feel complacent that the list of enzymes in Table 2 is the total number of copper enzymes in the human body. Many enzymes have not yet been studied adequately and some may yet prove to contain copper. Our own work suggests that phenylalanine hydroxylase (phenylalanine 4-monooxygenase, EC 1.14.16.1) contains copper in addition to the iron that is a generally accepted component (C. Yap, personal communication, 1977).

The defect in intestinal transport of copper and the renal copper wasting that occur in Menkes' steely-hair syndrome produce severe copper deficiency in this X-linked inherited disorder of humans (Danks 1977). The homologous allelic mutants, brindled (Mo^{br}/y) and blotchy (Mo^{blo}/y), in the mouse provide useful models of this disease (Hunt 1974, Danks 1977). The clinical effects of Menkes' syndrome and the mutant mice (Table 3) therefore comprise an important guide to the effects that may occur in human nutritional copper deficiency. However, in these mutations, copper deficiency is complicated by a defect in the intracellular transport of copper, which may alter the effects to some degree. The notable absence of anaemia, the effect most widely believed to occur in nutritional copper deficiency, must encourage us to proceed cautiously.

All these sources of information should be used to guide clinicians and research workers to the effects of nutritional copper deficiency in humans but

TABLE 1

Some effects of copper deficiency in different species

Effect	Present in	Absent in
Anaemia	Sheep, rat, rabbit, pig, chicken, cattle, dog, humans	Humans[a], mouse[a]
Neutropenia	Sheep, rat, humans	Humans[a]
Abnormal hair	Sheep, humans[a], rat, rabbit, guinea-pig, dog, cattle, mouse[a]	
Arterial disease (elastic laminae)	Pig, chicken, turkey, rat(F), humans, (mouse[a])	Sheep
Myocardial disease (necrosis & fibrosis)	Cattle, pig, rat	Sheep, horse
Bone changes	Rabbit, pig, dog, chicken, foal, humans, humans[a], mouse[a], (cattle, sheep)	
Cerebellar ataxia	Sheep(F), goat(F), pig(F)	
Other brain damage	Humans[a], rat(F), guinea-pig(F), mouse[a]	
Impaired growth	Mouse[a], cattle, (humans[a])	
Emphysema	Rat, mouse[a], humans[a]	
Diarrhoea	Cattle	
Hypercholesterolaemia	Rat	
Retinal dystrophy	Humans[a]	

[a]Effect occurring as a feature of Menkes' syndrome (humans) or mottled mutants (mouse)
(F) Fetal deficiency necessary to produce the effect.
More severely affected species are listed first. Brackets indicate mildly or variably affected species.

none should be allowed to constrain thinking. Concurrence of two or more of these features should particularly suggest copper deficiency, as in the discovery of copper deficiency in Menkes' syndrome.

The list of effects that have been attributed to copper deficiency as a result of direct observations in humans is much smaller:

Anaemia has attracted the most attention and there is no doubt that copper is important in the hypochromic normocytic anaemia that has been seen in

TABLE 2

Copper enzymes in humans

Common name	EC number	Functional role	Known or expected consequence of deficiency
Cytochrome c oxidase	1.9.3.1	Electron-transport chain	Uncertain
Superoxide dismutase	1.15.1.1	Free-radical detoxification	Uncertain
Tyrosinase (monophenol monooxygenase)	1.14.18.1	Melanin production	Failure of pigmentation
Dopamine β-hydroxylase (dopamine β-monooxygenase)	1.14.17.1	Catecholamine production	Neurological effects, type uncertain
Lysyl oxidase	a	Cross-linking of collagen and elastin	Vascular rupture
Caeruloplasmin	1.16.3.1	Ferroxidase and ? other roles	Anaemia
Enzyme not known		Cross-linking of keratin (disulphide bonds)	Pili torti

^aNo EC number yet allocated

Peru in malnourished children, aged six months to three years, who have chronic diarrhoea (Graham & Cordano 1969). The chronic diarrhoea is probably a cause of the development of the copper deficiency rather than a symptom of it. Careful studies showed that this anaemia was not corrected by the supply of all other haematinics but it was corrected when copper was added to the treatment regime. Anaemia has also been related to copper deficiency at three or four months of age in premature babies and babies of low birthweight (al-Rashid & Spangler 1971, Ashkenazi et al 1973), and in young babies receiving prolonged parenteral alimentation after bowel resection (Karpel & Peden 1972) or for other reasons (Heller et al 1978). Similar experience has been reported in older children (McCarthy et al 1978) and in adults (Dunlap et al 1974, Vilter et al 1974) who were receiving parenteral alimentation, usually after gut resection or in chronic gut disease. The regimes have always proved to be copper-deficient, and addition of copper has corrected the anaemia. The absence of anaemia in Menkes' syndrome, brindled mice (Camakaris et al 1979) and blotchy mice (J. Mann, personal communication, 1980) is curious, and research workers studying these conditions must obviously seek a special

TABLE 3

Features of Menkes' syndrome and mottled mouse mutants

Feature	Menkes' syndrome	Brindled (Mo^{br}/y)	Blotchy (Mo^{blo}/y)
X-linked inheritance	+	+	+
Premature birth	+	−	−
Poor growth	±	+ +	+
Early death	6–36 months	10–15 days	±
Hypothermia	+	0	0
Abnormal hair	+	+	+
Depigmentation	+	+	+
Brain damage	+ + +	+	±
Arterial disease	+ +	±	+
Osteoporosis and metaphyseal changes	+	+	+
Follicular hyperkeratosis	+	0	0
Retinal dystrophy	+	0	0
Emphysema	±	+	±
Low hepatic copper	+	+	+
High intestinal copper	+	+ +	+
High renal copper	+	+	+ +
Low catecholamines	+	+	+

− = absent
± = variably present
+ to + + + = degree of severity
0 = not studied

explanation of why haemopoiesis is excepted from the effects of copper deficiency. Alternatively this exception may indicate that an additional factor is necessary if copper deficiency is to cause anaemia.

Neutropenia usually accompanies the anaemia of copper deficiency in humans or in animals but it is not seen in Menkes' syndrome or in mutant mice. Vacuolation of bone marrow and cells and of circulating neutrophils is also described in most reports.

Bone changes were described in babies and children with copper deficiency in Peru (Graham & Cordano 1969) but were more pronounced in the young babies described in the 1970s (Griscom et al 1971, al-Rashid & Spangler 1971, Karpel & Peden 1972, Ashkenazi et al 1973, Heller et al 1978) and in Menkes' syndrome (Danks 1977). The splaying of metaphyses, accompanied by corner fractures, are similar to the changes seen in scurvy, but in copper deficiency and Menkes' syndrome osteoporosis is much more prominent and the extensive subperiosteal haemorrhage of scurvy is missing. The Wormian bones seen in the skull in Menkes' syndrome may be related to prenatal copper deficiency; they have not been described in the nutritional deficiency of young babies.

Changes in hair, skin and pigmentation are pronounced in Menkes' syndrome and in brindled and blotchy mice, but have received little comment in descriptions of nutritional copper deficiency. The gross pili torti (twisted hair) of Menkes' syndrome does not seem to be encountered in the nutritional conditions, but neither microscopy nor scanning electronmicroscopy have been used to look for minor degrees of this abnormality. Deficient pigmentation would not be noticed in white babies who need parenteral alimentation but it might be noticed in black babies with similar illnesses or in the rather older babies that develop anaemia in Peru. Of course, the general protein–calorie malnutrition seen in Peru may also have effects upon hair, skin and pigmentation. It would seem reasonable to study these aspects more carefully in patients who are believed to have copper deficiency. The patient described in detail by Ashkenazi et al (1973) is an exception, showing severe depigmentation of hair ('almost albino') and seborrhoeic dermatitis. Seborrhoea does occur in Menkes' syndrome, but follicular hyperkeratosis is the more frequent accompanying skin condition.

Vascular changes and *brain disease* are each prominent in Menkes' syndrome but have not been reported in nutritional copper deficiency, except in the severe case observed by Ashkenazi et al (1973). This baby showed profound hypotonia and lethargy and a degree of venous dilatation on the scalp which the authors attributed to an elastin defect but without any direct evidence for this claim. Perhaps nutritional deficiency is rarely sufficiently severe and prolonged to produce these effects. Prenatal deficiency may be necessary for their development, as it is for vascular disease in rats and neurological defects in most species. On the other hand comparison of Menkes' syndrome and the mutant mice shows that vascular disease is more prominent in the human disease than in either the brindled or the blotchy mice. It seems wiser to consider the possibility that humans are particularly susceptible to the effects of copper deficiency upon arterial elastin (as are pigs and turkeys) and to establish whether mild prolonged deficiency is linked to the development of the vascular diseases which are so common in western society.

Myocardial haemorrhage, necrosis and fibrosis have been reported in some animals (Table 1) but not in humans. However, post-mortem examinations do not seem to have been reported in humans suffering from nutritional copper deficiency. Myocardial changes have not been described as prominent in Menkes' syndrome.

CIRCUMSTANCES LIKELY TO CAUSE COPPER DEFICIENCY

Prematurity, especially of an extreme degree, seems to predispose to early onset of copper deficiency in babies if their intake of copper is low. This probably relates more to the rapid doubling in weight of small babies than to their lower copper stores, since their hepatic copper concentrations reach the high levels seen in the full-term newborn as early as 20 weeks after birth (Widdowson et al 1972). Consequently, small-for-dates babies are also at risk.

Cow's milk has a much lower copper content (150 µg/l) than human milk (450 µg/l) (Hambidge 1978), so artificial milk formulas need copper supplementation if fed to small babies. The same applies, of course, to parenteral alimentation regimes for all ages, but especially for premature babies. Excessive supply of amino acids with overflow aminoaciduria may contribute, since some amino acids carry copper with them (Henkin 1974).

Studies performed by Wilson & Lahey (1960) suggested that copper stores are sufficient to last at least five to six months, even after premature birth, despite a low copper intake of 15 μg kg^{-1} day^{-1}. This does not contradict the evidence just discussed or the possibility of copper deficiency occurring in babies of about this age who have excessive copper loss or requirement.

In recent years breast milk has been shown to have a beneficial effect upon zinc absorption, quite apart from any differences in its zinc content (Hurley et al 1978). After various claims have been made about polypeptides causing this effect, citrate has emerged as the agent responsible (Hurley et al 1979). No comparable phenomenon has been reported for copper, nor has a thorough search for such an effect been described, except that breast milk has been shown not to correct the absorptive defect in Menkes' syndrome (Williams et al 1979). Excessive zinc intake has been known for many years to interfere with copper absorption and has been studied carefully in rats (Hall et al 1979). This effect has been encountered in clinical practice in humans in recent years since zinc administration has become fashionable in treatment of various problems of wound healing and of sickle cell disease (Porter et al 1977, Prasad et al 1978).

Molybdenum excess has long been known to cause copper deficiency in sheep, and thiomolybdate complexes are now considered to be responsible. This phenomenon applies to some other species but the relevance to humans is unknown.

Chronic diarrhoea has generally been present in Peruvian children with copper deficiency and anaemia. Copper supplements have not corrected the diarrhoea which has therefore been considered as a cause of the deficiency, rather than a symptom. A number of the other patients described have had gut lesions or resection. Impaired absorption when there is damage to the small intestine is certainly possible, since absorption is an active process, but no definitive studies have been reported, except in the specific defect seen in Menkes' syndrome (Lucky & Hsia 1979). Enterohepatic circulation of copper was considered to be substantial in this case, but more recent studies suggest that copper excreted in bile is bound in a form that is poorly available for absorption (Frommer 1971). If this is true, illnesses that change the chemical form of copper in bile may be likely to enhance reabsorption rather than to prevent it.

The effect of gut resection on copper absorption is also ill documented. Studies of several animal species indicate that absorption is most active high in the small intestine, but absorptive ability is present elsewhere so resection may have to be extensive to cause deficiency. Little is known of the normal mechanism of absorption in humans. Copper deficiency has usually been described during phases of parenteral alimentation after gut resection rather than at later stages when oral feeding has been re-established.

Any disease leading to chronic loss of proteins of relatively high molecular weight (e.g. nephrotic syndrome, protein-losing enteropathy) may cause copper deficiency by loss of large amounts of caeruloplasmin and of copper bound to albumin.

Inadequate copper intake in healthy persons who simply eat average Western diets needs serious consideration. Copper is present at high concentrations in shellfish, animal livers and kidneys, at moderate concentrations in nuts, legumes and stone fruits, and at lower concentrations in meat and cereals. Concentrations of copper in dairy products are low. Adults who eat diets prepared from fresh foods, even with only an occasional high-content item, will be likely to ingest the 2–5 mg needed daily. However, those who rely heavily upon prepared foods from supermarket shelves are at real risk. Food processors are designed to eliminate copper because its oxidant properties limit shelf-life. Studies of diets eaten by urban New Zealanders (Guthrie et al 1978), by US college students and by various US hospital patients (Klevay et al 1979) indicate that intakes below 1.5 mg/day are common. Under these circumstances the type of water used may influence intake greatly. Some copper-lined hot-water systems deliver up to 1 mg copper/l in areas with very soft water supplies.

DIAGNOSIS OF COPPER DEFICIENCY

Lack of a simple, well-accepted method of establishing copper deficiency has been a great hindrance to diagnosis. Extreme copper deficiency, with serum copper levels less than 8 μmol/l (50 μg/l), presents no problem, but mild degrees of deficiency are hard to demonstrate.

Caeruloplasmin makes up 95% of serum copper and the range of concentrations encountered in healthy persons is wide, being partly determined genetically but also by oestrogen levels and by intercurrent infections that produce 'acute-phase reactions'. Consequently, wide ranges of serum copper are encountered among healthy people.

Urinary excretion of copper is low because copper is efficiently reabsorbed by the renal tubules. Consequently, minor changes in renal tubular efficiency cause major changes in urinary copper, rendering it a bad indicator of copper status. Hair content of copper is claimed to reflect copper status over the previous few months but is also affected by hair-washing and hair sprays. Hepatic copper estimation can be justified only when suspicion of deficiency is high and proof is vital.

There is one simple method of assessing copper status that seems to be worthy of trial. Caeruloplasmin levels could be measured twice, the second sample being taken after three or four days of copper administration at physiological replacement levels (10–20 mg daily in adults). In patients with copper deficiency the second result will be well above the first; patients who have low caeruloplasmin for any other reasons will not respond in this way. The logic behind the method is that copper deficiency limits the production of copper proteins such as caeruloplasmin, whereas moderate copper excess does not superinduce their production. The value of this approach has been demonstrated for zinc assessment in rats by the use of alkaline phosphatase (EC 3.1.3.1) or red cell carbonic anhydrase (EC 4.2.1.1) (Kirchgessner et al 1977). There is abundant evidence of rapid response of serum caeruloplasmin to copper repletion in rats (Paynter et al 1979) and in humans (Matsuda et al 1974). Lack of superinduction by excess copper is also established in rats (Linder et al 1979) and, less stringently, in humans. The suggested method may fail in protein-losing diseases in which a cellular copper enzyme may be a better index — e.g. red cell superoxide dismutase (EC 1.15.1.1).

MECHANISM OF PRODUCTION OF EFFECTS OF COPPER DEFICIENCY

The effects most easily explained are those not firmly identified in nutritional deficiency — depigmentation due to tyrosine deficiency, defective uncross-linked arterial elastin due to lysyl oxidase deficiency and hair changes due to defective disulphide bridging of keratin. These keratin changes probably also interfere with epidermal shedding in the skin, causing follicular hyperkeratosis, and faulty cross-linking in collagen may cause the bone changes.

Anaemia may be partly explained by lack of the ferroxidase effects of caeruloplasmin that are needed to release iron from ferritin in gut mucosa and in other tissues. However, lack of iron availability does not seem a sufficient explanation for the anaemia or neutropenia, because cross-over experiments suggest that there is deficiency of another circulating factor that is necessary for production of red cells and neutrophils (Zidar et al 1977).

Inefficiency of the terminal electron-transport chain due to diminished cytochrome c oxidase activity can be put forward to explain almost any ill-effect, but if we are honest we should admit that we really have no idea what effects this reduced enzyme activity may have. It is also fashionable to blame all sorts of cell damage onto superoxide radicals and hence onto a deficiency of superoxide dismutase. However, it is only within the last year that the first systematic attempt at studying the effects of copper deficiency on cytochrome c oxidase and superoxide dismutase in many different tissues has been published (Paynter et al 1979). Comparisons of the interesting differences between these two enzymes in different tissues from brindled and blotchy mice, and from mice with nutritional copper deficiency, plus correlation of these differences with the disease effects, may give a better idea of the consequences of interfering with these important enzymes.

'Copper is essential for normal myelination' is an impressive-sounding generalization repeated through many articles and texts. However, evidence for the statement is scanty. Myelination is defective in swayback (or enzootic ataxia) in sheep, and in various severe forms of experimental copper depletion in the rat or guinea-pig fetus. However, severe neuronal loss is also seen in these conditions and the myelin reduction could be secondary to brain cell damage. Studies of Menkes' syndrome show marked abnormalities of the development of neurons, especially of Purkinje cells in the cerebellum. Morphological changes in the brain of the brindled mouse are much less obvious and no quantitative or qualitative change in myelin was detected in chemical studies (D.M. Danks and J.M. Matthieu, unpublished results).

In my opinion the failure in dopamine β-monooxygenase activity and the consequent deficiency in catecholamine neurotransmitters may prove to be responsible for the brain lesions. These compounds are likely to function in brain development and organization before they start to function in their better understood transmitter roles.

Careful studies, in many laboratories around the world, of Menkes' syndrome and of brindled and blotchy mice have already added a great deal to our knowledge of human copper deficiency. These mutations affect a vital process in the normal intracellular transport of copper. When the nature of this process is eventually defined, it will provide an important landmark in the understanding of copper deficiency. The transport of copper from the gut lumen to the copper enzymes in the various body cells must involve a series of specific transport processes, which need to be defined before we can understand copper deficiency completely. Identification of mutants that affect each step may be the only way to define the normal process fully (as was done in the study of bacterial metabolism). The mouse is probably the best species to use because so many mutants are available, but it is unfortunate that little is known about copper nutrition in the mouse. We have examined two other mouse mutants – crinkled and quaking – without finding convincing evidence of a further copper mutant (see p 240-244, this volume).

Use of mutagens in cell culture systems may be another valuable approach, and cell hybridization will help to distinguish whether different processes are affected in different mutants.

A full knowledge of the relative affinities and capacities of various steps in copper transport, and of the apoenzymes concerned, will be needed eventually to explain such findings as the wide variation in the effects of copper deficiency on the activities of superoxide dismutase and cytochrome c oxidase in rats (Paynter et al 1979). Such knowledge may also explain the different effects seen in different species.

For the present, the human species has the honour of being the only species in which mutants that affect two different processes in copper transport are definitely known – those that occur in Menkes' syndrome and in Wilson's disease – so perhaps I should finish by pointing out that studies in humans may continue to provide new knowledge that is of interest to basic scientists and to veterinary research workers!

References

al-Rashid RA, Spangler J 1971 Neonatal copper deficiency. N Engl J Med 285:841-843
Ashkenazi A, Levin S, Djaldetti M, Fishel E, Benvenisti D 1973 The syndrome of neonatal copper deficiency. Pediatrics 52:525-533
Camakaris J, Mann JR, Danks DM 1979 Copper metabolism in mottled mouse mutants: copper concentrations in tissues during development. Biochem J 180:597-604
Danks DM 1977 Copper transport and utilisation in Menkes' syndrome and in mottled mice. Inorg Perspect Biol Med 1:73-100

Dunlap WM, James GW, Hume DM 1974 Anemia and neutropenia caused by copper deficiency. Ann Intern Med 80:470-476

Frommer D 1971 The binding of copper by bile and serum. Clin Sci (Oxf) 41:485-493

Graham GG, Cordano A 1969 Copper depletion and deficiency in the malnourished infant. Johns Hopkins Med J 124:139-150

Griscom NT, Craig JN, Neuhauser EBD 1971 Systemic bone disease developing in small premature infants. Pediatrics 48:883-895

Guthrie BE, McKenzie JM, Casey CC 1978 Copper status of New Zealanders. In: Kirchgessner M (ed) Trace element metabolism in man and animals. Arbeitskreis für Tierernährungsforschung, Freising-Weihenstephan, vol 3, p 304-306

Hall AC, Young BW, Bremner I 1979 Intestinal metallothionein and the mutual antagonism between copper and zinc in the rat. J Inorgan Biochem 11:57-66

Hambidge KM 1978 Trace elements in pediatric nutrition. Adv Pediatr 24:191-231

Heller RM, Kirchner SG, O'Neill JA, Hough AJ, Howard L, Kramer SS, Green HL 1978 Skeletal changes of copper deficiency in infants receiving prolonged total parenteral alimentation. J Pediatr 92:947-949

Henkin RI 1974 Metal-albumin-amino acid interactions: chemical and physiological interrelationships. In: Friedman M (ed) Protein-metal interactions. Plenum Press, New York, p 299-320

Hunt DM 1974 Primary defect in copper transport underlies mottled mutants in the mouse. Nature (Lond) 249:852-854

Hurley LS, Duncan JR, Eckhert CD, Sloan MV 1978 Zinc binding ligands in milk and their relationship to neonatal nutrition. In: Kirchgessner M (ed) Trace element metabolism in man and animals. Arbeitskreis für Tierernährungsforschung, Freising-Weihenstephan, vol 3, p 449-451

Hurley LS, Lönnerdal B, Stanislowski AG 1979 Zinc citrate, human milk and acrodermatitis enteropathica. Lancet 1:677-678

Karpel JT, Peden VH 1972 Copper deficiency in long-term parenteral nutrition. J Pediatr 80:32-36

Kirchgessner M, Roth H-P, Spoerl R, Schnegg A, Kellner RJ, Weigand E 1977 A comparative view on trace elements and growth. Nutr Metab 21:119-143

Klevay LM, Reck SJ, Barcome DF 1979 Evidence of dietary copper and zinc deficiencies. J Am Med Assoc 241:1916-1918

Linder MC, Houle PA, Isaacs E, Moon JR, Scott JE 1979 Copper regulation of ceruloplasmin in copper-deficient rats. Enzyme (Basel) 24:23-35

Lucky AW, Hsia YE 1979 Distribution of ingested and injected radiocopper in two patients with Menkes' kinky hair disease. Pediatr Res 13:1280-1284

Matsuda I, Pearson T, Holtzman NA 1974 Determination of apoceruloplasmin by radioimmunoassay in nutritional copper deficiency, Menkes' kinky hair syndrome, Wilson's disease and umbilical cord blood. Pediatr Res 8:821-824

McCarthy DM, May RJ, Mather M, Brennan MP 1978 Trace metals and essential fatty acid deficiency during total parenteral nutrition. Am J Dig Dis 23:1009-1016

Paynter DI, Moir RJ, Underwood EJ 1979 Changes in activity of the Cu-Zn superoxide dismutase enzyme in tissues of the rat with changes in dietary copper. J Nutr 109:1570-1576

Porter KL, McMaster D, Elmes ME, Love AHG 1977 Anaemia and low serum-copper during zinc therapy. Lancet 2:774

Prasad AG, Brewer GJ, Schoomaker EB, Rabboni P 1978 Hypocupremia induced by zinc therapy in adults. J Am Med Assoc 240:2166-2168

Underwood EJ 1977 Trace elements in human and animal nutrition, 4th edn. Academic Press, New York

Vilter RW, Bozian RC, Hess EV, Zellner DC, Petering HG 1974 Manifestations of copper deficiency in a patient with systemic sclerosis on intravenous alimentation. N Engl J Med 291:188-191

Widdowson EM, Chan H, Harrison GE, Milner RDG 1972 Accumulation of Cu, Zn, Mn, Cr and Co in the human liver before birth. Biol Neonate 20:360-367

Williams DM, Atkin CL, Seay AR, Bray PF 1979 Failure of human milk therapy in Menkes' kinky hair disease. Am J Dis Child 133:218-219

Wilson JF, Lahey ME 1960 Failure to induce dietary deficiency of copper in premature infants. Pediatrics 25:40-49

Zidar BL, Shadduck RK, Zeigler Z, Winkelstein A 1977 Observations on the anemia and neutropenia of human copper deficiency. Am J Hematol 3:177-185

Discussion

Shaw: Is it true that patients with Menkes' syndrome develop hypothermia and disorders of thermoregulation?

Danks: Yes. The degree of hypothermia can be extreme. These babies sometimes present with temperatures as low as 28 or even 25°C.

Shaw: Is that related to levels of cytochrome *c* oxidase activity, as has been suggested, and has metabolic response to cold been examined?

Danks: Yes; when I was in Lausanne on sabbatical leave a group of people expert at total body calorimetry examined a patient from Berne who had Menkes' syndrome, and to our amazement this child responded normally, in terms of respiratory quotient and energy production, to the stresses of feeding and of warming.

Shaw: Do disorders of thermoregulation arise in nutritional copper deficiency?

Danks: I have no direct experience of that.

Mills: Young copper-deficient rats appear to be hypothermic when they are handled but I am not aware of any measurements of their body temperature or heat output.

Graham: We have never seen hypothermia in nutritional copper deficiency.

Perhaps I could elaborate on our observations in these children. Now that we are aware of the problem with copper, we do not let our patients become deficient; we wait for the occasional patient to be admitted in a state of severe deficiency to enable us to make limited observations. All our patients have been marasmic, have usually been born at term and have had a short period of breast-feeding. They had diarrhoeal episodes almost continuously from an early age until they were finally admitted. All are depleted of many nutrients – copper, iron, magnesium, zinc, vitamin A – and in refeeding if we omit or fail to provide adequate amounts of any one of these nutrients the infants become overtly deficient. The first deficiency we discovered that way was one of magnesium.

We have never seen nutritional copper deficiency without the history of prolonged diarrhoea, so we have to assume that these babies have lost a lot of their original stores of body copper. That question deserves closer examination. I do not think the answer is simply that they fail to absorb dietary copper; they actually do not have sufficient remaining copper stores when we refeed them to allow them to increase their weight from 3 kg to 5 kg without becoming overtly deficient. Normal infants would have sufficient copper stores at birth to allow them to grow to 10 kg without any signs of deficiency. We believe that prolonged copper deficiency in association with diarrhoea is due to failure of the infants to reabsorb copper from the bile, but we have no proof of that. Occasionally the children were overtly deficient on admission but most deficiencies resulted from rehabilitation on a diet of modified cow's milk with added sugar and oil. Instead of the 42 μg copper $kg^{-1}\,day^{-1}$, which would have been provided by the milk formula, the babies received about 28 μg $kg^{-1}\,day^{-1}$ from the diluted version of the diet. We also used distilled and deionized water in preparation of the formulas, because we were interested in sodium balances. The background of depletion is there, and whether they become overtly deficient depends on a variety of circumstances.

In one extreme case, an infant was admitted with undetectable amounts of copper and caeruloplasmin in serum and with no neutrophils; it took four days of high doses of copper to produce a response, which normally would have occurred within a few hours. Interestingly, that child had no bone lesions, which seem to relate to the duration rather than to the severity of deficiency. The child's hair was examined by Dr Bradfield in California (see Bradfield et al 1980) and the copper content was not particularly low.

Danks: Copper content of the hair does not fall in Menkes' syndrome, incidentally, despite the other changes that occur in the hair (B.J. Stevens & D.M. Danks, unpublished work).

Graham: This child's hair had only the typical growth arrest of marasmus.

Anaemia is not present in all cases of nutritional copper deficiency. There may be poor or no iron absorption and some anaemic children do not develop reticulocytosis in response to oral iron but only after intramuscular iron. In prolonged deficiency there may be no response even to intramuscular iron, and the low serum concentration of iron cannot be increased.

We have measured the changes in serum concentrations of copper and in number of neutrophils after the first oral dose of copper, and in the first six hours there is a sharp drop in the copper concentration, which begins to rise again after a further six or eight hours. The first dose of copper may induce something that lowers serum copper concentrations. Neutropenia is the easiest diagnostic sign to use but it can be deceptive because a copper-deficient child can usually respond to an infection by an increase in the number of circulating neutrophils. The bone marrow reveals a neutrophil maturation arrest in copper deficiency.

Therefore, we might ask, why and how does copper deficiency produce a new set-point for circulating neutrophils? The answer may help us to understand how the body assigns priorities to whatever copper is available and why the body curtails the utilization of copper by neutrophils until it is needed. As Dr Danks mentioned, there has been only one case of copper deficiency (lasting five years) associated with *intractable* diarrhoea, which was stopped when the child was given copper. In that child there was no disaccharidase activity and only glucose could be tolerated. In none of the other cases did we believe that diarrhoea persisted or recurred because of the copper deficiency.

John Dobbing (personal communication) examined the effects of extreme malnutrition on brain development of a child who had severe copper deficiency in late infancy (ten months of age) and who died at seven years of age from typhoid fever. He found no detectable alteration in lipids or myelin.

We have followed the progress of most of the children who had severe nutritional copper deficiency in infancy. We have seen no specific behaviour or poor school performance that we could attribute to the deficiency rather than to the child's originally poor environment or general malnutrition. Therefore the manifestations of late (postnatal) copper deficiency are obviously not as devastating as those due to prenatal deficiency.

As far as growth is concerned, the nutritionally copper-deficient child gains weight and retains nitrogen, but in chronic deficiency, once linear growth of the bones is affected, the bone age is greatly retarded. After a few weeks of copper administration the bone age advances rapidly and the centres begin to mineralize. We believe that the copper-deficient infant may be more susceptible to neurotropic viruses.

Danks: In 1973, at a meeting on Wilson's disease in Paris, I was brash enough to suggest that within two or three years we would have worked out exactly how the disease works! I believe that copper has some specific destinations that it must reach in sufficient amounts. It is also a highly toxic ion which could not possibly be left to drift towards whatever protein might pick it up. It must therefore have a well organized system of carriers that relay it from the point of entry to the body to the point of utilization and eventually to the point of exit. There must be a rather neatly balanced hierarchical relationship between these carrier proteins. The shuffling between oxidation states may provide one mechanism by which the balance is achieved. Study of the Menkes' mutation should enable us to identify at least one of those carrier systems; (see discussion after Professor Hurley's paper, p 239-245). The metal seems to be diverted into a rather non-specific 'dump' of low-molecular-weight protein where it can be stored without ill-effects when it has failed to reach its original destination. We would like to identify why the copper is not transported normally.

Bremner: Have you any evidence of decreased enzyme activity in tissues that accumulate copper?

Danks: Yes; we find five times the normal concentration of copper and one fifth of the normal lysyl oxidase activity in the same cells in culture (P.M. Royce, J. Camakaris & D.M. Danks, unpublished work).

Hurley: Have cultured cells from patients with Menkes' syndrome been used?

Danks: Yes. Cells from Menkes' patients and brindled mice (P.M. Royce, J. Camakaris & D.M. Danks, unpublished work) produce similar results.

Mills: There is a chance that this phenomenon may not be as unique as we would think. When there are individual cases of enzootic ataxia among groups of lambs, I believe that brain lesions develop in animals that do not necessarily have the lowest copper status indicated by hepatic or blood concentrations of copper.

Lewis: The brain lesions in lambs probably arise when the liver content of copper is under 15 mg/kg on a dry matter basis, and the brain content of copper is about 3 mg/kg (Lewis et al 1974). We do not see many cases of swayback (enzootic ataxia) with brain concentrations of copper over 5 mg/kg dry matter. The relation between the brain and liver concentrations of copper is very variable; there is not a two-way relationship.

Graham: On the question of the diagnosis of copper deficiency I would like to draw an analogy with iron deficiency. Despite enormous effort over many years, the only accepted proof of iron deficiency is the presence of a response to iron administration; there is much overlap between the normal individual and the iron-deficient individual in ferritin concentration and transferrin saturation. I would suggest that in human copper deficiency the response of neutrophils to copper might be the quickest way to diagnose deficiency if there is any suspicion that it is present. The response occurs in 24 h and therefore might be a useful diagnostic tool.

Dormandy: If, on the other hand, we are to use small changes in concentrations of caeruloplasmin to make diagnoses, we should remember that there are two entirely different ways of measuring caeruloplasmin in clinical laboratory practice and that the results do not necessarily coincide. In at least one family with Wilson's disease there was a striking discrepancy between caeruloplasmin as measured by its oxidase function and as measured by immunodiffusion, which is now the current method (Gollan et al 1977). I think that we should therefore make it clear how measurements of caeruloplasmin have been made, at least for the time being, until the immunodiffusion method becomes universally adopted.

Mills: Under what circumstances do the results not agree?

Dormandy: It is quite common for there to be a small discrepancy, but diagnostic confusion arose in treatment of the family with Wilson's disease (Gollan et al 1977). The oxidase activity was low but the immunological measurement of caeruloplasmin was within the normal range.

Danks: For each element, one should study the particular metalloprotein that has the characteristics that are required: it needs to be accessible and to have a simple method of measurement, and one should work out the time-scale of response to replenishment. Caeruloplasmin is widely used because it is so accessible but superoxide dismutase may be a better index and it is, after all, equally accessible.

Riordan: In terms of the biochemical lesion in Menkes' disease, the apparent paradoxes within the cultured cells are revealing, in that there is a high concentration of copper and metallothionein (J.R. Riordan, unpublished work) (or whatever the low molecular weight binding ligands are), and yet the enzymes don't take up metal (D.M. Danks, B.C. Starcher, personal communications, 1980), and the cells are killed at the same time. So there are certain biochemical entities that are associated normally with copper and there are certain ones that are not. It is striking that copper can combine with one protein that responds to the concentration of the metal and then that there are other proteins that require copper for their function but are unable to combine with it at all. I am sure that if we could focus on that aspect a little more sharply we might begin to understand the biochemical lesion.

References

Bradfield RB, Cordano A, Baertl J, Graham GG 1980 Hair copper in copper deficiency. Lancet 2:343-344

Gollan JL, Stocks J, Dormandy TL, Sherlock S 1977 Reduced oxidase activity in the caeruloplasmin of two families with Wilson's disease. J Clin Pathol (Lond) 30:81-83

Lewis G, Terlecki S, Parker BNJ 1974 Observations on the pathogenesis of delayed swayback. Vet Rec 95:313-316

Copper in fetal and neonatal development

LUCILLE S. HURLEY, CARL L. KEEN and BO LÖNNERDAL

Department of Nutrition, University of California, Davis, California 95616, USA

Abstract The essentiality of copper for normal fetal and neonatal development has been well documented, although copper metabolism during this period is poorly understood. The dietary requirement for copper is influenced by genetic background. The neurological phenotypic characteristics of the mutant gene *quaking (qk)* in mice resemble in part those of copper-deficient animals. Supplementation of the maternal diet with copper during pregnancy and lactation, or during lactation alone, greatly reduced the frequency of tremors characteristic of these mutants, and brought the otherwise low copper concentrations in the brain to normal. Prenatal copper supplementation of *crinkled (cr)* mice increased neonatal survival and produced nearly normal development of skin and hair. Non-supplemented *cr/cr* mice showed anaemia at 21 days of age which disappeared later. Similarly, copper concentration in liver and hair was low in young but normal in old *cr/cr* mice. However, activity of copper–zinc-superoxide dismutase (Cu-ZnSOD) remained low even at 60 days of age. Copper supplementation brought both SOD activity and copper concentration of liver and hair to normal. The errors in copper metabolism produced by *qk* and *cr* appear to be expressed at different periods of development. The hypothesis that there are rapid changes in the metabolism of copper is supported by the observation that molecular distribution of copper in rat intestine changes drastically during the neonatal period.

CHARACTERISTICS OF PRENATAL COPPER DEFICIENCY

The presence of copper in animals and plants was known during the nineteenth century (Underwood 1977), but no evidence that it might be essential for animals was published until the third decade of the twentieth century. However, nutritional knowledge during the early 1920s was so limited that the purified diets used were incomplete in other nutrients as well as in copper (Underwood 1977). By 1928, however, Hart et al were able to demonstrate the essentiality of copper by reporting that this element, in addition to iron, was necessary for haemoglobin formation in rats. The importance of copper in the fetal development of mammals was first indicated by the work of Bennetts and his co-workers (1937, 1942) in their demonstration that enzootic ataxia of lambs, a disease affecting the developing fetus, could be prevented by giving the ewe additional copper during pregnancy. Subsequent experimental studies, as well as additional clinical or field work, showed that similar effects could be produced in several species by copper deficiency.

The characteristics of animals subjected to copper deficiency during prenatal life are similar in most species (Table 1), and appear to involve the same mechanisms that produce manifestations of copper deficiency postnatally (Underwood 1977, Hurley & Keen 1979), although they may occur at different stages of development. This is certainly due, at least in part, to differences in the pattern of development in various species. Thus, a particular maturational event may occur prenatally in one species, but neonatally in another. Myelination of the central nervous system is a good example; it is largely prenatal in the lamb and postnatal in the mouse or rat. Thus, one of the neurological abnormalities in enzootic ataxia and in other examples of either prenatal or neonatal copper deficiency is a paucity of normal myelin, which in turn is related to impaired biosynthesis of phospholipids. In addition, there may be cardiovascular lesions, abnormal lung development, and abnormal skin (all deriving from impaired cross-linking leading to formation of abnormal elastin), and skeletal abnormalities caused by insufficient cross-linking of collagen and defective matrix formation.

Biochemical signs of copper deficiency in prenatal, as in postnatal, life may include low levels of copper in liver, brain, or blood, and low activities of cop-

TABLE 1

Characteristic effects of prenatal copper deficiency

Fetal death	Abnormal lung development
Early neonatal death	Impaired cross-linking of elastin
Neurological abnormalities	Skin and hair abnormalities
Spastic paralysis	Skeletal abnormalities
Incoordination	Impaired cross-linking of collagen
Convulsive seizures	Low copper content of tissues
Collapsed cerebral hemispheres	Low activity of copper metalloenzymes
Small cerebellar lobes	Cytochrome c oxidase
Paucity of normal myelin	Copper–zinc-superoxide dismutase
Impaired phospholipid synthesis	Preventable by copper supplementation
Cardiovascular lesions	
Aneurysms of aortic arch	
Impaired cross-linking of elastin	

per enzymes such as cytochrome c oxidase (EC 1.9.3.1) and copper–zinc-superoxide dismutase (Cu-ZnSOD; EC 1.15.1.1). Perhaps the most crucial evidence for prenatal or neonatal copper deficiency is the prevention of these characteristics by copper supplementation.

GENETIC INTERACTIONS

Strain differences

The influence of genetic factors in susceptibility to copper deficiency is readily apparent from the studies of Wiener and his associates, who observed that sheep of the Welsh genotype had a lower incidence of swayback than did Blackface sheep; these differences were correlated with blood concentration of copper (Wiener 1966, Wiener & Field 1969). From embryo transfer studies, it was learned that the Blackface sheep had an abnormality in maternal copper metabolism, and that hepatic concentration of copper in the newborn was determined by the genotype of the ewe and not of the fetus (Wiener et al 1978).

Mutants

Several mutant genes that affect copper metabolism have been identified in animals and humans. Other papers in this symposium deal with the mutation in humans, Menkes' disease (Danks 1980, this volume), and the similarly sex-linked *mottled* mutant genes in mice (Hunt 1980, this volume).

Another mutant gene in mice that affects copper metabolism is *quaking (qk)*. Phenotypic characteristics of mutant mice that are homozygous for this gene include an intermittent axial body tremor, which is first seen on about day 10 of postnatal life, abnormalities of myelin, and reduced levels of brain cerebrosides and sulphatides (Sidman et al 1964, Baumann et al 1968). We have shown that *quaking* mice also have a low concentration of copper in the brain (Keen & Hurley 1976). We found that supplementation of the maternal diet with a high level of copper during pregnancy and lactation greatly reduced the frequency of tremors in the mutant offspring, and produced a normal concentration of brain copper (Keen & Hurley 1976). The critical period for copper supplementation in order to ameliorate the tremor of *qk/qk* mice was in the neonatal (suckling) period rather than the prenatal period (Table 2). Mutant mice from females fed a high copper diet only during lactation (Group III) showed a frequency of tremor as low as that of mice from females supplemented during both pregnancy and lactation (Groups II and V).

TABLE 2

Effect of different time periods of supplementation on frequency of tremors in *quaking (qk/qk)* mice

Dietary copper levels[a]		No. of mice	Age (days)		
Group Prenatal	Postnatal		21	23	25
mg Cu/kg diet			No. of tremors/5 seconds[b]		
I 6	6	4	52.0 ± 1.1[d,e,f]	54.5 + 1.3[d,e,f]	58.7 ± 0.8[d,e,f]
II 250	250	4	31.7 ± 1.9	39.7 ± 2.9	34.7 ± 2.8
III 6	250	5	38.0 ± 3.3	42.0 ± 2.3	38.0 ± 2.6[c]
IV 250	6	5	50.6 ± 1.1[d,e,f]	54.8 ± 2.3[d,e,f]	59.2 ± 1.9[c,d,e,f]
V 500	500	4	33.0 ± 2.9	35.2 ± 2.6	36.2 ± 3.7

[a]All diets were purified; control diet contained 6 mg Cu/kg; high copper diet contained 250 or 500 mg/kg; for experimental details see Keen & Hurley (1976)
[b]Values are expressed as mean ± SEM
[c]Value is for four animals
[d]Significantly higher than group II ($P < 0.01$); [e]group III ($P < 0.01$); and [f]group V ($P < 0.01$) (Student's *t*-test).

Although these findings clearly indicated an error in copper metabolism as a result of the *qk* gene, many other parameters often found to be abnormal during copper deficiency proved to be normal in 21-day old *qk/qk* mice. These measurements included the copper concentration of liver, kidney, and plasma; various indicators of anaemia (packed cell volume, haemoglobin, and red blood cell count); and activity of Cu-ZnSOD in liver and brain. In addition, the zinc concentrations of brain, liver, and kidney were also normal. These observations suggest that the gene *qk* does not produce a general deficiency of copper in the body, but an error in copper metabolism that is reflected specifically in neural tissue and is partially reversible during the suckling period.

The gene *crinkled (cr)* in mice is an autosomal recessive whose phenotypic characteristics in the homozygous animal include a smooth coat with thin skin, delayed pigmentation, retarded and abnormal development of hair bulbs and early mortality. Only straight hairs are formed, rather than the bent or crimped as well as straight hairs found in normal mice (Falconer et al 1952). The similarity of the hair abnormality to that in copper-deficient sheep (Underwood 1977), and of the early mortality and the thin skin to that in copper-deficient rats (O'Dell et al 1961), led us to study the possible relationship between this mutant gene and copper metabolism (Hurley & Bell 1975). When we fed a purified diet, containing a high level of copper, during pregnancy and lactation to female mice heterozygous for the gene cr, the diet ameliorated the effects of the mutant gene in homozygous mutant *(cr/cr)* offspring. Survival of *cr/cr* mice was significantly higher in the high copper group than in the normally fed controls (Table 3). In order to be effective, copper supplementation was required during the prenatal period; postnatal supplementation (during lactation) was without effect (Keen et al 1980b). Thickness of the skin and the epidermis, and development of the hair bulbs was nearly normal in offspring of dams supplemented during pregnancy and lactation.

Crinkled mutant mice also showed gross congenital malformations of the brain, including abnormal shape, cavitation, and small cerebellum, which were largely prevented by copper supplementation (Table 4). Histologically, there was disorganization and rarefaction of myelin, especially in the cerebellum (Theriault et al 1977). Biochemical evidence of cholesterol esters in the brains of one-year-old crinkled mice, but not in normal mice, was consistent with the view that the myelin disruption was secondary to axonal degeneration, since presence of this compound in adult neural tissue indicates active demyelination (Morell et al 1972). Brains contained a higher concentration of sulphatides in young *cr/cr* mice, and of cerebrosides in adult *cr/cr* mice, than did brains from non-crinkled controls (Theriault et al 1977). In young mutant mice whose dams were supplemented with copper during pregnancy and lactation, cerebroside concentration of the brain was normal (Table 4). Other brain lipids, phosphatidyl ethanolamine, phosphatidyl

TABLE 3

Effect of dietary supplementation with copper on survival of *crinkled (cr/cr)* and non-crinkled *(+/?)* mice

Diet	Genotype	Litters	Mice	Survival (Day 0–14)
		No.	No.	%
Stock diet, control[a] (12 mg Cu/kg)	+/?	11	81	72
	cr/cr		25	32[b]
Purified diet, control (6 mg Cu/kg)	+/?	15	101	75
	cr/cr		36	22[b,c]
Purified diet, copper-supplemented (500 mg Cu/kg)	+/?	15	115	80
	cr/cr		46	52[b]

[a]*Purina Laboratory Chow* (Ralston Purina Co., St. Louis)
[b]Significantly different from littermate controls by Chi-square test ($P < 0.01$)
[c]Significantly different from high-copper-fed mutants by Chi-square test ($P < 0.01$)

TABLE 4

Brain lipids in crinkled (cr/cr) and non-crinkled (+/?) mice[a]

Dietary group	Genotype	Brain (g, wet weight)	Brain lipids (μmole/g, wet weight)			Grossly malformed brains
			Cholesterol	Cerebrosides	Sulphatides	%
21-Days old						
Stock diet, control (12 mg Cu/kg)	+/?	0.34 ± 0.08(10)[b]	35.39 ± 1.17(5)	6.44 ± 0.77(5)	2.12 ± 0.52(5)	—
	cr/cr	0.32 ± 0.03(10)	32.90 ± 0.75(5)	6.26 ± 0.15(5)	3.87 ± 0.66(5)[c]	—
Purified diet, control (6 mg Cu/kg)	+/?	0.38 ± 0.04(7)	33.76 ± 0.90(5)	6.34 ± 0.68(8)	2.09 ± 0.17(7)	0(16)
	cr/cr	0.35 ± 0.01(8)	35.28 ± 0.82(5)	5.88 ± 0.62(8)	3.42 ± 0.57(6)[c]	33(18)
Purified diet, copper-supplemented (500 mg Cu/kg)	+/?	0.30 ± 0.02(6)	34.93 ± 1.33(3)	5.16 ± 0.72(3)	2.34 ± 0.35(3)	0(13)
	cr/cr	0.34 ± 0.03(6)	35.98 ± 1.90(3)	4.92 ± 0.38(3)	2.43 ± 0.38(3)	8(13)
Adult (17 months)						
Stock diet, control (12 mg Cu/kg)	+/?	0.40 ± 0.02(2)	54.17 ± 0.14(2)	8.73 ± 0.47(2)	4.25 ± 0.55(2)	—
	cr/cr	0.35 ± 0.05(4)	66.43 ± 7.52(4)	14.73 ± 0.59(4)[d]	4.18 ± 0.14(4)	—

[a]Values are expressed as mean ± SEM; for experimental details see Hurley & Bell (1975), and Theriault et al (1977)
[b]Number of animals tested is shown in parentheses
[c]Significantly different ($P < 0.05$) from non-crinkled littermate by Student's t-test
[d]Significantly different ($P < 0.01$) from non-crinkled littermate by Student's t-test
—, not examined

choline, phosphatidyl serine, sphingomyelin and phosphatidyl inositol, did not differ in concentration between mutants and non-mutants (Table 5).

Young cr/cr mice also showed the microcytic hypochromic anaemia that is characteristic of copper-deficient animals (Table 6). Haematocrit, haemoglobin, red blood cell count, mean corpuscular volume and mean corpuscular haemoglobin were all lower in mutant mice than in their non-mutant littermates. These differences were not seen at one year of age (Table 6).

TABLE 5

Brain phospholipid composition of 21-day old *crinkled (cr/cr)* and non-crinkled *(+/?)* mice[a]

Dietary group	Genotype	Brain lipids (μmole/g, wet weight)			
		Phosphatidyl ethanolamine	Phosphatidyl choline	Phosphatidyl serine	Sphingomyelin & phosphatidyl inositol
Stock diet, control (12 mg Cu/kg)	+/?	15.45 ± 2.10	18.35 ± 2.65	5.17 ± 0.91	3.20 ± 0.63
	cr/cr	15.53 ± 0.31	21.40 ± 4.03	5.60 ± 0.64	4.39 ± 0.49
Purified diet, control (6 mg Cu/kg)	+/?	16.64 ± 1.38	19.66 ± 1.34	5.96 ± 0.48	4.00 ± 0.18
	cr/cr	18.49 ± 1.79	18.65 ± 1.62	5.89 ± 0.43	3.87 ± 0.14
Purified diet, copper-supplemented (500 mg Cu/kg)	+/?	16.54 ± 2.12	18.44 ± 0.59	5.71 ± 0.39	4.12 ± 0.60
	cr/cr	16.09 ± 1.09	16.79 ± 4.00	5.56 ± 0.10	3.49 ± 0.93

[a]Values are expressed as mean ± SEM with 4 animals per group; for experimental details see Theriault et al (1977)

TABLE 6

Haematological characteristics of 21-day old and one-year old *crinkled (cr/cr)* and non-crinkled *(+/?)* mice[a]

Genotype	Age	Plasma copper[b] (μg/dl)	PCV[c] (ml/100 ml)	Haemoglobin (g/dl)	RBC[c] ($10^6/\mu$l)	MCV[c] (μm^3)	MCH[c] (pg)
+/?	21 days	70.5 ± 2.0	43.4 ± 0.5	13.3 ± 0.4	8.3 ± 0.2	51.7 ± 1.2	15.5 ± 0.5
cr/cr		73.1 ± 2.5	30.3 ± 1.0[d]	9.6 ± 0.3[d]	7.4 ± 0.3[d]	39.8 ± 1.9[d]	13.1 ± 0.5[d]
+/?	1 year	78.2 ± 3.7	41.3 ± 0.9	13.1 ± 0.3	10.7 ± 0.4	38.6 ± 1.6	12.2 ± 0.3
cr/cr		77.1 ± 3.9	41.7 ± 1.4	13.1 ± 0.7	10.3 ± 0.6	40.2 ± 1.9	12.5 ± 0.7

[a]All values are expressed as mean ± SEM with 5 animals per group. All animals were fed purified diets containing 6 mg Cu/kg
[b]Copper concentration was determined by atomic absorption spectrophotometry after wet ashing
[c]PCV, packed cell volume; RBC, red blood cells; MCV, mean corpuscular volume; MCH, mean corpuscular haemoglobin
[d]Crinkled mutants are significantly different from non-crinkled littermates ($P < 0.01$) by Student's t-test

Similar differences between crinkled and non-crinkled mice, and between young and old mutants, were found in the copper concentration of the hair (Table 7). Hair copper concentration was lower in *cr/cr* mice than in littermate controls, even at 60 days of age, and was normal with copper supplementation, but by 12 months of age, when the hair copper concentrations of the controls had declined, there was no difference between mutants and non-mutants.

At 14 days of age, both hepatic copper concentration and Cu-ZnSOD activity were lower in *cr/cr* mice than in their non-mutant littermates. However, at 60 days of age, hepatic copper concentration was the same in both mutants and non-mutants, while activity of Cu-ZnSOD remained low in *cr/cr* mice, suggesting that the activity of this enzyme may be permanently affected by the early nutritional environment (Keen & Hurley 1979a). Supplementation during pregnancy and lactation with a high level of dietary copper brought both Cu-ZnSOD activity and copper concentration in the liver of mutant offspring to levels similar to those of non-mutant littermates (Keen & Hurley 1979a).

Copper concentration in the brain, on the other hand, did not differ between the crinkled and the non-crinkled mice, and liver and brain concentrations of zinc also showed no differences (Keen & Hurley 1979b). In addition, activities of hepatic manganese-SOD and glutathione peroxidase (EC

TABLE 7

Copper and zinc in hair of *crinkled (cr/cr)* and non-crinkled *(+/?)*[a,b] mice

Dietary group	Age	Hair copper concentration (+/?)	Hair copper concentration (cr/cr)	Hair zinc concentration (+/?)	Hair zinc concentration (cr/cr)
Stock diet, control (12 mg Cu/kg)	2 months	9.3 ± 1.4	5.0 ± 0.9[c]	135 ± 6	130 ± 7
Purified diet, control (6 mg Cu/kg)	2 months	11.2 ± 1.5	6.2 ± 1.0[c]	118 ± 20	125 ± 9
Purified diet, copper-supplemented (500 mg Cu/kg)	2 months	8.1 ± 1.8	10.7 ± 2.2	140 ± 23	135 ± 6
Stock diet, control (12 mg Cu/kg)	1 year	5.5 ± 2.0	6.7 ± 2.1	162 ± 10	147 ± 12

[a]Copper and zinc concentrations are expressed as µg/g dry weight. Values were obtained by atomic absorption spectrophotometry after wet ashing
[b]All values are the mean ± SEM with 4 animals per group
[c]Crinkled mutants are significantly different from non-crinkled littermates ($P < 0.01$) by Student's *t*-test

1.11.1.9) (both the selenium-dependent and independent forms of the enzyme) showed no differences between the two groups.

The changes with age in the differences between crinkled mutant mice and their littermate controls led to a study of the developmental pattern of copper concentration in various tissues. The abnormal copper metabolism resulting from the mutant gene *cr* thereby became evident. Although hepatic copper concentration in young mice varied considerably among litters, hepatic copper concentration *within* each litter from a heterozygous female was significantly lower in mutant than in non-mutant littermates, from day 18 of gestation to 20 days after birth. Crinkled mice older than 20 days of age had hepatic copper concentrations similar to those of their littermates (Fig. 1) (Keen & Hurley 1979b). On day 21, the hepatic copper concentration was even higher in mutants than in non-mutants. This rather abrupt change at 20 to 21 days of age in crinkled mice, coupled with the changes in copper concentration of the liver during the early postnatal period in normal mice (see Figure 1), suggested that at approximately this time a new mechanism developed for intestinal transport and absorption of copper. The report of Bronner et al (1978) is consistent with this hypothesis. These workers found that intestinal transport of copper was lower than normal in crinkled mice early in life, and that it then increased above normal about day 21. After this time, it was similar in mutants and non-mutants.

We tested in rats the hypothesis that a shift in the mechanism of intestinal copper transport occurred during the early postnatal period. We chose the rat instead of the mouse in order to have larger amounts of tissue. Previous studies had demonstrated that the developmental curves of copper concentra-

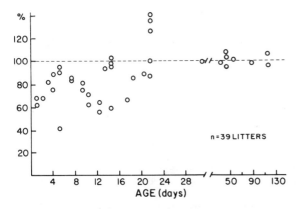

FIG. 1. Developmental pattern of hepatic copper (ordinate) in crinkled mice as a percentage of the concentration in control mice. Reprinted with permission from Keen & Hurley (1979b).

tion in rat liver (Fig. 2), brain, and kidney were similar to those of the mouse (Keen & Hurley 1980). Therefore the hypothesis was tested by study of the molecular distribution of copper in rat intestine. Gel filtration was used for separation of the copper-binding ligands; the gels were modified by reduction with sodium borohydride to remove surface charges and thereby to eliminate

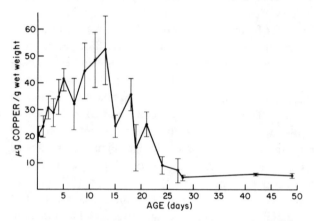

FIG. 2. Developmental pattern of hepatic copper concentration in the Sprague–Dawley rat. Each point represents the mean of measurements from 4 to 6 animals. A similar pattern is seen in the mouse. For experimental details see Keen & Hurley (1980).

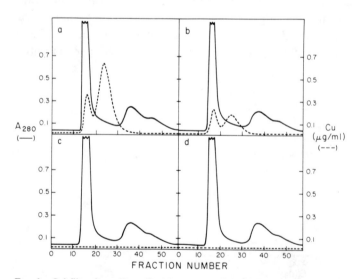

FIG. 3. Gel filtration of intestinal homogenates from (a) one-day, (b) 10-day, (c) 30-day and (d) 60-day old rats on Sephadex G-50 (1.6 × 90 cm) in 0.1 M ammonium acetate buffer (pH 6.5). Fraction volume: 4 ml. Sample volume: 2 ml. Flow rate: 60 ml/h. A_{280}, absorbance at 280 nm.

anomalous elution behaviour. Recovery of copper from the column under these conditions varied from 95–100% (Lönnerdal et al 1980). In the intestine of one- and 10-day old rats, copper was found in two peaks, one with a high molecular weight (greater than 30 000 daltons) and a second, containing most of the copper, with a low molecular weight (6000–8000) (Fig. 3). By 30 days of age, however (and also at 60 days of age), no copper peak was seen in the homogenates of intestine (Keen et al 1980a). These results support the hypothesis that there is a change in the mechanism or rate of intestinal absorption of copper during the period between 10 and 30 days of postnatal life in the rat and in the mouse. It should be emphasized that in these studies we have used conditions and amounts of copper-binding compounds that are physiological, and not the maximal amounts that can be demonstrated by the addition of copper to the system.

CONCLUDING REMARKS

The concentration of copper in the maternal diet is critical for normal development of the embryo and fetus. An inadequate amount of this element in either prenatal or neonatal life results in several developmental abnormalities. The requirement for copper is influenced by the genetic background; in some genetic disorders very high amounts may be needed in the prenatal or neonatal period.

A salient feature of copper metabolism in the fetus and neonate is the series of sizeable alterations that occur in the concentration of copper in various body tissues during the course of development. These changes may reflect developmental shifts in copper metabolism, and their significance may be elucidated by investigations of copper binding during development. Studies of normal animals, as well as the use of copper mutants as probes, may help to identify the biochemical parameters of genetic lesions and lead to better understanding of copper metabolism.

ACKNOWLEDGEMENTS

The authors wish to thank Donna D. Dungan for technical assistance.
This investigation was supported by NIH Research Grants No. HD-02355 and HD-01743 from the National Institute of Child Health and Human Development and by National Research Service Award DE-07001 from the National Institute of Dental Research.

References

Baumann NA, Jacque CM, Pollet SA, Harpin ML 1968 Fatty acid and lipid composition of the brain of a myelin deficient mutant, the 'quaking' mouse. Eur J Biochem 5:340-344

Bennetts HW, Beck AB 1942 Enzootic ataxia and copper deficiency of sheep in Western Australia. Aust CSIRO Bull 147:1-52

Bennetts HW, Chapman FE 1937 Copper deficiency in sheep in Western Australia: A preliminary account of the aetiology of enzootic ataxia of ewes. Aust Vet J 13:138-149

Bronner F, Golub EE, Veng TH, Bossak C 1978 Intestinal copper transport and copper-binding protein (CuBP) in crinkled (cr/cr) mice. Fed Proc 3:721

Falconer DS, Fraser AS, King JWB 1952 The genetics and development of 'crinkled', a new mutant in the house mouse. J Genet 50:324-346

Hart EB, Steenbock H, Waddell J, Elvehjem CA 1928 Iron in nutrition. VII: Copper as a supplement to iron for hemoglobin building in the rat. J Biol Chem 77:797-812

Hurley LS, Bell LT 1975 Amelioration by copper supplementation of mutant gene effects in the crinkled mouse. Proc Soc Exp Biol Med 149:830-834

Hurley LS, Keen CL 1979 Teratogenic effects of copper. Part II. In: Nriagu J (ed) Copper in the environment. John Wiley, New York, p 35-56

Keen CL, Hurley LS 1976 Copper supplementation in quaking mutant mice: Reduced tremors and increased brain copper. Science (Wash DC) 193:244-246

Keen CL, Hurley LS 1979a Superoxide dismutase activity in the crinkled mutant mouse: Ameliorative effects of dietary copper supplementation. Proc Soc Exp Biol Med 162:152-156

Keen CL, Hurley LS 1979b Developmental patterns of copper and zinc concentrations in mouse liver and brain: Evidence that the gene *crinkled (cr)* is associated with an abnormality in copper metabolism. J Inorg Biochem 11:269-277

Keen CL, Hurley LS 1980 Developmental changes in concentrations of iron, copper, and zinc in mouse tissues. Mech Ageing Dev 13:161-177

Keen CL, Lönnerdal B, Hurley LS 1980a Developmental changes in the molecular distribution of copper and zinc in rat tissues. Fed Proc 39:1749

Keen CL, Saltman P, Hurley LS 1980b Copper nitrilotriacetate: a potent therapeutic agent in the treatment of a genetic disorder of copper metabolism. Am J Clin Nutr 33:1789-1801

Lönnerdal B, Clegg M, Keen CL, Hurley LS 1980 Effects of wet ashing techniques on the determination of trace element concentrations in biological samples. In: Schramel P (ed) Trace element analytical chemistry in medicine and biology. Walter de Gruyter & Co, Berlin (Proc Int Workshop, Neuherberg, April 27-29 1980), in press

Morell P, Bornstein MB, Norton WT 1972 Diseases of myelin. In: Albers RW et al (eds) Basic neurochemistry. Little, Brown and Co, Boston, p 497-515

O'Dell BL, Hardwick BC, Reynolds G 1961 Mineral deficiencies of milk and congenital malformations in rats. J Nutr 73:151-157

Sidman RL, Dickie MM, Appel SH 1964 Mutant mice (quaking and jimpy) with deficient myelination in the central nervous system. Science (Wash DC) 144:309-311

Theriault LL, Dungan DD, Simons S, Keen CL, Hurley LS 1977 Lipid and myelin abnormalities of brain in the crinkled mouse. Proc Soc Exp Biol Med 155:549-553

Underwood EJ 1977 Trace elements in human and animal nutrition, 4th edn. Academic Press, New York (see Chapter 3, p 56-108)

Wiener G 1966 Genetic and other factors in the occurrence of swayback in sheep. J Comp Pathol 76:435-447

Wiener G, Field AC 1969 Copper concentrations in the liver and blood of sheep of different breeds in relationship to swayback history. J Comp Pathol 79:7-14

Wiener G, Wilmut I, Field AC 1978 Maternal and lamb breed interactions in the concentration of copper in tissues and plasma of sheep. In: Kirchgessner M (ed) Trace element metabolism in man and animals. Arbeitskreis für Tierernährungsforschung, Freising-Weihenstephan, vol 3, p 469-472

Discussion

Lewis: At what stage did you wean the group of rats in which you studied copper binding, Professor Hurley, or were they on a milk diet all the way through the tests?

Hurley: The rats in which we studied copper binding were weaned at 21 days onto a purified complete diet, but rats start nibbling their mother's food at about 14 or 15 days of age so we don't know whether the changes in copper binding in the liver were induced by the solid food or by another factor. Neither do we know whether these differences in copper binding between days 1 and 10 simply represent the lower amount of copper in the mother's milk: on day 1 the copper concentration in the milk was quite high; on day 10 it was significantly lower (Keen et al 1980b).

Bremner: There must have been tremendous differences in copper concentrations in the gut mucosa of the rats in the early stages after birth. For example, you showed, Professor Hurley, that there was increased binding of copper to specific fractions in the mucosa during days 1–5 but no detectable copper in the later samples. What were the total copper concentrations in the mucosa?

Hurley: We have not yet measured that.

Bremner: Mistilis & Mearrick (1969) showed in their studies on copper absorption in the neonatal rat that the intestinal mucosa in the young animal had a tremendously high affinity for copper. This seems to be a general feature in suckling animals, because similar observations have been made with other metals.

Harris: Evans et al (1970) have pointed out that the rat, instead of having a high concentration of copper in the liver at birth, appears to accumulate hepatic copper gradually over the first 10 days or so of its life. This accumulation occurs before there are maximum caeruloplasmin concentrations in the serum. Could this explain your results, Professor Hurley?

Hurley: We have observed that the concentrations of copper in the liver of both rats and mice increased during the first few days after birth, but the concentrations at birth in both species are higher than those in the adult. At present we have results on copper binding in the intestine and liver of rats for days 1, 10, and 30; we do not yet know exactly when the change between day 10 and day 30 occurs (Fig. 3).

Harris: Some people believe that caeruloplasmin is responsible for removing copper from the liver. Do caeruloplasmin concentrations in the plasma of these animals increase between day 1 and day 10?

Hurley: We haven't measured that.

Terlecki: It is very difficult to compare as you did, Professor Hurley, the abnormalities of the central nervous system in mice and sheep, because sheep are born with a mature central nervous system whereas in mice the CNS is still at a developing stage. In contrast to the enzootic ataxia (swayback) observed in sheep in Australia (Bennetts 1932), we do not observe a small cerebellum in lambs with swayback in Britain. In quaking mice there is a myelin deficiency. Does copper deficiency act directly on the synthesis of myelin lipids, or indirectly on the metabolism of the oligodendroglial cells that are responsible for myelin formation?

Hurley: I don't know the answer to that but I should think that both could occur.

The differences observed between sheep and mice could indeed arise because the animals develop at different times. The prenatal and postnatal manifestations of deficiency are more similar for copper than for deficiencies of many other nutrients. The various conflicting observations reported on sheep with swayback illustrate that copper deficiency has many different forms even within the same species. It is easier to explain different effects of copper deficiencies when they are due to mutations, but different effects in non-mutant animals are not so easy to explain.

Lewis: The sheep is also different from many species (including the mouse) because the lamb is usually born with a lower copper status than that of its dam, whereas in most other species, including the pig and the calf, the young are born with a higher copper status. In fact, pigs behave like your mice, Professor Hurley – they are born with a high copper status that falls over the first few months of life.

Could I now ask Dr Shaw a question? When you found low levels of copper in the neonatal children that you mentioned in our earlier general discussion (p 160), did you measure plasma copper or caeruloplasmin in their mothers?

Shaw: There is no evidence that the children we examined had deficiencies at birth. These infants simply failed to absorb copper adequately after birth.

Danks: I was delighted when more mutants seemed to be identified as responsible for copper deficiency in mice because one day it might be possible to map the process of copper transport by the same techniques that are used in bacterial metabolism, i.e. by sequential addition of the effects of a number of mutants. We have spent about 12 months studying crinkled mice but we obtained different results from those discussed by Professor Hurley (J. Mann, J. Camakaris, P.M. Royce, J.M. Matthieu, J.M. Gillespie, D.M. Danks, unpublished work). When we studied the survival of crinkled mice up to 60 days after birth we found that there was poor survival among the crinkled mice, as

Lucille Hurley also found. However, one of the notable differences is that the normal littermates thrive, in our hands, whereas about 15–20% of the normal littermates in Professor Hurley's experiments fail to survive to 60 days. In our work, copper supplements to the mothers do not change the survival of crinkled mice or of normal littermates, whereas Professor Hurley found that both crinkled mice and normal littermates survived better when the mother had been given copper supplements before the birth. We measured copper and caeruloplasmin concentrations at 4, 5 and 16 days of age and found no deficiency in any of the organs that we examined. We also analysed both copper levels by litter pairs and unlike Professor Hurley we found that the levels in crinkled mice were not consistently higher or lower than in the normal littermates. In the skin of crinkled mice we found no histochemical alteration in the staining properties of collagen or elastin. We did, of course, observe the deficiency of hair follicles that Falconer's group (1952) showed originally. They described a rather specific absence of follicles for certain types of hair in certain regions, which is quite different to the abnormality in the brindled and blotchy mouse. We found an increased hydroxyproline content and a decreased fat content in the skin of the crinkled mouse. If hydroxyproline is expressed as a fraction of the total weight of skin, there is an increase, in crinkled mice compared with normals, but if the fat content in the skin is extracted completely, there is no difference. We have measured lysyl oxidase activity against both collagen and elastin substrates in the whole skin and also in cultured lung fibroblasts and it was normal in both circumstances. This result was in contrast to what is found in the brindled mutants.

We found that skin rather than hair pigmentations were different in crinkled mice, compared with normals. Although in blotchy and brindled mice replacement of copper produces rapid hair darkening, we did not observe this effect either after postnatal treatment or after intrauterine treatment of the pregnant mother.

We have found a massive increase in free sulphydryl content of hair keratin in brindled mice (J.M. Gillespie, J. Mann, D.M. Danks, unpublished work), and of course this occurs also in Menkes' syndrome in humans (Danks et al 1972). However, we could not find any change in the disulphide content of the keratin from crinkled mutants. We therefore doubt that the *cr/cr* mouse is of interest to those working on copper metabolism.

In mottled (blotchy or brindled) mutant mice, who have severe copper disturbance, there are no defects in myelination (D.M. Danks, J.M. Matthieu, B. Koellreutter, J. Mann, unpublished work). Quaking mutant mice have a very specific type of myelin disturbance quite unlike anything that has been described in copper deficiency. We found normal levels of copper in the brain of the quaking mouse.

Mills: Are you sure, Dr Danks, that you are studying the same defect as Lucille Hurley?

Danks: Well, the crinkled mice that we used came from the same source that Professor Hurley used. Her descriptions of the syndrome, and those of Falconer et al (1952) are exactly like the syndrome in our mice. I suppose it is possible that there is a difference in the way that we handle our mice, which modifies the copper levels, but nothing is altered about the clinical syndrome in our mice.

Hurley: I cannot explain the difference between our results and those of David Danks, although we have corresponded about this already. We have recently repeated the supplementation experiments, this time with copper-NTA, which we found to be even more effective than copper sulphate in increasing survival of the crinkled mutants (Keen et al 1980a). Dr Danks administered copper to the mice in their drinking water whereas we gave it in the diet; I do not know whether that might make a difference to the results that we each obtained. All our tissue analyses for copper content are done after perfusion of the animals with saline so that there is no blood left in the tissues. Since copper content in the tissues is low anyway, failure to perfuse could leave some blood in the tissues which would lead to artifactual values of copper in Dr Danks's experiments. Felix Bronner found that crinkled mice have a reduced ability to absorb copper before, but not after, they reach 21 days of age (Bronner et al 1978). That result is compatible with our findings. We have also found that some of the abnormal types of hair present in Menkes' disease also occur in the crinkled mice (Hurley & Bell 1975). I understand that some of these abnormal hairs do not occur in any condition other than in Menkes' syndrome. In work on the quaking mutant, Suzuki & Zagoren (1978) have reported that the nodes of Ranvier in the peripheral nervous system of this mutant bind less copper than those in littermate controls.

Danks: The morphological changes in hair are quite different in crinkled mice on the one hand and in brindled mice or in Menkes' syndrome on the other. The effect of alterations in sulphydryl content are striking in Menkes' syndrome and brindled or blotchy mice, but are absent in crinkled mice which suggests to me that the basic lesion in hair is quite different.

Hurley: I am not sure that those changes would be absent in *my* crinkled mice! Perhaps we should exchange mice and then see what results are produced!

Mills: Professor Hurley, you referred earlier to a defect in copper absorption in crinkled mice up to 21 days of age (Bronner et al 1978). In your studies of copper in intestinal mucosa you used Sephadex G-50 to fractionate cytosol and were thus looking at the transport of soluble low-molecular-weight species. Mistilis & Mearrick (1969) have suggested that copper absorption in the early postnatal stages takes place by a pinocytic process that is strongly inhibited by cortisone. Supporting evidence for this is now available for a number of transition elements. Presumably, in this instance, copper would be associated with a macromolecular species during its absorption. Could a defect in pinocytic absorption account for some of your findings?

Hurley: This is certainly one possibility. However, the work that Bronner et al (1978) did was with isolated cells.

Terlecki: There seems to be a similarity between quaking mice and Border disease (a viral disease of sheep) in terms of pathological changes in myelin. The changes are similar morphologically and neurochemically (Barlow & Storey 1977). Border disease is characterized by defective quantities of myelin and by the composition of various lipid fractions of the myelin. It seems, therefore, that myelin defects in quaking mice are rather similar to those in Border disease of sheep and *not* to those in swayback (enzootic ataxia).

McMurray: Professor Hurley, you mentioned reversibility of pathological changes during copper deficiency. O'Dell et al (1978) found that the changes induced in rat lung were not fully reversible. In fact all lesions are not necessarily reversible.

Hurley: I referred not to the reversibility but to the prevention of the symptoms, at least in the lung. In addition, enzootic ataxia can be *prevented* in lambs by administration of copper to ewes during pregnancy, but that effect is not the same as reversibility.

Mills: I would like to return briefly to the effects of copper in non-mutant animals, and particularly in rats and mice. N.T. Davis & B.F. Fell (unpublished results) at Aberdeen have been studying copper deficiency during reproduction in the rat. They found a high rate of fetal deaths and resorptions, but sometimes only one fetus in a litter was affected, while the neighbouring littermates in the same horn of the uterus developed normally. Is this found also in mice?

Hurley: Well we haven't worked with copper deficiency *per se* so I cannot answer that particular question. However, the same phenomenon occurs also in other deficiencies. There are local differences in blood supply and in the placenta, which might be relevant.

Danks: Among the normal littermates of blotchy and brindled mice, we find that the postnatal rise in the liver content of copper is several-fold greater and more prolonged than the 50% rise found by Professor Hurley.

Hunt: We also found a higher postnatal rise of copper in the liver than did Professor Hurley, but the rise was not particularly prolonged and showed a dramatic fall between three and four weeks of age (Hunt & Port 1979). Such developmental differences in tissue copper levels could of course be due to the use of different strains of mice.

Riordan: What is the explanation for the rise in liver copper after birth — is it because of redistribution, or is copper taken up from the milk?

Danks: We fed copper isotopes to the mother and measured the uptake of copper from her milk. We found that the extra copper does not come from redistribution but from absorption. Much of this absorption of copper occurs through pinocytosis in the lower ileum.

Mangan: Professor Hurley, was there any copper in the mucosa of the rats used in the study of copper binding?

Hurley: There was no detectable bound copper found in the rat intestine after 10 days of age.

Danks: Surely this is related to Ian Bremner's discussion about absorption (pp 23-36)? If there is a deficiency in a process that takes copper through the mucosa into the bloodstream, young mice will store the copper in some form in the mucosa. If the process of transporting copper right through into the bloodstream later becomes available, or 'matures', then one would not expect to find high concentrations of stored copper in the mucosa.

Hurley: The *rate* of absorption might be the most important factor in determining the amount present in the intestine. All we can do is to describe the changes that occur. We may then begin to understand more about the way an animal handles copper during this period of development.

Harris: Dr Danks, I think that your idea (p 240) of developing a whole series of mutant animals to aid in identification of the pathways for copper in metabolism is useful. You mentioned a similar approach in microorganisms but of course that system is probably much less complicated. It seems likely that the various organs have their own intrinsic ways of handling copper and we cannot assume that metabolism of copper in the liver, for example, is identical to that in the kidney. If you are going to develop a series of mutations the number required would surely be enormous?

Danks: The system that I envisage would simply help to distinguish between some of the component systems of copper transport. Obviously one would need to return to whole animal experiments in order to study the interplay between those components.

References

Barlow RM, Storey IJ 1977 Myelination of the ovine CNS with special reference to Border disease. II: Quantitative aspects. Neuropathol Appl Neurobiol 3:255-265

Bennetts HW 1932 Enzootic ataxia of lambs in Western Australia. Aust Vet J 8:137-142; and 8:183-184 (correction)

Bronner F, Golub EE, Veng TH, Bossak C 1978 Intestinal copper transport and copper-binding protein (CuBP) in crinkled (cr/cr) mice. Fed Proc 3:721

Danks DM, Stevens BJ, Campbell PE, Gillespie JM, Walker-Smith J, Blomfield J, Turner B 1972 Menkes' kinky-hair syndrome. Lancet 1:1100-1102

Evans GW, Myron DR, Cornatzer NF, Cornatzer WE 1970 Age-dependent alterations in hepatic subcellular copper distribution and plasma caeruloplasmin. Am J Physiol 218:298-300

Falconer DS, Fraser AS, King JWB 1952 The genetics and development of 'crinkled', a new mutant in the house mouse. J Genet 50:324-346

Hunt DM, Port AE 1979 Trace element binding in the copper deficient mottled mutants in the mouse. Life Sci 24:1453-1466

Hurley LS, Bell LT 1975 Amelioration by copper supplementation of mutant gene effects in the crinkled mouse. Proc Soc Exp Biol Med 149:830-834

Keen CL, Saltman P, Hurley LS 1980a Copper nitrilotriacetate: a potent therapeutic agent in the treatment of a genetic disorder of copper metabolism. Am J Clin Nutr, 33:1789-1801

Keen CL, Lönnerdal B, Clegg M, Hurley LS 1980b Developmental changes in nutrient composition of rat milk. Fed Proc 39:903

Mistilis SP, Mearrick PT 1969 The absorption of ionic, biliary and plasma radiocopper in neonatal rats. Scand J Gastroenterol 4:691-696

O'Dell BL, Kilburn KH, McKenzie N, Thurston RJ 1978 The lung of the copper deficient rat. Am J Pathol 91:413-430

Suzuki I, Zagoren JC 1978 Studies on the copper binding affinity of fibers in the peripheral nervous system of quaking mouse. Neurosciences 3:447-455

Copper and neurological function

D.M. HUNT

School of Biological Sciences, Queen Mary College, University of London, Mile End Road, London E1 4NS, UK

Abstract The role of copper in maintaining normal neurological function has been examined in animals made copper-deficient by dietary means, and in the genetic disorders of copper homeostasis — Menkes' kinky-hair disease in humans and the mottled (*Mo*) mutants in the mouse. With the exception of the disorder in *Mo* mice, reduced myelination is a constant feature of these copper diseases but there is otherwise a lack of conformity in the structural defects produced in different species. Dietary copper-deficient animals show a reduction in noradrenaline and dopamine concentrations, together with a depressed tyrosine 3-monooxygenase activity (EC 1.14.16.2). Noradrenaline concentrations are also reduced in brain tissue of *Mo* mice and this reduction is associated with a decrease in the *in vivo* activity of the copper metalloenzyme, dopamine β-monooxygenase (EC 1.14.17.1). Many tissues contain potent inhibitors of dopamine β-monooxygenase activity, and assays of this enzyme have utilized cupric ions to inactivate these inhibitors. The elevated *in vitro* activities of dopamine β-monooxygenase obtained for both *Mo* brain and adrenal tissue may therefore reflect either a reduced inactivation of these endogenous inhibitors in the intact animal or the activation *in vitro* of apoenzyme. Concentrations of dopamine and tyrosine 3-monooxygenase are unchanged in *Mo* mice. The reduction in dopamine and tyrosine 3-monooxygenase activity in dietary copper-deficient animals may therefore reflect neuronal loss rather than reduced catalytic activity of the catecholamine biosynthetic pathway. The possible effects of depressed activities of cytochrome *c* oxidase (EC 1.9.3.1) and superoxide dismutase (EC 1.15.1.1) in the development of neurological dysfunction are also discussed, and attention is drawn to the possible significance of the elevated uptake of neutral amino acids, especially tyrosine and tryptophan, by *Mo* brain tissue.

Neurological dysfunction is invariably apparent in copper-deficient mammals. The dietary copper-deficiency disease of swayback or enzootic ataxia in lambs is associated with incoordination of the hind quarters; ataxia, tremor and a swaying gait have also been reported in dietary copper-deficient guinea pigs (Everson et al 1968). In rats born from copper-deficient mothers and maintained on a copper-deficient diet thereafter, behavioural changes may include spasticity, hyperventilation, anorexia, seizures, tremor, and priapism in males (Carlton & Kelly 1969, Prohaska & Wells 1974, Zimmerman et al 1976).

STRUCTURAL CHANGES IN THE CNS

Brain growth is reduced in dietary copper-deficient animals and extensive tissue necrosis may be present. In guinea pigs, cerebellar folia may be missing or malformed (Everson et al 1968). In the rat degeneration of the olfactory lobes occurs, and necrotic lesions have been reported in the cerebral cortex (Morgan & O'Dell 1977), in the corpus striatum and in the thalamic nuclei of the diencephalon (Carlton & Kelly 1969). There is also evidence of demyelination in swayback lambs and of reduced myelination in dietary copper-deficient rats (Prohaska & Wells 1974). However, postnatal copper replacement by fostering of rat pups born to copper-deficient mothers is sufficient to reverse the effect on myelination but is unable to overcome the effect on brain growth or on behaviour (Zimmerman et al 1976).

A defect in copper metabolism has been established by Danks et al (1972) in the human X-linked genetic disorder, Menkes' kinky-hair disease. Affected male infants suffer convulsions and spasticity, and widespread neuronal loss throughout the CNS, although the cortex of the cerebellum is the most severely affected area (Ghatak et al 1972). Extensive demyelination is indicated by high tissue concentrations of cholesterol esters (Lou et al 1974). O'Brien & Sampson (1966) also reported a reduction in the unsaturated fatty acid content of the phospholipid fractions, which suggests that an increase in lipid peroxidation occurs, possibly as a result of the reduced activity of soluble superoxide dismutase (EC 1.15.1.1), a copper–zinc-metalloenzyme.

Although the activity of this enzyme is reduced in brain tissue of dietary-deficient rats (Morgan & O'Dell 1977, Prohaska & Wells 1974), an increased level of peroxidation in tissue homogenates was not found (Prohaska & Wells 1975), and the presence of increased levels of unsaturated fatty acids was not confirmed in Menkes' disease by Lou et al (1974).

The mottled *(Mo)* mutants in the mouse have a disorder homologous to Menkes' disease in humans (Hunt 1974). Ataxia, mild tremor, occasional seizures and, in older males, priapism are apparent in mutant animals, but severe degenerative changes are restricted to the cerebral cortex and to the thalamic nuclei and there is no evidence for hypomyelination or demyelination (Yajima & Suzuki 1979). The absence of myelin defects in *Mo* mice cannot be attributed to less reduction in brain content of copper since the deficiency in *Mo* mice is similar to that in the dietary copper-deficient rats described by Zimmerman et al (1976), in which hypomyelination was evident. Regional differences in the severity of the copper deficiency can also be ruled out by the data presented in Table 1, which shows approximately equal reductions in the copper content of four brain regions of *Mo* mice. *Mo* mice are produced from *Mo* heterozygous mothers maintained on a normal diet and, with the exception of accumulation in the kidney, the copper status of such adult animals is relatively normal (Hunt & Port 1979). Abnormal whisker development at birth indicates that *Mo* mice are already copper-deficient; the copper status of pre- and postnatal *Mo* mice may therefore be similar to that of dietary copper-deficient rats produced from deficient mothers. A species difference in the sensitivity of the developing brain to copper deficiency in rats and mice is therefore indicated and, at least in *Mo* mice, the behavioural changes must be attributed to neuronal loss rather than to hypomyelination.

The neuronal necrosis in *Mo* mice is associated with enlarged mitochondria that have prominent vesicular or tubular cristae (Yajima & Suzuki 1979). Mitochondrial degeneration may therefore precede neuronal degeneration in this disorder. Mitochondrial dysfunction may also play a significant part in neuronal degeneration in other forms of copper deficiency. Enlarged mitochondria have been reported in the brain tissue of infants with Menkes' disease (Ghatak et al 1972), in dietary copper-deficient rats (Prohaska & Wells 1975) and in swayback lambs (Cancilla & Barlow 1966), and the activity of cytochrome *c* oxidase (EC 1.9.3.1), a mitochondrial copper-metalloenzyme, is considerably reduced in copper-deficient animals (Hunt 1977, Morgan & O'Dell 1977, Prohaska & Wells 1975), although the reduction is perhaps less than that required to produce general respiratory failure (Gallagher et al 1956).

TABLE 1

Copper content of brain tissue from seven-day-old $Mo^{br}/-$ male mice and from normal (+) littermates.

	Normal littermates (+) (7-days) (n = 3)	Brindled littermates ($Mo^{br}/-$) (7 days) (n = 3)	% Difference from normal[a]
Cerebrum	1.22 ± 0.21	0.37 ± 0.04	−72.2
Cerebellum	1.41 ± 0.39	0.42 ± 0.01	−70.2
Anterior brain stem	1.38 ± 0.03	0.42 ± 0.04	−69.6
Posterior brain stem	2.25 ± 0.10	0.57 ± 0.11	−74.7

Results are expressed in µg/g wet weight ± SEM
[a]Significant at the 1% probability level

CATECHOLAMINE BIOSYNTHESIS AND FUNCTION IN COPPER DEFICIENCY

The catecholamines noradrenaline, adrenaline and dopamine are 3,4-dihydroxy derivatives of phenylethylamine. Catecholamines are found in the CNS, in the peripheral sympathetic nervous system, and in the adrenal medulla. In the CNS, noradrenaline (NA) and dopamine (DA) act as synaptic neurotransmitters.

In Parkinson's disease in humans, motor disorders of an extrapyramidal origin are apparent and the disease is characterized by akinesis, rigidity, tremor and reduced striatal concentrations of DA and reduced tyrosine 3-monooxygenase activity. Some related behavioural changes in *Mo* mice first led us to examine the central concentrations of DA and NA in these mice (Hunt & Johnson 1972a). Three alleles at the *Mo* locus have been extensively investigated: the lethal brindled (*Mo*^{br}) allele, which causes death in mutant males at about 14 days after birth; the viable brindled (*Mo*^{vbr}) allele, which is semi-viable; and the blotchy (*Mo*^{blo}) allele, which is fully viable and fertile (Hunt 1974). As shown in Fig. 1 the NA levels are depressed in mutant mice throughout development and the reduction is generally in excess of that reported in dietary copper-deficient animals (Morgan & O'Dell 1977, Prohaska & Wells 1974). A study of regional NA levels in adult *Mo*^{vbr} mice (Table 2) shows that the reduction in NA is not limited to a particular region and is consistent therefore with the equal reduction in distribution of copper in all regions of the brain in neonatal *Mo*^{br} mice (Table 1).

In contrast to results from dietary copper-deficient rats (Morgan & O'Dell 1977) and lambs (O'Dell et al 1976), DA levels are not reduced in 7- and 13-day-old *Mo*^{br} mice (Hunt & Johnson 1972a) and a study of regional DA levels in adult *Mo*^{vbr} mice (Table 2) confirms that DA levels are not affected in this disorder. The reduced activity of tyrosine 3-monooxygenase (EC 1.14.16.2) in dietary copper-deficient animals (Morgan & O'Dell 1977) is considered to be a major reason for the reduced levels of catecholamines in copper-deficient brain tissue, especially since tyrosine 3-monooxygenase is generally believed to be the rate-limiting enzyme in the catecholamine pathway (Levitt et al 1965). The rate-limiting role of tyrosine 3-monooxygenase has been questioned by Molinoff & Orcutt (1973) who point out that since dopamine β-monooxygenase (EC 1.14.17.1) is sequestered in the chromaffin granules and adrenergic vesicles, the enzymes and substrates of the pathway do not have free access to each other. A direct role for copper deficiency in the reduction of tyrosine 3-monooxygenase activity is also made doubtful by the action of diethyldithiocarbamate, a powerful copper chelator, which reduces NA levels but slightly increases DA levels in rat brain (Magos & Jarvis 1970).

COPPER AND NEUROLOGICAL FUNCTION 251

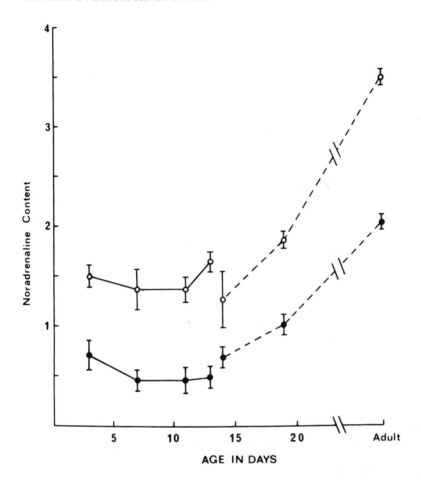

FIG. 1. Brain noradrenaline levels (nmoles/g) in *Mo* and normal mice. Normal (○); *Mo* (●); solid line *Mo^{br}* and normal littermates; broken line *Mo^{vbr}* and normal littermates; vertical lines indicate SEM.

There is little doubt that brain monoamine oxidase – MAO (EC 1.4.3.4) or amine oxidase (flavin-containing) – which is the major catecholamine degradative enzyme, is not copper-dependent (Nagatsu et al 1972). In *Mo* mice, the activity of this enzyme is entirely normal, as is the activity of catechol methyltransferase (EC 2.1.1.6, and see Table 3). Dopamine β-monooxygenase is therefore the only copper-metalloenzyme in the catecholamine biosynthetic pathway (Skotland & Ljones 1979). Assay of the activity of this enzyme *in vivo* (Hunt & Johnson 1972b) showed that 60 minutes after an intraperitoneal injection of [^{3}H]tyrosine, the quantity of

TABLE 2

Noradrenaline and dopamine concentrations in brain tissue of adult $Mo^{vbr}/-$ mice and normal (+) littermates.

	Noradrenaline nmoles/g ± SEM	% Difference from normal[a]	Dopamine nmoles/g ± SEM	% Difference from normal
Anterior brain stem				
(+) adult ($n = 3$)	6.55 ± 0.81	−53.3	10.29 ± 0.92	−9.2
($Mo^{vbr}/-$) adult ($n = 3$)	3.06 ± 0.18		9.34 ± 1.01	
Remainder[b]				
(+) adult ($n = 3$)	2.19 ± 0.11	−36.1	3.28 ± 0.23	+5.2
($Mo^{vbr}/-$) adult ($n = 3$)	1.40 ± 0.13		3.45 ± 0.47	

[a]Significant at 1% probability level.
[b]Remainder includes cerebrum, cerebellum and posterior brain stem.
Noradrenaline and dopamine were assayed by the method of Anton & Sayre (1962), as modified by Hunt & Johnson (1972b).

TABLE 3

Brain monoamine oxidase (MAO) and catechol methyl transferase (CMT) in neonatal $Mo^{br}/-$ male mice and normal (+) littermates.

	Normal littermates (neonates) (+)	Brindled littermates (neonates) ($Mo^{br}/-$)
MAO (μmoles g^{-1} hr^{-1} ± SEM)	3.04 ± 0.09 (3)	3.10 ± 0.08 (3)
CMT (μmoles g^{-1} hr^{-1} ± SEM)	212.0 ± 10.2 (3)	225.9 ± 4.4 (3)

MAO was assayed as described in Hunt & Johnson (1972b); CMT was assayed by the method of Jarrott (1971).
Numbers in parentheses refer to numbers of animals used.

[^{3}H]NA in the brains of *Mo* mice was severely reduced, while the quantity of [^{3}H]DA was significantly increased (Table 4). Therefore the effect of copper deficiency on the central concentrations of NA can be adequately accounted for by the reduced activity of dopamine β-monooxygenase and it is significant that the intraperitoneal injection of copper chloride into six-day-old Mo^{br} mice effectively eliminates the reduction in brain NA (Hunt 1976). An understanding of the precise mechanism of action of copper deficiency on enzyme function is complicated by the presence of potent low-molecular-weight inhibitors of dopamine β-monooxygenase (Molinoff & Orcutt 1973). These inhibitors are present in many tissues, including the brain and the adrenal medulla, and they can be inactivated *in vitro* either by cupric ions,

TABLE 4

In vivo and *in vitro* activities of dopamine β-monooxygenase in brain tissue from *Mo* male mice and normal (+) littermates.

	Age (days)	Normal mice (+)	Brindled mice ($Mo^{br}/-$)	% Difference from normal[a]
In vivo activity of dopamine β-monooxygenase (μCi/g ± SEM)	7	[³H]DA 0.511 ± 0.067 [³H]NA 0.374 ± 0.031	0.902 ± 0.045 0.047 ± 0.006	+76.4 −87.6
		(+)	($Mo^{br}/-$)	
In vitro activity of dopamine β-monooxygenase (nmoles g⁻¹ h⁻¹ ± SEM)	7	33.14 ± 2.39(13)	48.81 ± 2.92(12)	+47.3
		(+)	Blotchy mice ($Mo^{blo}/-$)	
	28	33.02 ± 3.20(6)	77.63 ± 5.26(6)	+135.1

The *in vivo* activity is estimated by the synthesis of [³H]noradrenaline and [³H]dopamine 60 min after an i.p. injection of 50 μCi [³H]tyrosine. *In vitro* activity was estimated as previously described in Hunt (1974).
[a]Significant at the 1% probability level.
Numbers in parentheses refer to numbers of animals used.

N-ethylmaleimide or by *p*-hydroxymercuribenzoate, which suggests that the inhibitors may be sulphydryl compounds that act by copper chelation (Skotland & Ljones 1979). The *in vitro* activity of dopamine β-monooxygenase was assayed in brain tissue from *Mo* and normal mice by the addition of copper sulphate to inactivate the inhibitors (Hunt 1974). Maximal activity was achieved in extracts of both normal and mutant tissue at a concentration of 3.25×10^{-4}M CuSO$_4$. At seven days postnatally, for Mo^{br} mice, and at 28 days, for Mo^{blo}, the activity of dopamine β-monooxygenase in mutant tissue *in vitro* significantly increased (Table 4). This raises a question about whether the reduced activity *in vivo* depends primarily on the presence of an inactive apoenzyme or on the non-inactivation of inhibitors.

A similar discrepancy between the *in vivo* and *in vitro* activities of dopamine β-monooxygenase is apparent in *Mo* adrenals (Hunt 1977). Adrenal tissue from 13-day-old Mo^{br} mice shows a reduced catecholamine content with a normal *in vitro* activity of dopamine β-monooxygenase, whereas in 28-day-old Mo^{blo} mice, the activity of the enzyme *in vitro* is increased, but this is not reflected in an altered catecholamine content (Table 5). Also, as found for the increased dopamine β-monooxygenase in the brain, this increase in activity is independent of the cupric ions required for inhibitor

TABLE 5

Concentrations of catecholamines, catecholamine biosynthetic enzymes and copper in *Mo* and normal adrenals.

	Age in days	Normal mice	Mutant mice	% Difference from normal
Catecholamines (nmoles/adrenal ± SEM)	13	+ 3.57 ± 0.17(5)	$Mo^{br}/-$ 2.07 ± 0.21(5)	-42.0^b
	28	+ 23.4 ± 3.1(4)	$Mo^{blo}/-$ 25.5 ± 2.0(4)	+9.0
Dopamine β-monooxygenase (nmoles/adrenal pair/h ± SEM)	9	+ 1.31 ± 0.21(9)	$Mo^{br}/-$ 1.29 ± 0.31(9)	−1.5
	28	+ 4.62 ± 1.02(5)	$Mo^{blo}/-$ 9.26 ± 1.74(5)	$+100.6^b$
Phenylethanolamine-N-methyltransferase (μmoles/adrenal pair/h ± SEM)	9	+ 39.7 ± 5.9(8)	$Mo^{br}/-$ 21.0 ± 5.3(8)	-47.1^a
Tyrosine 3-monooxygenase (nmoles/adrenal pair/h ± SEM)	13	+ 6.61 ± 0.36(13)	$Mo^{br}/-$ 5.77 ± 0.58(11)	−12.7
Amine oxidase (flavin-containing (MAO) (nmoles/adrenal pair/h ± SEM)	9	+ 5.14 ± 1.17(11)	$Mo^{br}/-$ 2.44 ± 0.34(12)	-52.9^a
Copper content (μg/g wet weight ± SEM)	9	+ 27.2 ± 10.3(4)	$Mo^{br}/-$ 19.4 ± 4.9(4)	−28.7
	Adult (i.e. >42)	+ 2.49 ± 0.43(3)	$Mo^{vbr}/-$ 2.31 ± 0.17(3)	−7.2

For details of analytical methods, see Hunt 1977.
Significant at the [a]5% and [b]1% probability levels.
Numbers in parentheses refer to numbers of animals used.

inactivation (Fig. 2). Copper concentrations, however, are not depressed in *Mo* adrenals, which suggests that either the copper present is not available for enzyme activation or for inhibitor inactivation, or that changes elsewhere in the animal are responsible for the altered dopamine β-monooxygenase activity. MAO and phenylethanolamine-N-methyltransferase (EC 2.1.1.28) (PNMT) activities are also depressed in *Mo* adrenals and a regulatory change in catecholamine biosynthesis may therefore be responsible for the reduced catecholamine levels in young *Mo* mice. Regulation of catecholamine biosyn-

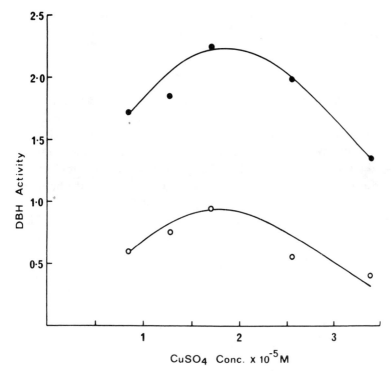

FIG. 2. Effect of cupric ion on dopamine β-monooxygenase (DBH) activity in adrenal tissue from adult Mo^{blo} male mice and normal littermates. The DBH activity (μmoles g^{-1} h^{-1}) in 1 : 400 (w/v) homogenates was assayed in the presence of different concentrations of copper sulphate as described in Hunt (1977). Each point is the mean of three separate determinations. (Mo^{blo}) (•); normal (○).

thesis is achieved centrally via the splanchnic nerve, and additional control is exercised by the glucocorticoids and pituitary hormones (Ciaranello 1979). The effect of copper deficiency may be to interfere with one or more of these control processes.

An alternative explanation for the reduced tyrosine 3-monooxygenase and DA concentrations reported in dietary copper-deficient animals (Morgan & O'Dell 1977, O'Dell et al 1976), is that both deficiencies arise from loss of dopaminergic neurons. The reported lesions in the corpus striatum (Carlton & Kelly 1969), a region rich in dopamine-containing neurons, in dietary copper-deficient rats supports this interpretation. The failure of Mo mice to undergo this loss may reflect their less severe copper deficiency in the brain compared to the deficiencies reported by Morgan & O'Dell (1977) in dietary copper-deficient rats. Of course, loss of adrenergic neurons may also occur in *Mo* mice and would contribute to the reduced levels of tyrosine 3-monooxygenase and NA.

NEUTRAL AMINO ACIDS

The concentrations of a number of neutral amino acids are increased in *Mo* brain tissue. In six-day-old Mo^{br} mice, free tyrosine, phenylalanine, methionine, threonine and histidine concentrations are raised (Hunt & Johnson 1972b) and by 11 days after birth, free tryptophan concentrations are also increased (Hunt & Johnson 1972a). Tyrosine levels in Mo^{br} brain show a maximal rise at day 7, whereas in Mo^{vbr} mice, tyrosine levels in the brain are high on days 9 and 13 but fall to normal by day 19 (Fig. 3). The neutral amino acids share a common carrier system across the blood–brain barrier (Oldendorf & Szabo 1976) and an increase in this transport system in *Mo* mice is in-

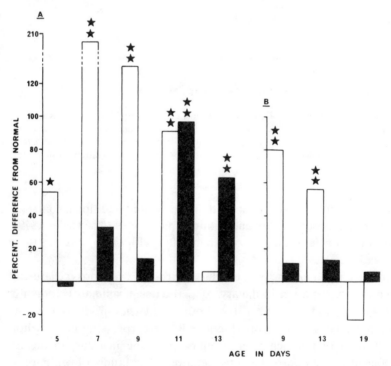

FIG. 3. Differences in brain tyrosine and tryptophan concentrations, in Mo^{br} and Mo^{vbr} male mice, compared to concentrations in normal littermates. A, Mo^{br}; B, Mo^{vbr}; solid bars, tryptophan; open bars, tyrosine; ★ and ★★ denote significance at the 5% and 1% probability levels respectively.

dicated by the competitive inhibition by isoleucine of tyrosine uptake and accumulation, and by the greater uptake of [C^{14}]methionine reported by Hunt & Johnson (1972b).

An increase in the neutral amino acid transport system is also seen in rats with a portacaval anastomosis (James et al 1978), an animal model of chronic liver disease in humans. Increased levels of free tryptophan in the brain have been implicated in the symptoms of neurological impairment termed hepatic encephalopathy, which frequently accompanies chronic liver disease (Young et al 1975). The rise in brain tryptophan may contribute to the development of the *Mo* syndrome, and hepatic failure may underlie the increased concentrations of tyrosine aminotransferase (EC 2.6.1.5) and the reduced levels of both phenylalanine 4-monooxygenase (EC 1.14.16.1) and tryptophan 2,3-dioxygenase (EC 1.13.11.11) in *Mo* liver tissue. It is tempting to speculate that a transient rise in the neutral amino acid transport system may also occur in Menkes' disease and in dietary copper-deficient animals.

COPPER-BINDING PROTEINS

Fractionation of the cytosolic copper-binding proteins by column chromatography on Sephadex G75 reveals 4 major copper-binding fractions (Fig. 4). Fraction 1 coincides with the void volume of the column and may represent a number of different protein species with molecular weights greater than 50 000. Fraction 2 is similar in separation characteristics to the fraction expected for superoxide dismutase (cerebrocuprein). Fraction 3, with a molecular weight of 14 000–15 000, is similar in size to the protein isolated from liver and kidney tissue of neonatal *Mo*^{br} and normal mice by Port & Hunt (1979). Fraction 4 represents either free copper or copper bound to low molecular weight compounds such as peptides and amino acids. A comparison of the two profiles shows that in *Mo*^{br} mice, copper is reduced in all 4 fractions but is proportionately more reduced in fraction 3. The copper deficiency in *Mo* brain tissue is not limited at the subcellular level to a particular protein species although there may be a slightly disproportionate reduction in the copper content of the different protein fractions. Until these fractions have been purified and characterized, it is not possible to comment further on the significance of these observations to neurological function.

CONCLUSION

Although hypomyelination or demyelination is widespread in copper-deficient animals, it is not present in *Mo* mice (Yajima & Suzuki 1979). The

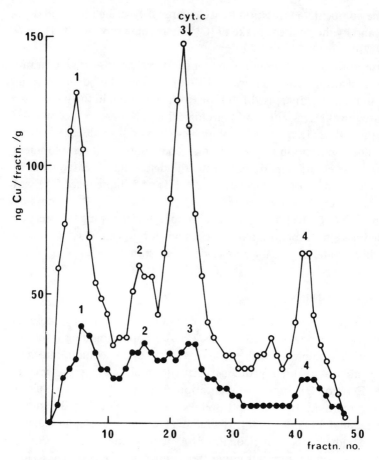

FIG. 4. Sephadex G75 separation of proteins in the supernatant fraction of brain from 7–8 day Mo^{br} male mice and normal littermates. The 80 000 g supernatant fraction in 10 mM Tris-HCl buffer (pH 7.4) was concentrated in an Amicon pressure cell equipped with a UM2 filter. The concentrated supernatant was applied to a Sephadex G75 column (2.2 × 40 cm) and eluted with 10 mM Tris-HCl buffer (pH 7.4). Fractions of 2.8 ml were collected. MO^{br} (●); normal (○); cyt. c, cytochrome c.

behavioural changes associated with copper deficiency must, at least for Mo mice, arise from tissue necrosis and neuronal loss in the CNS. Central catecholamine concentrations are consistently reduced in copper-deficient animals. The *in vivo* activity of dopamine β-monooxygenase, the only copper-metalloenzyme in the catecholamine biosynthetic pathway, is reduced in Mo mice (Hunt & Johnson 1972b) but has not been assayed in dietary copper-deficient animals. This reduced activity may be largely responsible for the reduced concentrations of noradrenaline reported in the brain tissue of *Mo* mice and dietary copper-deficient rats. Tyrosine 3-monooxygenase activity is not affected in neonatal *Mo*^{br} mice, but is reduced in dietary copper-deficient adult rats (Morgan & O'Dell 1977), which also have reduced concentrations of dopamine. Since a direct involvement of copper deficiency in the catalytic function of tyrosine 3-monooxygenase is unlikely, the changes in enzyme activity and in dopamine concentrations reported in older dietary copper-deficient animals may reflect the continuing tissue necrosis that occurs during development and the consequent loss of dopamine-containing neurons. Catecholamine concentrations may also be depressed as a result of increased autoxidation mediated by the superoxide radical that arises from reduced levels of superoxide dismutase in copper-deficient animals. There is at present, however, no evidence that the decreased activities of this enzyme are sufficient to change catecholamine levels significantly. Further defects in *Mo* mice which may contribute to neurological impairment are the increased uptake and accumulation of neutral amino acids that occur in brain tissue (Hunt & Johnson 1972a,b). The existence of a similar change in dietary copper-deficient animals or in patients with Menkes' disease has yet to be examined.

ACKNOWLEDGEMENTS

This work was largely supported by grants from the Medical Research Council of Great Britain. The assistance of Mrs. Ann Port is gratefully acknowledged.

References

Anton AH, Sayre DF 1962 A study of factors affecting the aluminium oxide-trihydroxyindole procedure for the analysis of catecholamines. J Pharmacol Exp Ther 138:360-375

Cancilla PA, Barlow RM 1966 Structural changes of the central nervous system in swayback (enzootic ataxia) of lambs. II. Electron microscopy of the lower motor neuron. Acta Neuropathol 6:251-259

Carlton WW, Kelly WA 1969 Neural lesions in the offspring of female rats fed a copper-deficient diet. J Nutr 97:42-52

Ciaranello R 1979 Genetic regulation of the catecholamine synthesizing enzymes. In: Shire JGM (ed) Genetic variation in hormone systems, Vol II.CRC Press, Florida

Danks DM, Campbell PE, Stevens BJ, Mayne V, Cartwright E 1972 Menkes' kinky hair syndrome. An inherited defect in copper absorption with widespread effects. Pediatrics 50:188-201

Everson GJ, Shrader RE, Wang T 1968 Chemical and morphological changes in the brains of copper-deficient guinea pigs. J Nutr 96:115-125

Gallagher CH, Judah JD, Rees KR 1956 The biochemistry of copper deficiency: I. Enzymological disturbances, blood chemistry and excretion of amino acids. Proc R Soc Lond Ser B 145:134-150

Ghatak NR, Hirano A, Poon TP, French JH 1972 Trichopoliodystrophy. II. Pathological changes in skeletal muscle and nervous system. Arch Neurol 26:60-72

Hunt DM 1974 Primary defect in copper transport underlies mottled mutants in the mouse. Nature (Lond) 249:852-854

Hunt DM 1976 A study of copper treatment and tissue copper levels in the murine congenital copper deficiency, mottled. Life Sci 19:1913-1920

Hunt DM 1977 Catecholamine biosynthesis and the activity of a number of copper-dependent enzymes in the copper deficient mottled mouse mutants. Comp Biochem Physiol 57C:79-83

Hunt DM, Johnson DR 1972a Aromatic amino acid metabolism in brindled (Mo^{br}) and viable-brindled (Mo^{vbr}), two alleles at the mottled locus in the mouse. Biochem Genet 6:31-40

Hunt DM, Johnson DR 1972b An inherited deficiency in noradrenaline biosynthesis in the brindled mouse. J Neurochem 19:2811-2819

Hunt DM, Port AE 1979 Trace element binding in the copper deficient mottled mutants in the mouse. Life Sci 24:1453-1466

James JH, Escourrou J, Fischer JE 1978 Blood-brain neutral amino acid transport activity is increased after portacaval anastomosis. Science (Wash DC) 200:1395-1397

Jarrott B 1971 Occurrence and properties of catechol-O-methyl transferase in adrenergic neurons. J Neurochem 18:17-27

Levitt M, Spector S, Sjoerdsma A, Udenfriend S 1965 Elucidation of the rate limiting step in norepinephrine biosynthesis using the perfused guinea pig heart. J Pharmacol Exp Ther 148:1-8

Lou HC, Holmer GK, Reske-Nielsen E, Vagn-Hansen P 1974 Lipid composition in gray and white matter of the brain in Menkes' disease. J Neurochem 22:377-381

Magos L, Jarvis JAE 1970 Effects of diethyldithiocarbonate and carbon disulphide on brain tyrosine. J Pharm Pharmacol 22:936-938

Molinoff PB, Orcutt JC 1973 Dopamine-β-hydroxylase and the regulation of catecholamine biosynthesis. In: Usdin E, Snyder SH (eds) Frontiers in catecholamine research. Pergamon Press, New York, p 195-200

Morgan RF, O'Dell BL 1977 Effect of copper deficiency on the concentrations of catecholamines and related enzyme activities in the rat brain. J Neurochem 28:207-213

Nagatsu T, Nakano G, Mitzutani K, Harada M 1972 Purification and properties of amine oxidase in brain and connective tissue (dental pulp). Adv Biochem Psychopharmacol 5:25-36

O'Brien JS, Sampson EL 1966 Kinky hair disease. II. Biochemical studies. J Neuropathol Exp Neurol 25:523-530

O'Dell BL, Smith RM, King RA 1976 Effect of copper status on brain neurotransmitter metabolism in the lamb. J Neurochem 26:451-455

Oldendorf WH, Szabo J 1976 Amino acid assignment to one of three blood-brain barrier amino acid carriers. Am J Physiol 230:94-98

Port AE, Hunt DM 1979 A study of the copper-binding proteins in liver and kidney tissue of neonatal normal and mottled mutant mice. Biochem J 183:721-730

Prohaska JR, Wells WW 1974 Copper deficiency in the developing rat brain: a possible model for Menkes' steely-hair disease. J Neurochem 23:91-98

Prohaska JR, Wells WW 1975 Copper deficiency in the developing rat brain: evidence for abnormal mitochondria. J Neurochem 25:221-228

Skotland T, Ljones T 1979 Dopamine β-mono-oxygenase: structure, mechanism and properties of the enzyme-bound copper. Inorg Perspect Biol Med 2:151-180

Yajima K, Suzuki K 1979 Neuronal degeneration in the brain of the brindled mouse. Acta Neuropathol 45:17-25

Young SN, Lal S, Sourkes TL, Feldmuller F, Aronoff A, Martin JB 1975 Relationships between tryptophan in serum and CSF, and 5-hydroxyindoleacetic acid in CSF in man: effect of cirrhosis of liver and probenecid administration. J Neurol Neurosurg Psychiatry 38:322-330

Zimmerman AW, Matthieu JM, Quarles RH, Brady RO, Hsu JM 1976 Hypomyelination in copper-deficient rats. Arch Neurol 33:111-119

Discussion

Riordan: Could you comment further, Dr Hunt, on the significance of a reduction in phenylethanolamine N-methyltransferase activity, and its effects on phospholipid metabolism? Does this reduction contribute to the neurological changes in Menkes' syndrome?

Hunt: The reduced activity of phenylethanolamine N-methyltransferase will affect the synthesis of adrenaline from noradrenaline. As far as I am aware, this reduction is restricted to the adrenal medulla.

Sourkes: I would like to ask about the increased absorption, or the increased transport, of the large neutral amino acids. There is much work already on tryptophan, but is it possible that the behavioural changes that you mentioned may be explained by increased brain concentrations of serotonin (5-HT)?

Hunt: We have found that serotonin concentrations are not abnormal in the mutant mice with Menkes' syndrome (Hunt & Johnson 1972), and therefore serotonin is probably not responsible for the behaviour of the mutants.

Harris: It would be interesting to find the primary and secondary causes for some of the effects you described in your paper. One primary cause may be related to the energy requirements for the whole process. If there was a reduction in activity of the cytochrome c oxidase or a failure to sustain the concentrations of ATP, there would be an impairment of active transport and perhaps some of the other symptoms you described would also be explained.

Hunt: We have not looked at ATP levels, but the reduction in cytochrome c oxidase activity of about 50% in the brain of brindled mice (Hunt 1977) is probably not sufficient to cause serious respiratory disturbances.

Mills: I think that your explanation might apply to copper deficiency in the rat brain but not in the sheep brain, which has very low levels of cytochrome c oxidase.

How easy is it to measure copper distribution in the adrenal gland, when you are dealing with such a local phenomenon?

Hunt: It is difficult to make these measurements because one is considering the adrenal gland as a whole rather than its individual parts. The concentrations of copper may be differentially altered but we cannot detect that.

Mills: Dr J. Hesketh (unpublished results) in Aberdeen has also found a reduction of noradrenaline concentrations in adrenal glands from copper-deficient cattle. Like you, he found no significant difference in copper content. The total adrenal copper therefore does not decline rapidly but nevertheless the difference in noradrenaline concentrations occurs even before overt signs of copper deficiency appear.

Hurley: Manganese is also necessary for dopamine metabolism, transport of tryptophan and brain function. Do you know whether this mutation might affect manganese metabolism as well as copper metabolism?

Hunt: We have not studied manganese metabolism in these mutants.

Lewis: Was there any difference in weights of the adrenal glands between mutants and non-mutants, and were there any histological changes?

Hunt: There is no difference between weights of adrenal glands from mottled mutants and from normal mice, but we have not examined the adrenal glands histologically.

Danks: I would like to comment on the relationship between myelination and copper disturbances in general. We found no changes in myelin in either blotchy or brindled mice compared with non-mutant mice (D.M. Danks, J.M. Matthieu, B. Koellreutter, J. Mann, unpublished work). (We have not looked at viable-brindled mice.) This finding compelled me to examine the published work on myelination in copper deficiency more closely and I concluded that there was no evidence that copper had any specific and essential effect on the process of myelination. Only *severe* copper deprivation during intrauterine life will cause widespread brain damage in rats (Hall & Howell 1973) and guinea pigs (Everson et al 1967). The animals undergo considerable neuronal loss, so it is not surprising that the overall brain content of myelin is reduced in those cases.

Terlecki: I agree that there is little evidence of specific effects of copper deficiency on myelination. Our study of copper-deficient ovine fetuses (S. Terlecki & G. Lewis, unpublished results) showed that the first changes, in the form of cerebral lesions, occurred in so-called 'myelination glia centres', which are chiefly composed of oligodendroglial cells responsible for myelin formation. If copper was responsible for those changes, its effect on myelin synthesis would therefore be only indirect. Our study also showed that the pathogenesis of cerebral lesions in swayback differs from that of lesions which occurred in the *brain stem* and *spinal cord* in this disease. These observations agree with those of Fell et al (1965) and suggest that the low copper content in the central nervous system which exists in swayback, and in the similar disease called enzootic ataxia in Australia, leads to a deficiency of cytochrome *c* oxidase. A low activity of this enzyme primarily affects the large neurons in the brain stem and secondarily produces degeneration of the associated axons and myelin in the spinal cord. An objection to this hypothesis is that the number of affected fibres in the spinal cord is disproportionately greater than the number of degenerating neurons observed at any one time.

Danks: Many text books still claim that copper is essential for lipid formation in myelin but there is little foundation for such a view.

Lewis: Another anomaly is that swayback occurs only in copper-deficient lambs yet very many equally copper-deficient lambs exhibit no symptoms of swayback and no changes in myelin formation.

Mills: Dr Danks, some years ago you suggested (Danks et al 1972) that those who were trying to find the cause of enzootic ataxia (swayback) might be looking in the wrong place, because cerebral vascular lesions may be induced by copper deficiency and because these lesions, by restricting blood and oxygen supply to the brain, may have a greater effect on oxidative processes than would the loss of cytochrome *c* oxidase activity.

Danks: You are quoting me correctly. When we examine the brain of a patient with Menkes' syndrome at death, many of the visible changes are the result of vascular accidents. However, these changes can become so gross during the late stages of the disease that they obscure any effects of the earlier changes. The comment that you quoted was made in an attempt to direct the attention of veterinary pathologists to the vascular supply of the brain. John Howell in Perth (personal communication, 1979) has examined the vascular supply and tells me he can find no arterial changes in sheep, but I would be glad to hear the results of other studies.

Terlecki: I agree with John Howell — there are no vascular changes!

Danks: Well there are some species differences to be considered. We have recently seen a case of what seems to be the human equivalent of the 'blotchy' phenotype — in other words, a mild Menkes' syndrome (P. Procopis, J. Camakaris, D.M. Danks, unpublished work). The outstanding neurological feature in this child was cerebellar ataxia which was greater than could be accounted for by the degree of mental retardation in the child. Several groups have recently reported that cerebellar Purkinje cells in Menkes' syndrome have an excessive number of dendritic connections with other cells (Iwata et al 1979). It seems therefore, that a rather specific lesion in the cerebellum is caused by the disease in humans.

Shaw: Is the blotchy phenotype in humans linked to the X chromosome?

Danks: Yes, probably.

Terlecki: Red deer farming is well established in New Zealand, and even in Britain there are now about 40 red deer farms. The spinal cord lesions in ataxic red deer are almost indistinguishable from those found in sheep with swayback (Terlecki et al 1964). Opinions about the presence of neuronal changes in the brain stem are divided: Barlow et al (1964) did not observe them but we did, and they were similar to those seen in swayback, although less frequent. An interesting observation was that even the normal red deer had low copper levels (about 14 mg/kg dry matter) in comparison with domestic ruminants. Copper supplementation by means of mineral licks appears to reduce the incidence of ataxia but its true effectiveness is difficult to assess because it is also usual to cull affected animals in order to avoid a possible spread of a genetic component which may be involved in the aetiology of the disease.

Mills: Perhaps Miss Lewis would like to comment on the sensitivity of the fetal lamb to small changes in copper supply during late development? I understand that the incidence of swayback (i.e. enzootic ataxia) can be dictated by weather conditions, but we don't know whether this is relevant to disease in humans.

Lewis: The earliest changes we have seen occurred at about 100 days of gestation (S. Terlecki and G. Lewis, unpublished work). The period between 100–120 days of gestation is the time when myelin is actively laid down in the fetal lamb. Even a small supplementation of the copper in the mother's diet at this critical period of fetal development may prevent the disease. Farmers in this country have always said that during a mild winter there is a high incidence of swayback, but when there is a lot of snow on the ground there is a low incidence of the disease. We have shown that there is a causal relationship between the weather and the incidence of swayback (C.B. Ollerenshaw, G. Lewis and L.P. Smith, unpublished results). When the weather is mild the sheep are left to eat grass, which may contain little copper and may also contain inhibitory factors such as molybdenum and sulphur. Furthermore, sheep left outside in winter may also consume about 20% of their intake of dry matter as soil, which has been shown to reduce copper absorption. In a mild winter, therefore, these factors tend to reduce the availability of copper to the ewe and the fetus.

However, during winters when there is snow on the ground the sheep are given supplementary feeds. In addition, the copper in hay is more available to the sheep than the copper in grass, and the other inhibitory factors are reduced. The lambing season is in March and April, so if there is snow in January or February then the extra copper becomes available at the time when the fetus is most susceptible to the effects of copper deficiency, and the incidence of swayback is thus reduced. This empirical approach has been used to forecast the incidence of swayback in England and Wales. Results show that the forecasts have been reasonably accurate and can make a useful contribution to the prevention of this disease.

Hurley: Miss Lewis, in view of what you have just said, how can it be stated dogmatically that copper deficiency does not reduce myelination? There may not be a direct cause and effect relationship, but reduced myelination may be one of the sequelae.

Lewis: We think that copper is involved in myelination, but it is harder to explain why some copper-deficient lambs have no symptoms of swayback although, as I mentioned before, a *pre-requisite* for swayback is copper deficiency. I believe that a familial factor may determine whether or not swayback occurs in a copper-deficient animal, because a ewe that has had a lamb with swayback is highly likely to produce subsequent swayback lambs, given the right conditions (G. Lewis & S. Terlecki, unpublished work).

Danks: I think that you may have misinterpreted my comments. I feel that the categoric statement that copper is essential for myelination is over-stated, and based on insufficient evidence, but I would not wish to imply that copper does not function at all in myelination.

Mills: There is still a conflict of opinion on this point. Although we (Mills & Fell 1960) did not obtain quantitative data there appeared to be a closer relationship of ataxia to motor neuronal lesions than to myelin defects in lambs depleted of copper during fetal and early postnatal development. On the other hand, I don't think that this view is universally accepted.

Terlecki: We have tried to compare the severity and incidence of ataxia with the severity of lesions and found no direct correlation (S. Terlecki & G. Lewis, unpublished results).

Riordan: Dr Hunt, I do not understand how copper deficiency in the central nervous system is propagated to the adrenal glands, or what the implications of this are. Why is it that in the absence of copper deficiency there can be a decreased enzyme activity in the mice?

Hunt: Since copper is deficient in the CNS of mottled mice, and since we have shown that the activity of a number of copper-dependent enzymes and the levels of the neurotransmitter, noradrenaline, are also reduced, it is possible that the altered enzyme activities arise from lesions in brain centres that control adrenal functions. Direct nervous control of enzyme activity is achieved via the splanchnic nerve and there is evidence for additional hormonal control. Copper deficiency could interfere with one or more of those control processes.

Riordan: But aren't those enzymes dependent on copper? If copper is still present, what causes the decrease in enzyme activity?

Hunt: Of those enzymes in the catecholamine pathway of synthesis and degradation, only dopamine β-monooxygenase is a copper-metalloenzyme.

References

Barlow RM, Butler EJ, Purves D 1964 An ataxic condition in red deer (*Cervus elaphus*). J Comp Pathol 74:519-529

Danks DM, Campbell PE, Stevens BJ, Mayne V, Cartwright E 1972 Menkes' kinky hair syndrome. An inherited defect in copper absorption with widespread effects. Pediatrics 50:188-201

Everson GJ, Tasi H-GC, Wang T-I 1967 Copper deficiency in the guinea pig. J Nutr 93:533-540

Fell BF, Mills CF, Boyne R 1965 Cytochrome oxidase deficiency in the motor neurones of copper deficient lambs: a histochemical study. Res Vet Sci 6:170-177

Hall GA, Howell JMcC 1973 Lesions produced by copper deficiency in neonate and older rats. Br J Nutr 29:95-104

Hunt DM 1977 Catecholamine biosynthesis and the activity of a number of copper-dependent enzymes in the copper deficient mottled mouse mutants. Comp Biochem Physiol 57C:79-83

Hunt DM, Johnson DR 1972 Aromatic amino acid metabolism in brindled (Mo^{br}) and viable-brindled (Mo^{vbr}), two alleles at the mottled locus in the mouse. Biochem Genet 6:31-40

Iwata M, Hirano A, French JH 1979 Degeneration of the cerebellar system in X-chromosome-linked copper malabsorption. Ann Neurol 5:542-549

Mills CF, Fell BF 1960 Demyelination in lambs born of ewes maintained on high intakes of sulphate and molybdate. Nature (Lond) 185:20-22

Terlecki S, Done JT, Clegg FG 1964 Enzootic ataxia of red deer. Br Vet J 120:311

Copper and hepatic function

CHARLES A. OWEN, Jr.

Mayo Clinic and Mayo Foundation, Rochester, Minnesota 55901, USA

Abstract When hepatic excretion of copper into bile is impaired, the amount of copper in the liver increases. This happens in extrahepatic cholestasis, primary biliary cirrhosis and in two inherited diseases of copper metabolism, Wilson's disease in humans and the Bedlington terrier copper disease. By six months of age the homozygously affected Bedlington terrier has already begun to accumulate copper in its liver. The trend continues, peaking at the age of five to eight years, when hepatic copper may exceed 10 000 µg/g dry weight (normal 90–400 µg/g in livers from mongrels). Despite these concentrations, which are several times higher than those found in any human disease, there is remarkably little evidence of hepatic inflammation or fibrosis in younger Bedlington terriers. The copper is condensed in lysosomes and is identified by X-ray emission spectroscopy. Hepatic cirrhosis eventually develops and death is often associated with ascites and jaundice. Despite characteristic histological differences between the livers in Wilson's disease and in the Bedlington disease, there is a striking general resemblance between the two conditions.

There is little question that the liver plays a central role in the metabolism of copper and, in turn, that abnormalities of copper metabolism have a direct effect on hepatic function.

Once absorbed from the intestine, ionic copper is transported in the blood as an albumin complex. The liver separates the copper from the albumin complex and complexes the metal with a low-molecular-weight protein of its own. From this point, copper enters a number of distinct pathways within the liver. Some of the copper is rapidly excreted into the bile. More slowly, copper is converted into caeruloplasmin and returned to the blood. If more copper reaches the liver than can readily be converted to caeruloplasmin or excreted in the bile, the excess is stored for future disposition.

Hazelrig et al (1966) studied the rates of conversion of radiolabelled copper into caeruloplasmin and its rate of excretion into the bile by isolated rat livers;

an example of these results is shown in Fig. 1. To the ^{64}Cu were added various amounts of non-radioactive copper to study the impact of excessive copper. Computer analysis of the data suggested three distinct hepatic copper compartments: bile, caeruloplasmin and storage. All three could be in two-way communication with the blood and with each other. However, the closest computer fitting of the experimental data was obtained when it was assumed that copper could pass from the blood into all three compartments but back to the blood from only two, the biliary and caeruloplasmin compartments. Further, there seemed to be no communication between the storage and biliary

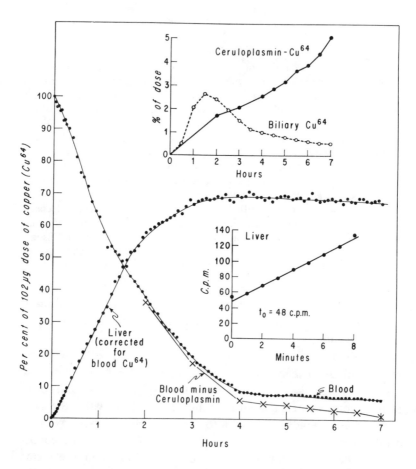

FIG. 1. Perfusion of the isolated liver of a normal rat with blood containing ^{64}Cu. Biliary excretion of the ^{64}Cu is rapid; release of labelled caeruloplasmin is considerably slower. (Reproduced from Owen & Hazelrig (1966), by permission of the publisher.) C.p.m., counts per minute.

pools and only one-way passage between storage and caeruloplasmin pools, from former to latter (Fig. 2).

Such a scheme is obviously an over-simplification because copper goes along other pathways within the liver. One prominent such pathway is that for formation of superoxide dismutase (EC 1.15.1.1) in the hepatic supernatant cytosol (Table 1). Other pathways that seem likely are the incorporation of copper into cytochrome c oxidase (EC 1.9.3.1) and into other copper enzymes.

The three proposed hepatic compartments are mathematical, not anatomical, ones and conversion between the two types is not simple. Caeruloplasmin synthesis is believed to be in the polyribosomes (Gaitskhoki et al 1975) and there is mounting evidence that the lysosome is the storage site

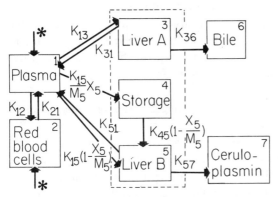

FIG. 2. Proposed model for the handling of copper by the isolated, perfused rat liver. X refers to the percentage of the administered dose of ^{64}Cu in a compartment, M is the maximal capacity of a compartment, and the K values are transfer rates. (Reproduced from Owen & Hazelrig (1968), by permission of the publisher.)

TABLE 1

Distribution of ^{67}Cu in supernatant cytosol of normal rat liver after intravenous injection of carrier-free ^{67}CuCl$_2$

Hours after injection of ^{67}Cu	% of dose/gram in:		% of ^{67}Cu in supernatant in mol. wt. fraction:		
	whole liver	supernatant	> 100 000	40 000	10 000
0.5	4.2	1.3	21	38	42
1	2.8	1.4	20	50	31
6	1.6	1.4	5	85	10
24	1.6	1.4	5	93	2

for excess copper (Goldfischer 1965). The lysosome might also be the biliary compartment in the scheme, since bile is believed to be derived from the lysosome (de Duve & Wattiaux 1966).

This hypothetical concept of liver compartments is subject to limited direct analysis. For example, the scheme indicates that ionic copper not only enters the liver from the bloodstream but leaves the liver to return to the blood. This was verified by injection of radiolabelled copper into intact rats. Then, at various intervals, the livers were removed and perfused with non-radioactive blood. Ionic copper was promptly released by these livers, but progressively less was released as the interval between injection of the ^{67}Cu and extirpation of the liver was increased (Owen & Orvis 1970).

The model suggests that in copper deficiency states copper going to the biliary compartment is penalized in favour of more going toward the synthesis of caeruloplasmin. This was found to be correct in rats on a low-copper diet (Owen & Hazelrig 1968). However, there is also a virtual cessation of biliary copper excretion by the pregnant rat despite her hepatic copper remaining normal. The compartment analysis does not explain the radical fall in her serum caeruloplasmin immediately *post partum* either (Terao & Owen 1977). The copper-laden rat conforms well to the model. Biliary copper increases to a maximum, synthesis of caeruloplasmin changes little, but there is progressively more storage of copper (Owen & Hazelrig 1968).

One can extrapolate from this mathematical model in the isolated perfused rat liver to disease states in humans and other species. In copper poisoning the expected significant accumulation of copper in the liver is also observed. If the biliary system is occluded there should also be a hepatic accumulation of the metal since other excretory pathways have very limited capabilities for ridding the body of copper. These secondary pathways include the kidney, the gastrointestinal tract and sweat glands.

The occlusive biliary diseases include: partial or complete agenesis of the biliary tree of infants; biliary cirrhosis of infancy and its adult counterpart, primary biliary cirrhosis; and obstruction of the biliary pathway by stone, cancer or scar. The representative data from published work in Table 2 show that hepatic copper rises sharply in all these diseases.

But hepatic copper also increases in various inflammatory–fibrous diseases of the liver (cirrhosis), as well as in Wilson's disease. One might attribute the former to scarring of bile canaliculi. However, it is reasonably certain that the accumulation of copper in the liver precedes the cirrhotic changes in the patient with Wilson's disease. The impaired excretion of copper in this patient (Frommer 1974) must be caused by a biochemical mutation rather than by a physical obstruction.

TABLE 2

Hepatic copper content in normal adults and in various disease states

Condition	Hepatic copper (μg/g dry matter)
Normal	10–30
Wilson's disease (Scheinberg & Sternlieb 1959)	158–2950
Primary biliary cirrhosis (Benson 1979): asymptomatic	~ 100
active	~ 300
advanced	~ 1100
Common duct obstruction (Smallwood et al 1968): < 6 months	< 150
> 12 months	200–950
Indian childhood cirrhosis (Dang & Somasundaram 1979)	53–5800
Biliary cirrhosis of infancy (Sternlieb et al 1966)	135–769
'Juvenile cirrhosis' (Bartók et al 1971)	2947–3564

The symptomatic patient with Wilson's disease also has excessive copper in the brain and kidney but virtually nowhere else. This could be because these organs, plus the liver, have an unusual avidity for ionic copper, which could accumulate throughout the lifetime of the patient (Uzman et al 1956).

A more popular hypothesis today is that the liver can accumulate copper up to only a certain point and that further copper spills over into the brain and kidney.

One could propose another hypothesis, based on the concept that caeruloplasmin functions primarily as a donor of copper (Broman 1964). In those patients having little caeruloplasmin in their serum, Wilson's disease would represent a state of functional copper deficiency in which liver, kidney and brain try to compensate by excessive, and ineffective, uptake of ionic copper.

The sole defect in Wilson's disease, on the other hand, could be an error in biliary excretion of copper. Perhaps the liver cannot form a complex between copper and the peculiar, low-molecular-weight biliary compound (Terao & Owen 1973) which is characterized by its poor absorbability from the intestinal tract (Owen 1964). Perhaps there is an anatomical defect in the lysosome. Perhaps, in most patients, there is an associated defect in the caeruloplasmin-synthesizing mechanism. Or perhaps there is a more fundamental block in a common pathway whereby copper is supplied to the biliary and caeruloplasmin compartments. Support for the concept that the liver is the sole organ responsible for Wilson's disease comes from reports of cure of the disease after transplantation of normal livers into these patients (Groth et al 1973).

One unexplained aspect of copper toxicity, the haemolytic crisis, occurs after ingestion of copper sulphate by humans, ingestion by sheep of feed contaminated with copper spray, or in Wilson's disease in humans and in Bedlington disease in dogs. Studies on sheep convincingly demonstrate that haemolytic episodes accompany the release of stored copper from the liver. It is widely held, therefore, that the resulting hypercupraemia is directly responsible for the haemolysis. However, the serum concentrations of copper are less than those required to induce haemolysis *in vitro*. Further, hepatic copper is virtually all protein-bound so that when it is released its effect might not be comparable to that of ionic copper. Finally, Thompson & Todd (1970) have suggested that the haemolytic agent in the liver may not be a copper compound at all but a phospholipid.

Elucidation of the fundamental pathological process in Wilson's disease is hindered by two things. First, the disease can exist for years before symptoms arise and the distinction between primary and secondary defects is therefore difficult. Secondly, humane constraints strictly limit human research.

A potential animal model for Wilson's disease has been discovered which may permit the investigator much more latitude than study of the human disease.

In 1975, Hardy et al at the University of Minnesota observed several Bedlington terriers dying of hepatic cirrhosis, ascites and jaundice. Hepatic tissue stained intensely for copper, and direct assays revealed much higher concentrations of copper than are observed in patients with Wilson's disease. A detailed report of the pathological process in these dogs has recently appeared (Twedt et al 1979).

Perhaps two-thirds of all the 6000 or 8000 Bedlington terriers in the United States are affected by this autosomal recessive disease. By the time the terrier is six months of age one can usually determine whether it is affected or not by the concentration of copper in the liver. During the next two years hepatic copper rises and rapidly tapers off to a peak when the affected dog is about five or six years old. Thereafter hepatic copper begins to fall moderately (Fig. 3).

During the period of rapid accumulation of copper the liver exhibits almost no histological abnormality other than intense deposits of copper that can even be distinguished by haematoxylin and eosin staining when the copper content reaches several thousand μg/g dry matter. Only very occasional in-

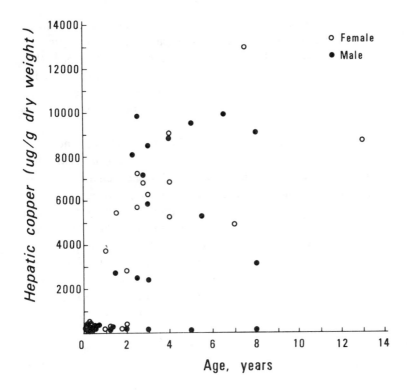

FIG. 3. Progressive rise in hepatic copper in certain Bedlington terriers as they age.

flammatory cells are seen and there is no fibrosis at this stage. Hepatic cirrhosis develops considerably later so that one cannot attribute the hepatic accumulation to a distortion of the biliary tree. There is also an accumulation of iron in the liver of the most affected Bedlington terriers. The iron is present diffusely in the cytoplasm as well as in a granular form in Kupffer cells. The latter form is seen in humans when there is excessive breakdown of erythrocytes.

The Bedlington terrier disease bears a certain resemblance to Wilson's disease but there are also distinct differences. In both diseases copper progressively accumulates in the liver and is primarily localized in the lysosomes; the accumulation of copper precedes clinical symptoms; both diseases are fatal if untreated; and penicillamine is an effective de-coppering agent for both.

In discussing differences between the Bedlington disease and Wilson's disease one must first point out the unusual handling of copper by normal dogs. In the United States, a 10 kg dog eating a diet of commercial chow ingests at least as much copper as a 70 kg person because commercial preparations contain so much liver. Whether dog livers contain more copper than they did 50 years ago (Table 3) is uncertain because of changes and improvements in measurements of tissue copper. Nevertheless, normal mongrel dogs in the United States have hepatic copper concentrations of 90–400 $\mu g/g$ dry matter. Mean values of 190–250 $\mu g/g$ have been reported (Twedt 1979).

One reason for these high levels of copper might be the unusual handling of copper in the blood of normal dogs. Serum albumin in the dog lacks histidine in position 3 from the N-terminus, and cannot bind copper (Dixon & Sarkar 1974). Whether related to this or not, canine serum copper is normally only about one-half that found in humans or rats. Further, 20–30% of the copper can readily be dialysed out of the serum in dogs, again unlike that in other mammals. The non-dialysable serum copper is presumably caeruloplasmin and is low by human standards. When dog serum is chromatographed through diethylaminoethylcellulose not one caeruloplasmin peak emerges from the column, as it does from human or rat serum, but three or four peaks (Fig. 4). Are there several caeruloplasmins in normal dog serum or are there non-caeruloplasminic copper proteins?

The diseased Bedlington terrier shares with the normal dog a high intake of copper and the inability of its serum albumin to bind copper. However, other differences distinguish the canine disease from Wilson's disease. Are they simply species differences or are the diseases fundamentally different?

The accumulation of excess copper in the liver is lysosomal in both dog and humans but only in selected hepatocytes. In Wilson's disease the affected hepatocytes are periportal and in the Bedlington terrier they are centrilobular. In Wilson's disease the electron-dense lysosomes are consistently pericanalicular; in primary biliary cirrhosis they are usually perinuclear and in the Bedlington disease they are randomly distributed. The dense canine lysosomes contain very few lipid deposits whereas round lipid inclusions

TABLE 3

Reported values for canine hepatic copper

Report	Hepatic copper ($\mu g/g$ dry matter)
Flinn et al 1929	7
Meyer & Eggert 1932	44
Beck 1956	80
Owen (unpublished) 1980	260

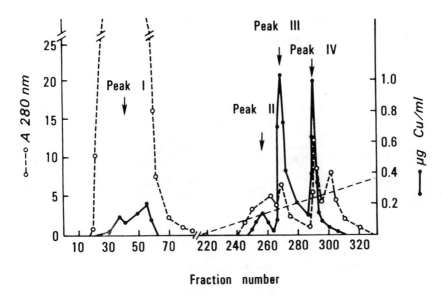

FIG. 4. Three copper-protein chromatographic peaks (Peaks II, III and IV) from normal dog serum passed through diethylaminoethylcellulose. (Only one caeruloplasmin peak is found in normal human or rat serum.) A, absorbance.

characterize Wilson's disease, and crescentic inclusions are typically found in primary biliary cirrhosis.

Serum caeruloplasmin concentration is normal in the Bedlington terrier, as it is in a small percentage of patients with Wilson's disease.

Copper concentration was found to be increased in the kidneys of most Bedlingtons (84–244 μg/g dry matter; normal 27.8 ± 4.5 SD) as it is in most patients with Wilson's disease. However, in only three of the 11 Bedlingtons that we studied was there a dramatic rise in *brain* concentrations of copper: 237, 331 and 2621 μg/g dry matter (normal 26.4 ± 11.8 SD). The range for the remainder was 11–50 μg/g. Perhaps our colony of dogs is too young – all under the age of eight years – to accumulate encephalic copper.

Studies with radiolabelled copper show a parallelism between Wilson's disease and the Bedlington disease. Hepatic uptake of copper, based upon external counting, is modestly increased in the dogs, biliary excretion is sharply diminished and the emergence of radiolabelled caeruloplasmin in the blood is both slower and more limited than in normal dogs.

Whether the copper-laden Bedlington terrier has true Wilson's disease or not, it has an hepatic block to biliary excretion of copper and an accumulation of the metal in the liver before the development of symptoms. Here, then, is an ideal model for trying to discover how one genetic abnormality can so markedly alter copper metabolism.

ACKNOWLEDGEMENTS

Dr. Juergen Ludwig carrier out the histology, Dr. Steven Barham the electron microscopy and X-ray emission spectrometry, Dr. John McCall the tissue copper assays, Dr. Shahnaz Ravanshad the serum chromatography and Dr. Le-Chu Su the radioactive studies. Expert technical assistance was furnished by Helen Sievers, Curtis Grabau, Steven Ziesmer and Joseph Standing. This work was supported in part by grant AM 21738 from the National Institutes of Health.

References

Bartók I, Szabó L, Horváth E, Ormos J 1971 Juvenile Zirrhose mit hochgradiger Kupferspeicherung in der Leber; klinische, pathologische, histochemische und elektronmikroskopische Untersuchungen. Acta Hepatosplenol 18:119-128
Beck AB 1956 The copper content of the liver and blood of some vertebrates. Aust J Zool 4:1-18
Benson GD 1979 Hepatic copper accumulation in primary biliary cirrhosis. Yale J Biol Med 52:83-88
Broman L 1964 Chromatographic and magnetic studies on human ceruloplasmin. Acta Soc Med Upsal 69 (suppl 7):1-85
Dang HS, Somasundaram S 1979 Copper levels in Indian childhood cirrhosis. Lancet 2:246
de Duve C, Wattiaux R 1966 Functions of lysosomes. Annu Rev Physiol 28:435-492
Dixon JW, Sarkar B 1974 Isolation, amino acid sequence and copper(II)-binding properties of peptide (1-24) of dog serum albumin. J Biol Chem 249:5872-5877
Flinn FB, Mouye JM 1929 Some physiological aspects of copper in the organism. J Biol Chem 84:101-104
Frommer DJ 1974 Defective biliary excretion of copper in Wilson's disease. Gut 15:125-219
Gaitskhoki VS, Kisselev OI, Moshkov KA, Puchkova LV, Shavlovski MM, Shulman VS et al 1975 On the defect of synthesis ceruloplasmin in the liver polyribosomes in Wilson's disease. Biochem Genet 13:533-550
Goldfischer S 1965 The localization of copper in the pericanalicular granules (lysosomes) of liver in Wilson's disease (hepatolenticular degeneration). Am J Pathol 46:977-983
Groth CG, Dubois RS, Corman J, Halgrimson C, Rodgerson DO, Starzl TE 1973 Hepatic transplantation in Wilson disease. Birth Defects: Orig Art Ser 9(2):106-108
Hardy RM, Stevens JB, Stowe CM 1975 Chronic progressive hepatitis in Bedlington terriers associated with elevated copper concentrations. Minn Vet 15:13-24
Hazelrig JB, Owen CA Jr, Ackerman E 1966 A mathematical model for copper metabolism and its relation to Wilson's disease. Am J Physiol 211:1075-1081
Meyer AE, Eggert C 1932 Iron and copper in liver and liver extracts. J Biol Chem 99:265-270
Owen CA Jr 1964 Absorption and excretion of Cu^{64}-labeled copper by the rat. Am J Physiol 209:900-904
Owen CA Jr, Hazelrig JB 1966 Metabolism of Cu^{64}-labeled copper by the isolated rat liver. Am J Physiol 210:1059-1064
Owen CA Jr, Hazelrig JB 1968a Copper deficiency and copper toxicity in the rat. Am J Physiol 215:334-338
Owen CA Jr, Hazelrig JB 1968b Copper metabolism in the rat. In: Bergsma D (ed) Wilson's disease. The National Foundation—March of Dimes, White Plains (Birth Defects: Orig Art Ser 4), p 1-7

Owen CA Jr, Orvis AL 1970 Release of copper by rat liver. Am J Physiol 218:88-91
Scheinberg IH, Sternlieb I 1959 The liver in Wilson's disease. Gastroenterology 37:550-564
Smallwood RA, Williams HA, Rosenoer VM, Sherlock S 1968 Liver copper levels in liver disease. Studies using neutron activation analysis. Lancet 2:1310-1313
Sternlieb I, Harris RC, Scheinberg IH 1966 Le cuivre dans la cirrhose biliaire de l'enfant. Rev Int Hepatol 16:1105-1110
Terao T, Owen CA Jr 1973 Nature of copper compounds in liver supernate and bile of rats; studies with ^{67}Cu. Am J Physiol 224:682-686
Terao T, Owen CA Jr 1977 Copper metabolism in pregnant and postpartum rat and pups. Am J Physiol 232:E172-E179
Thompson RH, Todd JR 1970 Chronic copper poisoning in sheep; biochemical studies of the haemolytic process. In: Mills CF (ed) Trace element metabolism in animals. E & S Livingstone, Edinburgh, p 120-123
Twedt DC, Sternlieb I, Gilbertson SR 1979 Clinical, morphologic, and chemical studies on copper toxicosis of Bedlington terriers. J Am Vet Med Assoc 175:269-275
Uzman LL, Iber FL, Chalmers TC 1956 The mechanism of copper deposition in the liver in hepatolenticular degeneration (Wilson's disease). Am J Med Sci 231:511-518

Discussion

Dormandy: Could your model (Fig. 2, and Hazelrig et al 1966) help us to distinguish between primary biliary cirrhosis and atypical Wilson's disease? In typical Wilson's disease one finds low caeruloplasmin levels yet this is not true in every case. Would analysis of bile samples give any indication of a difference between the two diseases?

Owen: One can almost make the distinction on clinical grounds alone. Primary biliary cirrhosis usually develops after the age of 40 and usually in women. Wilson's disease, on the other hand, usually becomes symptomatic before the age of 20. Histologically, the liver is distinctly different in the two diseases. The best way to distinguish definitely between the two is simple. If radiolabelled copper is injected into a patient with Wilson's disease virtually none of the copper will be incorporated into circulating caeruloplasmin. The patient with primary biliary cirrhosis, on the other hand, not only converts the labelled copper to caeruloplasmin but often does so at a considerably accelerated rate; the distinction is clear.

Danks: I agree. Children sometimes present with cirrhosis at the age of about 10 years, an age at which Wilson's disease can cause cirrhosis. We also see patients with cirrhosis that is due to causes other than Wilson's disease and these patients may have a high concentration of copper in their livers. In these cases, kinetic studies with radiolabelled copper leave no doubt about the origin of the disease.

Bremner: I have always been intrigued by the emphasis placed on the abnormal binding sites for copper in albumin from the dog and by the claim that this explains the apparent susceptibility of the dog to copper poisoning. I believe that albumin from pigs also contains the same abnormal copper-binding site (S.H. Lawrie, personal communication) and therefore one would also expect pigs to be susceptible to copper poisoning. However, pigs are extremely tolerant of copper and it is routine practice in this country to supplement their diets with about 250 mg copper per kg diet.

Owen: I am not sure what the lack of albumin binding does in the dog. But if the uptake patterns of copper for a rat and a dog are superimposed, the labelled copper injected into the bloodstream of the dog reaches the liver more rapidly and builds up to a higher level than it does in the rat. This could simply be a species difference or it could arise because albumin does not control the rate at which copper leaves the blood and enters the liver in the dog. But I have seen no clinical evidence that the lack of albumin has any effect on the dog.

Österberg: Some years ago I did some studies on this together with Dr Sarkar in Toronto (Österberg et al 1975). In humans, histidine is in the third position of the N-terminal sequence of albumin, and the histidyl residue increases the stability of the copper complex. The difference in stability between the strongest copper complex of two different amino acids and that of albumin is of the order of 10^5, i.e. the albumin complex is 10^5 times more stable than the mixed amino-acid complex. If we assume that the histidine residue instead were present in position two of the N-terminal albumin sequence, then the albumin complex would be only about 100 times more stable than the amino-acid complex. In dog albumin there is no histidine in the N-terminal sequence of the peptide chain and the stability of the albumin–copper complex is only about 0.01% of that for the amino-acid complex (Österberg 1974). Therefore, for dogs, one would expect that amino acids take over the binding of the labile copper in blood plasma.

Owen: I believe that they do that, because when one dialyses normal dog plasma, about a third of the copper diffuses out (C.A. Owen, unpublished results), and I suspect that the copper is in the form of histidine and other copper amino acids. One does not see this in animals in which the primary copper binding is to albumin.

Österberg: The fact that albumin is a large molecule might be important. One can compare albumin with transferrin, which can become attached to certain sites on the erythrocytes. In humans and other species who have copper bound to albumin, the albumin might also be attached to certain sites on the cells that require copper. Amino acids, which are small molecules, may not have the specific architecture for these binding sites on the cell surface. So, if this idea is correct there might be difficulty in delivery of copper to the cells via the amino-acid complexes; if albumin behaves similarly to transferrin then for dogs this might cause problems for the copper transport.

Harris: Dr Owen, I have always been intrigued by Dr Hazelrig's multicompartment model (Fig. 2), because I think it explains the kinetics very well. Does an equilibrium exist between some compartments so that there is a free flow of copper in any direction depending on where there is an accumulation of copper, or is energy required before copper can move between different compartments?

Owen: This is only a mathematical compartmentalization, not an anatomical one, as Dr Hazelrig, who formulated it, will admit. Where an arrow goes only in one direction in the scheme (Fig. 2) it simply means that the computer matched the experimental data better by using one arrow than two. In functional terms, one arrow means that copper can move only in one direction, whereas two arrows indicate an equilibrium. We disbelieved this scheme when Dr Hazelrig first presented the data to us because she was suggesting that ionic copper can leave the liver. So we labelled rats with one isotope of copper and labelled the blood with the other isotope. After removing the liver and putting it in a perfusion cabinet we measured the exchange. There was complete exchange within 60 minutes (Owen & Orvis 1970), so Dr Hazelrig was right about the free flow of ionic copper from liver back to blood.

Hill: The fact that Dr Hazelrig said that she could fit this scheme to a *minimum* number of compartments is a key point because if it could be fitted to that number of compartments it could also be fitted to more.

Sourkes: Is there, in the compartment, a stable component that corresponds to lysosomal copper, and is anything known about the chemical forms of the copper?

Owen: It is difficult to convert these mathematical compartments to anatomical ones. The excretory pathway in practically every organ of the body is assumed to be the lysosome. Therefore the lysosome takes copper (or any other waste product of the liver) to the bile for excretion. This is reasonable but, as I have shown, the copper that is sequestered is also in the lysosome, which is presumably its storage site. Does the lysosome therefore represent *two* of the mathematical compartments? If so, it is a curious little organelle! I assume that the third compartment, which synthesizes the protein, is the ribosome. Porter & Hills (1974) suggested that the copper-binding protein in the cytosol polymerizes, develops a high sulphur content and then actually *becomes* the lysosomal storage component of copper.

Tanner: Copper accumulation also occurs in the disease called Indian childhood cirrhosis, as we discovered by accident when we were looking for hepatitis B surface antigen in hepatic biopsy sections by use of the orcein stain

(Portmann et al 1978). Since that time, we have studied a larger group of Indian children; all those who had the classical histological features of Indian childhood cirrhosis also had granular orcein staining (unpublished results). The high copper levels therefore form an integral part of the disease. Now whether the disease results from copper poisoning (in which case it would be preventable), or whether it is another inherited abnormality of copper metabolism, remains to be determined. Also, I do not know whether copper is both a necessary *and* a sufficient cause for the disease. The course of Indian childhood cirrhosis resembles the terminal stage of the Bedlington terrier disease, as described by Dr Owen, in that after a period of asymptomatic hepatomegaly these children go into a very rapid phase of jaundice and hepatocellular failure, and they all die. We have some X-ray dispersive pictures that are similar to Dr Owen's, showing copper and sulphur in electron-dense bodies which we presume to be derived from lysosomes (unpublished results).

Riordan: I was surprised that only about 50% of the copper in the Wilson's-diseased liver is associated with the lysosomes. We have purified copper-metallothionein both from so-called 'normal' human fetal liver and from Wilson's-diseased liver (Riordan & Richards 1980). We found that in the fetus approximately 80% of the total tissue copper is associated with an insoluble fraction, and this fraction contains a lot of copper-metallothionein. But, in the Wilson's-diseased liver, where the copper concentration is about 5–10 times higher than that of the fetus, less than 40% of the total copper is associated with this insoluble fraction. Perhaps I shouldn't associate the insoluble fraction with the lysosomal fraction but one tends to do so. It is interesting that in Wilson's disease not all the copper is aggregated into an insoluble polymer that may associate with the lysosome.

Bremner: Goldfischer & Sternlieb (1968) found differences in the distribution of copper in the liver of patients with Wilson's disease. The differences depended on how far the disease had progressed. In young asymptomatic patients, copper staining was diffuse in the cytoplasm whereas in older patients, with advanced disease, copper was found exclusively in the lysosomes.

Riordan: The stage that I was referring to was the terminal phase of the disease, because the liver samples were taken at autopsy.

Bremner: I am interested in the differences that you reported, Dr Owen, in the localization of the residual bodies in liver in both Wilson's disease and the Bedlington terrier disease. In work I have been doing with Dr Timothy King (King & Bremner 1979) on the occurrence of these bodies in copper-poisoned sheep we found no evidence for excretion of copper from the residual bodies into the bile. Copper could not be detected in or near the bile canaliculi. The residual bodies were incorporated into macrophages which were then excreted into the portal tract. This may explain the massive liberation of copper from the liver of these animals into the blood in the terminal stages of copper toxicosis.

I think that the sheep therefore provides a useful woolly model for Wilson's disease! Sheep don't have low caeruloplasmin activity but they do have low rates of biliary excretion of copper. Although there is quite a good correlation in the pig between the levels of copper in the bile and in the liver (Skalicky et al 1978), there is no such correlation in the sheep (Norheim & Søli 1977).

Hill: Is it possible to reverse the progression of disease in Bedlington terriers by chelation therapy?

Owen: One of the champion Bedlington terriers in Chicago was treated for over a year with penicillamine. This treatment reduced the dog's liver content of copper from over 5000 μg/g dry weight to less than 1000 μg/g (Ludwig et al 1980). Because the dog has been asymptomatic since before the treatment was started, and because it is regularly winning prizes at shows, I assume that I haven't done it any harm by the treatment!

Shaw: Penicillamine therapy may cause zinc deficiency (Klingberg et al 1976) – did you therefore need to give zinc supplements to the dog?

Owen: No; our treatment of the dog did not include extra zinc.

Mangan: When D-penicillamine is used to treat Wilson's disease does the drug remove both types of copper, i.e. the 50% that is in the lysosome and the 50% that is outside, or is it incapable of removing the insoluble fraction?

Owen: I don't know.

Riordan: I would like to comment on my findings in three cases of Wilson's disease. In two of the three cases, treatment with penicillamine had continued for many years and the total concentration of copper in these patients was very high. There was no reduction like the one that you observed in the dog, Dr Owen.

Sourkes: That is consistent with the observation that penicillamine begins to reverse the symptoms even before there is much excretion of copper (Sass-Kortsak 1974). There are certainly at least two forms of copper; one which is actively toxic and the other which is deposited in tissue in some inert form.

Harris: Dr Owen, is it correct that a patient with Wilson's disease cannot synthesize caeruloplasmin, or can the patient make the protein but not incorporate the copper?

Owen: About 90% of patients with Wilson's disease have only traces of caeruloplasmin or apocaeruloplasmin in their circulation (Carrico & Deutsch 1969). They have less than 2–4 mg/100 ml, whereas the normal level is about 30 mg/100 ml. The patients have no more apocaeruloplasmin in their circulations than a normal patient, but the level of true caeruloplasmin is very low. The other 5–10% of patients have normal or only moderately reduced levels of caeruloplasmin. But even those with normal caeruloplasmin cannot convert labelled copper to labelled caeruloplasmin to a significant extent. This is usually assumed to be because labelled copper is extensively diluted by the copper stored in the liver. I would question this explanation, however, because if it were true, the same thing should be seen in primary biliary cirrhosis, when even higher levels of copper may be found in the liver. In biliary cirrhosis, ionic copper is converted to caeruloplasmin rapidly.

References

Carrico RJ, Deutsch HF 1969 Some properties of ceruloplasmin from patients with Wilson's disease. Biochem Med 3:117-129
Goldfischer S, Sternlieb I 1968 Changes in the distribution of hepatic copper in relation to the progression of Wilson's disease (hepatolenticular degeneration). Am J Pathol 53:883-899
Hazelrig JB, Owen CA Jr, Ackerman E 1966 A mathematical model for copper metabolism and its relation to Wilson's disease. Am J Physiol 211:1075-1081
King T, Bremner I 1979 Autophagy and apoptosis in liver during the prehaemolytic phase of chronic copper poisoning in sheep. J Comp Pathol 89:515-530
Klingberg WG, Prasad AS, Oberleas D 1976 Zinc deficiency following penicillamine therapy. In: Prasad AS, Oberleas D (eds) Trace elements in human health and disease. Vol 1: Zinc and copper. Academic Press, New York, p 51-65
Ludwig J, Owen CA Jr, Barham SS, McCall JT, Hardy RM 1980 The liver in the inherited copper disease of Bedlington terriers. Lab Invest 43:82-87
Norheim G, Søli NE 1977 Chronic copper poisoning in sheep. II: The distribution of soluble copper-, molybdenum- and zinc-binding proteins from liver and kidney. Acta Pharmacol Toxicol 40:178-187
Österberg R 1974 Models for copper protein interaction based on solution and crystal structure studies. Coord Chem Rev 12:309-347
Österberg R, Branegård B, Ligaarden R, Sarkar B 1975 Copper(II) induced polymerization of human albumin, and its depolymerization by diglycyl-L-histidine: a pH static and ultracentrifugation study. Bioinorg Chem 5:149-165
Owen CA Jr, Orvis AL 1970 Release of copper by rat liver. Am J Physiol 218:88-91
Porter H, Hills JR 1974 The half-cystine-rich copper protein of newborn liver. Probable relationship to metallothionein and subcellular localization in non-mitochondrial particles probably representing heavy lysosomes. In: Hoekstra WG et al (eds) Trace element metabolism in animals. University Park Press, Baltimore, p 482-485
Portmann B, Tanner MS, Mowat AP, Williams R 1978 Orcein-positive liver deposits in Indian childhood cirrhosis. Lancet 1:1338-1340
Riordan JR, Richards V 1980 Human fetal liver contains both zinc- and copper-rich forms of metallothionein. J Biol Chem 255:5380-5383
Sass-Kortsak A 1974 Hepatolenticular degeneration (Kinnear Wilson's disease). In: Schwiegk H (ed) Handbuch der inneren Medizin. Springer-Verlag, Berlin, p 627-665
Skalicky M, Kment A, Halder I, Leibetseder J 1978 Effects of low and high copper intake on copper metabolism in pigs. In: Kirchgessner M (ed) Trace element metabolism in man and animals. Arbeitskreis für Tierernährungsforschung, Freising-Weihenstephan, vol 3:163-166

Therapeutic uses of copper-chelating agents

RAGNAR ÖSTERBERG

Department of Medical Biochemistry, University of Göteborg, Box 33031, S-400 33 Göteborg, Sweden

Abstract Equilibrium analysis of a model system for the *in vivo* reactions between penicillamine and copper (I), the penicillamine–glutathione–Cu(I) system, indicates that in a certain concentration range the therapeutic use of penicillamine will not disturb the normal Cu(I) metabolism. The equilibrium data required for this analysis were obtained by e.m.f. titrations on the Cu(I)-glutathione (H_3A) and on the Cu(I)-penicillamine (H_2A) systems at 25 °C, in 0.5 M $NaClO_4$ medium, using glass and copper amalgam electrodes; the data were analysed first by various graphical methods and then by a general least-squares computer program. The results show that mononuclear Cu(I) species, $Cu(HA)_2$, form in both systems; in addition, the polynuclear $Cu_5A_4^{3-}$ species forms in the penicillamine system and the mononuclear $CuHA^-$ species might form in the glutathione system. The results are discussed in relation to the therapeutic use of penicillamine.

In Wilson's disease copper accumulates as a result of a congenital disorder, yet it is one of the few diseases that are effectively controlled by therapy with complex-forming agents (Walshe 1966). Nowadays, the treatment most commonly involves D-penicillamine; although, due to serious side-effects of penicilliamine in some patients, other drugs are successively being developed (Sarkar 1976).

Penicillamine therapy is also used to treat many other disorders, such as metal intoxication, liver disorders, and crippling arthritis (Multi-Centre Trial Group 1973). Since penicillamine forms strong complexes with many metal ions it is possible that some of its effects are due to disturbances in metal-ion metabolism, particularly in copper metabolism. The possible interaction of penicillamine with copper ions and with other physiologically important metal ions may, however, be balanced by ligands present within the organism. Therefore, in our work involving equilibrium analysis we have considered penicillamine not only in relation to the copper ions but also in relation to an important ligand present *in vivo*, glutathione (Österberg et al 1979). In the first step Cu(I) complexes, rather than Cu(II) complexes, have been studied.

The univalent state is supposed to be important in the cell membrane and within the cell, where the local redox potential is supposed to be considerably lower than in blood plasma. A support for this idea is that Cu(I) rather than Cu(II) complexes are efficient as anti-inflammatory agents (Whitehouse & Walker 1978). The special character of Cu(I) indicates that *in vivo,* thiol groups will be the most important ligands for this particular valency (Ahrland et al 1958, Österberg 1974). Because of this fact, glutathione rather than amino acids or proteins is used in this study; glutathione is present in all cells in concentrations varying from 1 to 6 mM (Long 1961, Tietze 1969). The structural formulae of the ligands are shown in Fig. 1.

The formation of complexes between Cu(I) ions and the ligands was studied by e.m.f. methods using constant-current coulometry (Österberg 1970). The experimental approach employed to prevent Cu(I) from disproportionating into Cu(II) and into solid copper was to generate the Cu(I) ions into the ligand solution by constant-current electrolysis of a two-phase amalgam and to measure the amount of electricity used with coulometry (Österberg 1970).

METHODS

The complex formation was measured by varying the total copper ion concentration and recording the free concentrations of copper and hydrogen ions by means of copper amalgam and glass electrodes. In each titration the total concentrations of ligand, A, and hydrogen ions, H, were kept constant. The copper ions were successively generated into the titration solution by constant-current electrolysis of a two-phase copper amalgam (Österberg 1970).

The electromotive forces were measured for electrochemical cells having a copper amalgam electrode or a glass electrode. The equilibrium solution, the

$$H_2A = \begin{matrix} (CH_3)_2CSH \\ | \\ \overset{+}{H_3N}-CH-COO^- \end{matrix} \qquad (I)$$

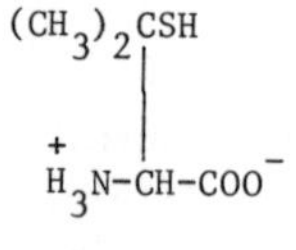

$$H_3A = \begin{matrix} COO^- & H_2CSH \\ | & | \\ \overset{+}{H_3N}-CHCH_2CH_2CONHCHCONHCH_2-COOH \end{matrix} \qquad (II)$$

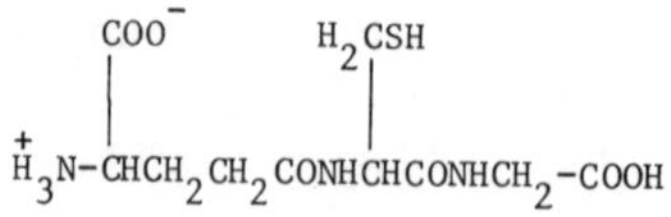

FIG. 1. Structural formulae of D-penicillamine (I) and glutathione (II) and their corresponding abbreviations.

reference half cell, and the salt bridge all had concentrations of 0.5 M NaClO$_4$ and it is therefore assumed that variations in activity factors are effectively reduced so that all the species in solution may be considered in terms of their concentrations.

The titrations were carried out in two steps. First, we determined the cell constant, E_{0G}, from a titration in which OH$^-$ was successively added to a solution containing HClO$_4$ (Österberg 1965). Then the ligand was introduced and the pH was adjusted by addition of OH$^-$ ions which were generated by constant-current electrolysis at a platinum gauze immersed in the solution (Biedermann & Ciavatta 1964), so that redox impurities and other impurities from sodium hydroxide could be avoided. In the second part of the titration the copper amalgam was added and copper ions were generated in the solution by constant-current electrolysis at successive intervals. After each generation the electromotive forces were measured. This gave the data (E_{Cu}, pH, μ)$_{A,H}$, where μ is moles electrons generated per litre.

At the end of each titration the solution was separated from the amalgam and the total concentrations of copper ions ($B+D$) were determined gravimetrically after electrodeposition on a platinum gauze. Thus, the total concentrations for both copper(II), B, and copper(I), D, were obtained, since $\mu = 2B + D$. Some of these solutions were also subjected to paper chromatography, which gave just one ninhydrin-positive component.

The E_0 constants for the copper amalgam and glass electrode cells were determined in a separate titration of a solution of Cu(II) and H$^+$ by means of constant-current electrolysis and a Gran plot (Gran 1952). The majority of the measurements were carried out in the pH range 6.5 to 7.6; in this range the pH could be easily varied by the generation of copper ions into the solution.

ANALYSIS OF DATA

The copper analysis indicates that in these concentration ranges ($\mu = 0.04 - 0.6$ mM, $A = 2$–40 mM) only Cu(I) complexes are formed in the solution. The coulometer reading, denoted μ (in moles electrons generated per litre), is then given by the equation

$$\mu = D = d + \Sigma \Sigma \Sigma p [\mathrm{Cu}_p \mathrm{H}_q \mathrm{A}_r], \tag{1}$$

where D and d are the total and free Cu(I) concentrations, respectively. It is assumed that the Cu$_p$H$_q$A$_r$ complexes formed can be described by the general reaction

$$p\mathrm{Cu}^+ + q\mathrm{H}^+ + r\mathrm{A}^{y-} \rightleftharpoons \mathrm{Cu}_p^\mathrm{I} \mathrm{H}_q \mathrm{A}_r^{(p+q-ry)+} \tag{2}$$

and by the equilibrium constant β_{pqr}, where p, q, and r are integers ($p, q, r = 0, 1, 2, \ldots$) and y is the charge of the anion, A^{y-}. In order to find the predominating species $Cu_pH_qA_r$ and their equilibrium constants of formation the data were analysed both by graphical methods and by computer methods.

The calculation of the free concentration of Cu(I) and ligand

From the primary data (E_{Cu}, pH, μ, A) we first determined d, the free concentration of Cu(I) ions. Then the data μ/d were calculated for each point measured. In the calculation of the free ligand concentration, a, as a first approximation, we neglected the contribution of Cu(I) complexes to the quantity A, the total concentration of ligand, since $A \gg D$. Thus,

$$A = a + a \sum h^q \beta_{0q1} + \sum \sum \sum r \, [Cu_p^I H_q A_r] \approx a(1 + \sum h^q \beta_{0q1}) \quad (3)$$

where h is the free concentration of H^+ ions. Furthermore, in the pH range studied, the ligand mainly exists in the form of the species H_2A. Therefore

$$A \approx [H_2A] = a \cdot h^2 \beta_{021} \quad (4)$$

Thus, for constant A, the free ligand concentration, a, varies mainly with h.

Acid–base equilibria

Analysis of data obtained by acid–base titration of glutathione in 0.5 M $NaClO_4$, using a glass electrode, yielded four β_{0q1} values; they correspond to the following four pK values:

$$pK_{H_4A} = 2.34, \; pK_{H_3A} = 3.48, \; pK_{H_2A} = 8.625, \; pK_{HA} = 9.43.$$

The first two pK values, pK_{H_4A} and pK_{H_3A}, may essentially correspond to the carboxyl groups; pK_{H_2A} and pK_{HA} may correspond to the α-amino and the thiol group, respectively.

Analysis of the data obtained by acid–base titrations of penicillamine in 0.5 M $NaClO_4$ medium yielded three β_{0q1} values; they correspond to the following three pK values:

$$pK_{H_3A} = 1.95, \; pK_{H_2A} = 7.92, \; pK_{HA} = 10.50.$$

The pK values may correspond to those of the carboxyl, amino and thiol groups, respectively.

Copper(I)-glutathione

When log μ/d of equation (1) was plotted against pH, the data varied both with A and with pH; however, when log μ/d was plotted against (log A + pH), most of the data fell on the same curve. The data may then be explained by a series of mononuclear complexes of the type $CuHA$, $Cu(HA)_2$, $Cu(HA)_3$, ... For these species the function μ/d can be rewritten:

$$\mu/d = 1 + \Sigma \, (ha)^r \beta_{1rr} \approx \Sigma \, (ha)^r \beta_{1rr} \quad (r = 1, 2, ...) \tag{5}$$

By substituting $a = Ah^{-2}\beta_{021}^{-1}$ into equation 5, and assuming that the species $CuHA$ and $Cu(HA)_2$ predominate, we obtain

$$\mu/d = D/d = Ah^{-1}\beta_{111}\beta_{021}^{-1} + A^2 h^{-2} \beta_{122}\beta_{021}^{-2} \tag{6}$$

This equation can be normalized into

$$Y = Av + A^2 v^2 \tag{7}$$

where

$$Y = \mu/d(\beta_{122}\beta_{111}^{-2}) \tag{8}$$

and

$$v = h^{-1}(\beta_{021}^{-1}\beta_{122}\beta_{111}^{-1}) \tag{9}$$

Fig. 2 illustrates the comparison between the experimental data, log (D/d) versus pH at constant A, and the model function log $Y(\log v)_A$ (Sillén 1956). From the position of the best fit we obtained log $\beta_{111} = 25$ and log $\beta_{122} = 38.7$. These constants were refined by using the general least-squares computer program, Letagropvrid (Ingri & Sillén (1965). The result was:

$$\log \beta_{111} = 24.9 \, (<25.3); \quad \log \beta_{122} = 38.8 \pm 0.1.$$

Copper(I)-penicillamine

When log μ/d was plotted against (log A + pH), the data for $A \geq 20$ mM fell on the same curve with a slope equal to 2; from the intercept we obtained log $\beta_{122} = 38.9$. Thus, in this range, it is indicated that a single $CuH_2A_2^-$ species exists.

In the next step the data for $A \leq 20$ mM were analysed. The term $[Cu(HA)_2]$ was eliminated from μ (equation 1), and the resulting quantity μ^* was divided by d. When log μ^*/d was plotted against the quantity (log d + log A + 2pH), the plotted data fell on one single curve with the limiting slopes 3 and 6. Thus,

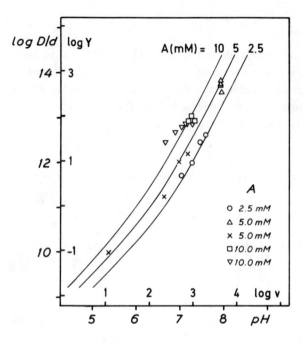

FIG. 2. The Cu(I) complexes of glutathione. (Österberg et al 1979). Experimental data log D/d versus pH for constant A compared with normalized graphs, log $Y(\log v)_A$. The fully drawn curves have been calculated assuming $CuHA^-$ and $Cu(HA)_2^{3-}$ species using equation 1. See text for definition of terms.

μ^*/d may be a function of $(da)^n = (dAh^{-2}\beta_{021}^{-1})^n$ for $n = 3, 4, 5$ or 6; and one or more species of the series, $Cu(CuA)_n^{(n-1)-}$, may exist in the solution.

In order to analyse which $Cu(CuA)_n^{(n-1)-}$ species exist, the primary experimental data were compared with normalized model functions (Sillén 1956) calculated on the assumption that the species $CuH_2A_2^-$ and one single particle $Cu(CuA)_n^{(n-1)-}$ exist simultaneously. The best fit was obtained for the combination $CuH_2A_2^-$ and $Cu_5A_4^{3-}$; and for this combination the experimental data were plotted on the form $(\log D/d - 2pH)$ versus $(4\log d + 6pH)$. These data were compared with the model function log $Y(\log v)_A$ defined as

$$Y = A^2 + A^4 v \tag{10}$$

where

$$Y = \mu d^{-1} h^2 \beta_{021}^2 \beta_{122}^{-1} \tag{11}$$

and

$$v = 5d^4 h^{-6} \beta_{504} \beta_{021}^{-2} \beta_{122}^{-1} \tag{12}$$

THERAPEUTIC USES OF COPPER-CHELATING AGENTS

The comparison of the experimental data with normalized graphs calculated from equation 10 is illustrated in Fig. 3. From the differences between the calculated and experimental readings on each axis, we obtained from the position of the best fit: log $\beta_{122} = 39.1$ and log $\beta_{504} = 101.5$. These constants were refined by using the general least-squares computer program, Letagropvrid. The result was:

$$\log \beta_{122} = 39.18 \pm 0.03 \qquad \log \beta_{504} = 101.50 \pm 0.05.$$

Distribution of Cu(I) ions among complexes

The calculations of the distribution of Cu(I) ions among the predominating species in each system indicated that, at the physiological concentration of glutathione (1–6 mM), the species $CuH_2A_2^{3-}$ predominates over CuHA and, at the therapeutic concentration of penicillamine, both $CuH_2A_2^-$ and $Cu_5A_4^{3-}$ predominate.

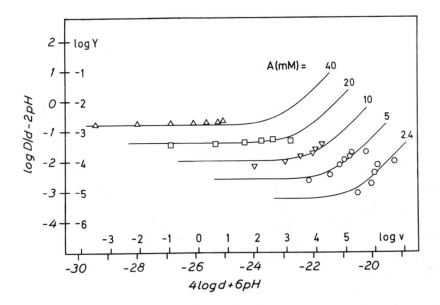

FIG. 3. The Cu(I) complexes of penicillamine (Österberg et al 1979). Experimental data (log D/d − 2pH) *versus* (4 log d + 6pH) for constant A, compared with normalized graphs, log Y(log v)$_A$. The fully drawn curves have been calculated assuming $CuH_2A_2^-$ and $Cu_5A_4^{3-}$ species using equation 10. See text for definition of terms.

DISCUSSION

The results described in the previous section provide evidence for the formation of diligandic mononuclear species, of the type Cu(I) (HA)$_2$, in both the glutathione and the penicillamine systems. In addition, a polynuclear species of the type Cu$_5$A$_4^{3-}$ forms in the penicillamine system; this species might be the most important member of a series of polynuclear species of the type Cu(CuA)$_n^{(n-1)-}$. Such polynuclear species do not seem to form in the glutathione system, where a possible minor species might be the mononuclear CuHA$^-$ species. It seems possible that under more oxidizing conditions the penicillamine-Cu(I) species Cu$_5$A$_4^{3-}$ may have its counterpart in a mixed valency Cu(I), Cu(II) species, perhaps with a composition similar to that present in the crystals of copper and penicillamine, i.e. Cu(I)$_8$Cu(II)$_6$A$_{12}^{4-}$ (Birker & Freeman 1976) (see also Schugar et al 1976, Wright & Frieden 1975).

The distribution of the Cu(I) ions among the complexes at a physiological pH (pH 7.45) indicates that in the penicillamine system the polynuclear species predominates at relatively low concentrations of penicillamine and at high concentrations of copper. These ranges are within the concentrations of penicillamine in blood plasma, when the drug is used against Wilson's disease (Walshe 1966). In the glutathione system, on the other hand, the species CuH$_2$A$_2^{3-}$ predominates in the physiological range of glutathione concentrations (1-6 mM). As a result, under conditions in which univalent copper is important, e.g. in the cell membrane and probably within the cells, CuH$_2$A$_2^{3-}$ would be the predominating glutathione species and CuH$_2$A$_2^-$ and Cu$_5$A$_4^{3-}$ would be the predominating penicillamine species. When the physiological and medical implications of these results are discussed, it must be kept in mind that the present medium, 0.5 M NaClO$_4$, differs from the physiological one, 0.15 M NaCl: the relatively high concentration of background electrolyte used in this study is required in order to keep the activity factors effectively constant. However, this difference in ionic strength may not be of great importance, because many low-molecular-weight systems show no major change in their stability constants when the medium changes from 0.15 to 0.5 M ionic strength (Sillén & Martell 1964). Therefore, as a first approximation, our data may be discussed in relation to their possible medical implications.

The results of the equilibrium analysis of the present systems were used to determine the concentrations of the various Cu(I) species in a system of Cu(I), penicillamine, and glutathione, i.e. a system that may simulate the conditions supposed to prevail *in vivo*. For this purpose a computer program, Haltafall (Ingri et al 1967), was used. The results indicate that, for the concentration of glutathione supposed to exist in blood (1.0 mM), penicillamine can successively compete with glutathione for the Cu(I) ions. When the labile copper concentration is very high, 500 μM, the main Cu(I) species in solution is generally the penicillamine species $Cu_5A_4^{3-}$. On the other hand, when the labile copper concentration is about the same as the physiological concentration of copper in blood, about 1 μM, then species of the type $Cu(HA)_2$ predominate in both systems; however, the glutathione species is much more important at a physiological pH than that of penicillamine. Thus, if the penicillamine concentration is kept equal to or lower than the concentration of glutathione, our results indicate that penicillamine is efficient for the treatment of diseases of excess copper and metal intoxications. By this treatment the excess of metal ions can be eliminated, thus leaving the physiological Cu(I) metabolism essentially undisturbed. As illustrated by Fig. 4, a well controlled dose of penicillamine that yields a blood concentration of 0.2 to 1.0 mM seems to be optimal, because 0.2 - 1.0 mM penicillamine will bind only a minor fraction of total labile copper at its physiological concentration (1 μM); then $CuH_2A_2^-$ is the main penicillamine species which does not compete as well with the corresponding glutathione species at physiological pH. It should be noted that these data are based on the assumption that Cu(I), rather than Cu(II) or a mixed valency Cu(I)–Cu(II) species, is important locally where penicillamine reacts *in vivo*. A recent study supports this idea since complexes of Cu(I) rather than Cu(II) are efficient agents against inflammation (Whitehouse & Walker 1978).

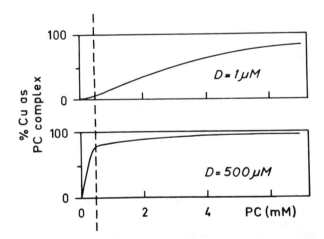

FIG. 4. Moles per 100 moles of copper ions bound to penicillamine (PC) at pH 7.45 in the Cu(I)-glutathione-penicillamine system for a physiological (1.0 μM) and a toxic level (500 μM) of labile copper(I) ions (Österberg et al 1979). The concentration of glutathione is the same as that in human blood (1.0 mM). The vertical dashed line indicates optimal penicillamine concentration (~ 0.5 mM). D is the total concentration of free Cu(I) ions.

That Cu(I) rather than Cu(II) species are efficient is also in agreement with the suggested mechanism of action of copper-penicillamine: Cu(I) ions act as scavengers for superoxide (O_2^-) radicals present in the tissues (Younes & Weser 1977).

In humans and many animals, a toxic reaction develops usually as a result of copper being introduced by injection or by haemodialysis. Oral doses of copper are generally tolerated without harmful effects. If copper ions are introduced by injection or haemodialysis, then as Fig. 4 illustrates, penicillamine at 0.2 to 1.0 mM concentration is a useful antidote to the toxicity. For sensitive subjects, a concentration of 0.5 mM penicillamine in the blood would be an excellent preventive agent for suspected intoxication from copper ions during haemodialysis (Fig. 4). As far as copper-sensitive animals such as sheep (Süveges et al 1971) and dogs (Goresky et al 1968) are concerned, great precautions should be exercised if copper ions are administered to the animal, especially when they are injected intravenously, due to the possibility of haemolysis. This study indicates that if the animals developed signs of intoxication, penicillamine is a useful antidote, especially at carefully controlled doses.

In conclusion, this study shows how penicillamine interacts with Cu(I) ions when it is present *in vivo*. Penicillamine has proved efficient in the treatment of Wilson's disease, metal intoxication, and arthritis. In carefully controlled doses, penicillamine can safely be used both therapeutically and for eliminating toxic metal ions (including copper ions), without disturbing the normal concentration of copper(I) ions.

ACKNOWLEDGEMENTS

This work was supported by grants from the Swedish Natural Science Research Council and the International Copper Research Association.

References

Ahrland S, Chatt J, Davies NR 1958 The relative affinities of ligand atoms for acceptor molecules and ions. Q Rev Chem Soc Lond 12:265-276
Biedermann G, Ciavatta L 1964 Studies on the hydrolysis of metal ions. Ark Kemi 22:253-279
Birker PJMWL, Freeman HC 1976 Metal-binding in chelation therapy: X-ray crystal structure of a copper(I)-copper(II) complex of D-penicillamine. Chem Common 1976(9):312-313
Goresky CA, Holmes TH, Sass-Kortsak A 1968 The initial uptake of copper by the liver in the dog. Can J Physiol Pharmacol 46:771-784
Gran G 1952 Determination of the equivalence point in potentiometric titrations. Part II. Analyst Lond 77:661-671

Ingri N, Sillén LG 1965 High-speed computers as a supplement to graphic methods. IV. An Algol version of Letagropvrid. Ark Kemi 23:97-121
Ingri N, Kakolowics W, Sillén LG, Warnqvist B 1967 High-speed computers as a supplement to graphical methods. V. Talanta 14:1261-1286
Long C 1961 In: King EJ, Sperry WM (eds) Biochemists' handbook. E & FN Spon, London, p 682
Multi-Centre Trial Group 1973 Controlled trial of D-penicillamine in severe rheumatoid arthritis. Lancet 1:275-280
Österberg R 1965 The metal complexes of peptides and related compounds. 1. Copper(II) complexes of O-phosphorylserylglutamic acid in aqueous solution. Acta Chem Scand 19:1445-1468
Österberg R 1970 The metal complexes of peptides and related compounds. 4. A mixed copper(I)-copper(II) complex of glycylglycylglycine. Eur J Biochem 13:493-503
Österberg R 1974 Models for copper protein interaction based on solution and crystal structure studies. Coord Chem Rev 12:309-347
Österberg R, Ligaarden R, Persson D 1979 Copper(I) complexes of penicillamine and glutathione. J Inorg Biochem 10:341-355
Sarkar B 1976 Miscoordination in a metal-dependent disease: Wilson's disease. In: Williams DR (ed) An introduction to bio-inorganic chemistry. Thomas, Springfield USA, p 318-333
Schugar HJ, Ou C, Thich JA, Potenza JA, Lalanzette RA, Furey W Jr 1976 Molecular structure and copper(II)-mercaptide charge-transfer spectra of a novel $Cu_{14}[SC(CH_3)_2CH_2NH_2]_{12}Cl$ cluster. J Am Chem Soc 98:3047-3048
Sillén LG 1956 Some graphical methods for determining equilibrium constants. II. On 'curve-fitting' methods for two-variable data. Acta Chem Scand 10:186-202
Sillén LG, Martell AE 1964; 1971 Stability constants, 2nd edn; Suppl No 1, Stability constants of metal-ion complexes. The Chemical Society, London (Chem Soc Spec Publ Nos 17;25)
Süveges T, Rátz F, Sályi G 1971 Pathogenesis of chronic copper poisoning in lambs. Acta Vet Acad Sci Hung 21:383-391
Tietze F 1969 Enzymic method for quantitative determination of nanogram amounts of total and oxidized glutathione. Anal Biochem 27:502-522
Walshe JM 1966 Wilson's disease, a review. In: Peisach J et al (eds) The biochemistry of copper. Academic Press, New York, p 475-498
Whitehouse MW, Walker WR 1978 Copper and inflammation. Agents Actions 8:85-90
Wright JR, Frieden E 1975 Properties of the red-violet complex of copper and penicillamine and further insight into its formation reaction. Bioinorg Chem 4:163-175
Younes M, Weser U 1977 Superoxide dismutase activity of copper-penicillamine: possible involvement of Cu(I) stabilized sulphur radical. Biochem Biophys Res Commun 78:1247-1253

Discussion

Dormandy: Presumably, when a patient with rheumatoid arthritis is given penicillamine, the drug is converted in the body to the chelate, copper-penicillamine. After several weeks or months the drug often loses its effectiveness and has to be replaced by one of a number of other drugs. Perhaps the initial beneficial effect is due to the chelate of copper-penicillamine, and once the mobilizable supply of copper in the body has been exhausted the drug ceases to be effective. If one were to supplement penicillamine treatment with extra copper in these patients (who have no excess copper initially) the beneficial effect of the penicillamine might be prolonged.

Österberg: There are consistent reports that copper is involved in the therapeutic effect of penicillamine in rheumatoid arthritis. Generally, when penicillamine alone is administered, low doses are given initially in order to produce a particular response but, later, higher doses are required to evoke the same response. But as you saw from the data that I just presented, this would not be advisable because it would seriously disturb the physiological metabolism of copper; the essential copper might no longer be present at the sites where it exercises its biological functions. Therefore, in rheumatoid arthritis, in which copper is believed to have a specific effect, it is advantageous to give copper together with the penicillamine.

Hill: When you did the computer fit for the distribution of the copper between glutathione and penicillamine, Dr Österberg, did you consider the possibility of a mixed complex, i.e. one of copper-glutathione-penicillamine?

Österberg: No, we did not consider a mixed complex, but it is possible that such species exist. So far, we have not studied the effects of copper and the two ligands together, but I agree with you that this should be examined.

Hill: You said it was likely that these ligands could interfere with copper metabolism. If they form the copper complexes then it is also possible that they could interfere with the metabolic processes that require oxygen. At the Ciba Foundation meeting two years ago we showed preliminary data on the vitamin K-dependent and oxygen-dependent synthesis of prothrombin (Esnouf et al 1979a). Since then, we have shown (Esnouf et al 1979b) that *in vitro,* in a liver microsomal fraction, the synthesis of prothrombin is markedly inhibited by a variety of copper complexes including copper-penicillamine. The copper complexes are more effective inhibitors than the dismutase and we believe that this is related to the ease of access of the enzyme to this rather complex microsomal apparatus: i.e. access is less restricted for the copper complexes. We have also shown that dismutase is a very effective inhibitor of the production of hydroxyl radicals by neutrophils during the respiratory burst associated with phagocytosis. There is probably no problem of access in that case and consequently, although the copper complexes are effective inhibitors of hydroxyl radical production, they are less effective inhibitors than the dismutase.

Hurley: I have seen two reports of women with arthritis who were given large amounts of penicillamine during pregnancy and who gave birth to babies with abnormal connective tissue. There was no attempt to explain the abnormality but it seems likely that it was due to a copper-induced deficiency (Hurley & Keen 1979). Would you care to comment about that?

Österberg: Well, it seems likely that your interpretation is correct. One of my main points of emphasis is that doctors who give penicillamine to patients should carefully avoid overdosages, which interfere with the physiological level of copper ions.

Hurley: Scheinberg & Sternlieb (1975) reported that in a large number of patients with Wilson's disease there were no teratogenic effects of penicillamine. But those patients have high levels of copper in their bodies, so the results may not be relevant to detrimental effects of penicillamine in patients without Wilson's disease (Hurley & Keen 1979).

Österberg: I believe that large doses of penicillamine should generally be accompanied by administration of copper ions, or at least some tests should be made on the excretion of copper ions in order to ensure that large amounts of copper are not eliminated.

So far we have discussed the possible induction of copper deficiency as a result of penicillamine treatment. In other circumstances, the toxic effects of excess copper can be prevented by penicillamine. For instance, during haemodialysis, even small amounts of copper in the solution have caused sudden deaths (Matter et al 1969), and for these patients, low doses of penicillamine, which would not alter the physiological metabolism of copper ions, would prevent any harmful effects of copper.

Mills: In vivo, the cellular environment is quite rich in zinc. I appreciate that the stability constants of zinc complexes are about half those of the copper complexes, but is the effect of penicillamine related in any way to the presence of the zinc-rich medium? If it is not, then why does administration of penicillamine often increase zinc excretion even in patients with Wilson's disease? In many arthritic patients treated with penicillamine I believe that several of its side-effects could be related to induced defects in zinc metabolism; for example, loss of smell and taste acuity (Klingberg et al 1976).

Österberg: As I indicated in my paper, low doses of penicillamine will not interfere with the copper ions at the physiological concentration of labile copper, about 1 μM. Therefore, the penicillamine is indeed free to bind any other metal ions, and the Zn^{2+} ion would be a likely candidate.

Frieden: You studied these reactions aerobically, Dr Österberg, but have you considered the possible role of hydrogen peroxide, which is a by-product of the re-oxidation of Cu(I) to Cu(II)? Hydrogen peroxide can be removed by catalase but nevertheless it is an active molecule for producing other oxidation products that might concern you.

Österberg: I agree. In your own studies, Professor Frieden, I believe that you isolated from the urine a polymer of Cu(I), Cu(II) and penicillamine (Wright & Frieden 1975). In the presence of oxygen this polymer is always likely to be formed. When penicillamine is present, Cu(II) is generally reduced to Cu(I) and this redox reaction is highly favoured. We have done some preliminary studies on amino acids, copper ions and sulphydryl compounds under aerobic and anaerobic conditions and these studies do not generally invalidate the results obtained under anaerobic conditions (R. Österberg, unpublished results). In the blood plasma and especially in the urine, Cu(I) may be oxidized to Cu(II) due to the aerobic conditions, whereas in the tissues, where reduced glutathione is always available, Cu(II) would be reduced to Cu(I). Thus, if the Cu(I) species is removed from the anaerobic environment into an aerobic one, then it is likely that the mixed complex that you found, Professor Frieden, will be formed.

Mills: Are you suggesting that this happens *in vivo* as well?

Österberg: Well, most of the data are based on model studies and the true *in vivo* conditions remain to be proven.

Mills: Hill & Matrone (1970) showed that metabolic interactions between copper and zinc or between copper and cadmium are more readily explicable on structural grounds if we assume that copper is present as Cu(I) and not as Cu(II).

In contrast, the metabolic antagonism exhibited by silver against copper is relatively weak, as might be expected if Cu(II) plays a less significant part than Cu(I) in copper transport and in tissue storage.

Owen: We must be cautious about translating data from *in vitro* studies to conditions *in vivo*. Copper will form a complex with penicillamine in the test tube. Furthermore, the administration of penicillamine to normal people or diseased patients greatly increases the urinary excretion of copper, zinc and other metals. The assumption has been made that the urinary copper is present as a copper-penicillamine complex but the work of McCall et al (1969, 1970) shows that the urinary copper is bound to a protein or polypeptide, and not directly to penicillamine. Zinc, however, may form a complex with penicillamine. The chemical mechanism by which penicillamine removes copper from the body is still not clear.

Frieden: We were nevertheless able to isolate a mixed-valence complex of Cu(I), Cu(II) and penicillamine from the urine several years ago (Wright & Frieden 1975).

Hill: That mixed-valence complex is easy to recognize because it turns the urine purple.

Mangan: I have never been certain whether the beneficial effects of D-penicillamine in rheumatoid arthritis are due directly to the D-penicillamine or to its copper complex. If the effect of D-penicillamine is to mobilize copper from sites in the body where it is causing damage, then I would not expect copper-D-penicillamine to have any therapeutic effect. I am interested to know whether patients with rheumatoid arthritis have been given copper-D-penicillamine.

Animal models of inflammation shed little light on this problem; in standard models, e.g. in adjuvant arthritis in the rat (Liyanage & Currey 1972), D-penicillamine itself is not active but, as I mentioned earlier in the meeting, many simple copper-containing compounds *are* active. When we administer copper-D-penicillamine subcutaneously to rats, the urine rapidly becomes purple, indicating that the complex passes through the body unchanged (F.R. Mangan, unpublished work).

Österberg: Dr Owen, did you say that copper in the urine is not in the form of copper-penicillamine?

Owen: If the copper chloride-penicillamine complex is given intravenously to an animal, the violet complex is excreted directly in the urine (Wright & Frieden 1975). However, only a small amount of that violet complex is found in *human* urine, and on other grounds (Laurie & Prime 1979) it seems unlikely that penicillamine directly chelates copper in the bloodstream.

Österberg: When penicillamine is administered alone it takes up very little copper under physiological conditions and then copper may be excreted in the form of a complex with peptides, amino acids or perhaps low-molecular-weight proteins (if the kidneys are not damaged). But when copper-penicillamine is administered, as far as I know, the copper *is* excreted in the form of a penicillamine complex.

Owen: That may be so.

Österberg: What you suggested earlier, Dr Owen, is however in good agreement with our data.

Owen: As penicillamine becomes less effective in the treatment of Wilson's disease, the dose of penicillamine must be raised; otherwise the amount of copper in the urine progressively falls.

Österberg: That is also related to the copper concentration in blood plasma. In a patient with Wilson's disease, treated with penicillamine, there is a fairly low concentration of copper in the blood plasma. As penicillamine is excreted, the plasma concentration of copper ions decreases, and if the rate of excretion of copper ions is to be maintained, the plasma concentration of penicillamine must be increased.

Dormandy: There are many reports of the toxic effects of penicillamine in patients with rheumatoid arthritis and rheumatologists therefore regard the drug as potentially dangerous. By contrast, in most patients with Wilson's disease, penicillamine is given for many years and with no ill-effects whatever.

Mills: In how many of these cases of rheumatoid arthritis has zinc replacement been attempted?

Dormandy: I am not aware of any, but patients with rheumatoid arthritis who have been given penicillamine show no evidence of zinc depletion, at least as reflected by serum measurements of zinc.

Hill: But isn't penicillamine the wrong drug to give in rheumatoid arthritis? In Wilson's disease the drug is used as a chelating agent. I suggest that in rheumatoid arthritis one should be using the metal chelate and not the chelating agent, whose properties are quite different.

Dormandy: That's right, and the same is true of penicillamine treatment for biliary cirrhosis.

Harris: In Wilson's disease one is dealing with accumulation of copper intracellularly and so the effectiveness of any therapeutic agent depends upon its ability to enter the cell. The status of the copper within the cell is also a problem, whether it is bound tightly to proteins or to smaller organic ligands. If a substantial fraction of the copper is bound to glutathione in the cell, will penicillamine be able to release that copper?

Österberg: Whether or not penicillamine can release copper from the cells will depend on the copper concentration within the cells. If the physiological concentration of labile copper is about 1 μM, then penicillamine would not pick up much copper at all since it cannot compete with glutathione or with other strong copper-complex-forming agents such as copper proteins. On the other hand, if there is a large concentration of labile copper then penicillamine would be very efficient in removing copper. That is what our data indicate. The reason for this is that the polynuclear penicillamine complex becomes important at high copper concentrations. Penicillamine is an interesting drug because of its ability to form more than one complex with both Cu(I) and Cu(II).

Harris: Am I correct to assume that in Wilson's disease one would expect to see a large amount of copper bound to glutathione?

Österberg: I am using glutathione as a model ligand for the thiol ligands present *in vivo*, but that does not necessarily mean that copper is bound to glutathione *in vivo*; it might be bound to stronger complex-forming compounds than glutathione, such as metallothionein. What I was saying was that under the conditions in which glutathione successfully competes with penicillamine for copper ions, penicillamine will not interfere with binding of the copper ions.

Hill: Penicillamine is an ideal reagent to remove copper from the cells because it will coordinate to Cu(I); it may reduce Cu(II) to Cu(I) and then the excess penicillamine would bind to the Cu(I) that is formed, thus giving *solu-*

ble Cu(I) complexes. On the other hand, at least one of the forms identified by Dr Österberg does not carry a large charge and so it may be quite mobile across cell membranes. Penicillamine is therefore an ideal stripping reagent for copper.

References

Esnouf MP, Green MR, Hill HAO, Irvine GB, Walter SJ 1979a Dioxygen and the vitamin K-dependent synthesis of prothrombin. In: Oxygen free radicals and tissue damage. Excerpta Medica, Amsterdam (Ciba Found Symp 65), p 187-197

Esnouf MP, Green MR, Hill HAO, Walter SJ 1979b The inhibition of the vitamin K-dependent carboxylation of glutamyl residues in prothrombin by some copper complexes. FEBS (Fed Eur Biochem Soc) Lett 107:146-150

Hill CH, Matrone G 1970 Chemical parameters in the study of in vivo and in vitro interactions of transition elements. Fed Proc 29:1474-1481

Hurley LS, Keen CL 1979 Teratogenic effects of copper. In: Nriagu JO (ed) Copper in the environment. Part II: Health effects. John Wiley, New York, p 35-56

Klingberg WG, Prasad AS, Oberleas D 1976 Zinc deficiency following penicillamine therapy. In: Prasad AS, Oberleas D (eds) Trace elements in human health and disease. Vol 1: Zinc and copper. Academic Press, New York, p 51-66

Laurie SH, Prime DM 1979 The formation and nature of a mixed valence copper-D-penicillamine-chloride cluster in aqueous solution and its relevance to the treatment of Wilson's disease. J Inorg Biochem 11:229-239

Liyanage SP, Currey HLF 1972 Failure of D-penicillamine to modify adjuvant arthritis or immune response in the rat. Ann Rheum Dis 31:521

Matter BJ, Pedersen J, Psimenos G, Lindeman RD 1969 Lethal copper intoxication in hemodialysis. Trans Am Soc Artif Intern Organs 15:309-315

McCall JT, McLennan KG, Goldstein NP, Randall RV 1969 Copper and zinc homeostasis during chelation therapy. In: Hemphill DD (ed) Trace elements in environmental health. University of Missouri, Columbia, vol 2:127-140

McCall JT, Goldstein NP, Randall RV 1970 Metabolism of zinc and copper in patients with metal poisoning. In: Hemphill DD (ed) Trace substances in environmental health. University of Missouri, Columbia, vol 3:91-105

Scheinberg HI, Sternlieb I 1975 Pregnancy in penicillamine treated patients with Wilson's disease. N Engl J Med 293:1300-1302

Wright JR, Frieden E 1975 Properties of the red-violet complex of copper and penicillamine and further insight into its formation reaction. Bioinorg Chem 4:163-175

Anaerobic potentiation of copper toxicity and some environmental considerations

DAVID C.H. McBRIEN

Biochemistry Department, School of Biological Sciences, Brunel University, Uxbridge, Middlesex UB8 3PH, UK

Abstract The toxicity of copper is substantially greater when the metal ion is applied to cells under conditions of anoxia than under aerobic conditions. The increase in toxicity occurs because Cu(II) is reduced to Cu(I) which is stable under anoxia and is more toxic than the oxidized species. This effect has been observed in mammalian cells in tissue culture and in bacteria. Anoxic potentiation of copper toxicity has also been observed in algal, fungal and yeast cells. The reduction of Cu(II) has been demonstrated by electron paramagnetic resonance spectrometry. The appearance of Cu(I) has been observed by means of a specific colorimetric reagent. The reduction of Cu(II) to Cu(I) is not dependent upon concomitant metabolism and requires only a supply of sulphydryl groups.

Microorganisms that are responsible for two anaerobic processes of economic importance to humans, the digestion of sewage sludge in effluent treatment plants and the digestion of food in ruminant animals, may be exposed to high concentrations of copper. The environmental consequences of this exposure are discussed.

It is possible to argue that in a symposium devoted to 'Biological roles of copper' it is inappropriate to consider copper toxicity. Toxicity is a biological property but not a biological function. However, the topic is sufficiently important for semantic arguments to be set aside, even though few people would agree 'that its widespread use in industry and its high toxicity distinguish copper as the most potentially hazardous metal released to the environment' (Schmidt 1978). Overt toxicity caused by environmental exposure to copper is rare in humans. Copper is an essential element for living organisms and as a result of evolution, humans and other higher animals have elaborated mechanisms for the absorption, compartmentation and metabolism of copper that minimize its toxic effects. However, simpler organisms lack such sophisticated machinery and are more susceptible to copper intoxication. We use this sensitivity to our own advantage and disseminate much copper, deliberately, into the environment, as a toxic agent to control fungi on crops, molluscs in fields and algal blooms on reservoirs. The effect of these and our other uses of copper adds to the natural geochemical mobilization of the metal an environmental burden that may produce unwanted toxic effects on the more susceptible simple organisms. I propose first to describe some laboratory investigations into copper toxicity and then to consider the possible environmental relevance of the findings.

COPPER TOXICITY UNDER ANOXIA

Copper ions normally exist in one of two oxidation states, Cu(I) which is diamagnetic and Cu(II) which is paramagnetic. Of these two species Cu(I) is unstable in aqueous solutions in the presence of oxygen or oxidizing metal ions such as Fe^{3+} and is rapidly oxidized to Cu(II). For this reason, when the toxicity of copper to isolated cells is investigated, salts or compounds of Cu(II) are the usual sources of the applied copper. In a variety of different types of cell – yeast (Murayama 1961), algae (McBrien & Hassall 1965, 1967) and bacteria (Beswick et al 1976) – the toxic effects of Cu(II) are greater when the ions are applied to cells under anoxia rather than under aerobic conditions. This remains true, after the addition of copper, even if aerobic conditions are restored (Beswick et al 1976, McBrien & Hassall 1967). When we compared the toxicity of Cu(I) and Cu(II) solutions under anoxic conditions Cu(I) was substantially more toxic than Cu(II) (Fig. 1) even though electron paramagnetic resonance (e.p.r.) spectroscopy revealed (Fig. 2) that most of the Cu(II) was converted to a non-paramagnetic species, probably Cu(I) (Beswick et al 1976).

EPR SPECTRA OF FROZEN CELL SUSPENSIONS

In our earlier study (Beswick et al 1976) e.p.r. spectra of *Escherichia coli B* exposed to Cu(II) under anoxia were recorded at ambient temperatures. We were unable to obtain for comparison spectra from cells incubated with Cu(II) under aerobic conditions. A culture of *Escherichia coli B*, washed and suspended in 0.85% w/v sodium chloride with 2.8×10^9 cells ml^{-1}, continues endogenous respiration for several hours at ambient temperatures. When the culture is transferred to an e.p.r. cuvette, this respiration removes all the dissolved oxygen from the incubation medium within 10 min, whereas an e.p.r. spectrum often takes up to 30 min to be recorded. More recently we have extended our e.p.r. measurements to cultures incubated with copper under aerobic conditions, by freezing the cultures rapidly in liquid nitrogen, after placing them in the sample-tubes. This freezing can be accomplished

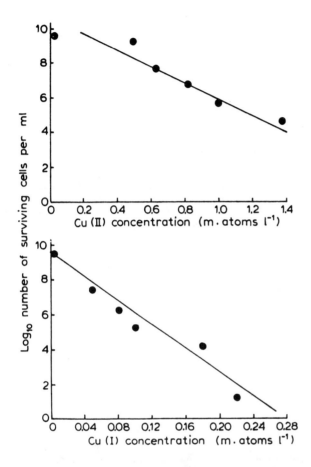

FIG. 1. The relationship between concentration of copper and the number of cells surviving a 30 min exposure to the metal. In both cases the log initial cell concentration was 9.7. Reproduced from Beswick et al (1976), with permission from Elsevier/North-Holland.

within 30 s of removing the sample from the culture. The e.p.r. spectrum is then recorded using a temperature-controlled cavity at $-160°C$. Some results obtained using murine leukaemic L5178Y cells have already been reported (Hesslewood et al 1978). We showed that when a suspension of 10^8 cells ml^{-1} was incubated under anoxic conditions with cupric chloride (0.066 m atoms Cu(II) l^{-1}), the e.p.r. signal obtained from Cu(II) had almost completely disappeared within 2.5 min of addition of the copper (Fig. 3) whereas under aerobic conditions the Cu(II) signal remained at its original intensity for at least 70 min.

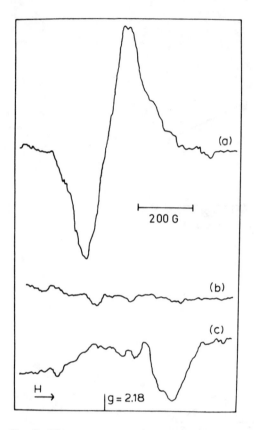

FIG. 2. Electron paramagnetic resonance spectra of (a) an anoxic solution of 0.5 m atoms l^{-1} copper as cupric chloride in 0.85% w/v sodium chloride; (b) an anoxic suspension of *E. coli* (4.8 × 10^9 cells ml^{-1}); (c) an anoxic mixture containing the same concentration of copper used for curve (a) and the same concentration of cells used for curve (b), recorded 5 min after mixing. Reproduced from Beswick et al (1976), with permission from Elsevier/North-Holland.

Using the technique of recording e.p.r. spectra of frozen cell suspensions we have now made a series of observations using *Escherichia coli B* as the test organism. Fig. 4 shows some spectra obtained when suspensions of 6.0 × 10^9 cells ml^{-1} in 0.85% w/v sodium chloride were treated with 0.066 m atoms Cu(II) l^{-1} and incubated under aerobic or anoxic conditions. Because a high proportion of tubes cracked when introduced into liquid nitrogen we were unable to use expensive fixed-volume e.p.r. tubes. For this reason, and because the high gain which we were obliged to use reveals distortion of the baseline due to a slight cavity contamination, it is not possible to make precise quantitative comparisons between the spectra. However, the aerobic suspension shows a strong e.p.r. signal ascribable to Cu(II) even 10 min after addition of the copper salt, whereas the signals from the anoxic suspension have almost disappeared at that time. Figs. 3 and 4 show that in all the samples in which both cells and copper are present, the strongest e.p.r. signal is obtained at a higher field strength than the strongest signal obtained with copper alone. In both these examples the high-field signal virtually disappears within a few minutes of addition of the copper when incubation is under anoxia. In our earlier experiments (Beswick et al 1976), using 4.8×10^9 cells ml^{-1} and a much higher copper concentration (0.5 m atoms l^{-1}), we found that on com-

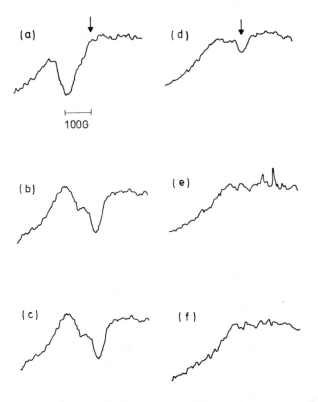

FIG. 3. Electron paramagnetic resonance spectra showing the loss of Cu(II) signal during incubation of cupric chloride with L5178Y cells under anoxia. The magnetic field increases from left to right. The arrows indicate g (spectroscopic splitting factor) = 2.00. Cell concentration was 10^8 ml^{-1} and copper concentration 6.6×10^{-2} m atoms l^{-1}. (a) Copper in saline; (b) copper plus medium; (c) copper plus cells ($t = 0$ min); (d) copper plus cells ($t = 2.5$ min); (e) copper plus cells ($t = 60$ min); (f) cells alone. Residual signal is background due to cavity contamination. Reproduced from Hesslewood et al (1978) by permission of H.K. Lewis and Co.

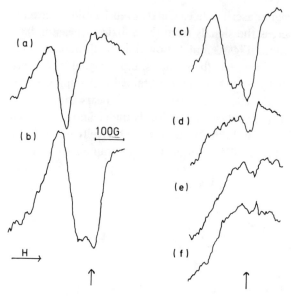

FIG. 4. Electron paramagnetic resonance spectra showing the loss of Cu(II) signal during incubation with *E. coli B* under anoxia. Copper concentration was 6.6×10^{-2} m atoms l^{-1} and cell concentration 6.0×10^9 cells ml^{-1}. The arrows indicate $g = 2.00$. (a) Copper alone; (b) copper plus cells, aerobic incubation ($t = 10$ min); (c) copper plus cells, anoxic incubation ($t = 0$ min); (d) copper plus cells, anoxic incubation ($t = 2$ min); (e) copper plus cells, anoxic incubation ($t = 10$ min); (f) cells alone. Residual signal is from the cavity.

pletion of reduction there was a substantial residual high-field signal which we attributed to Cu(II) that was bound to cell constituents and therefore resistant to reduction. However, we now know that the amount of copper represented by the high-field signal varies with the initial concentration of copper. We have observed, using a cell suspension of 6.0×10^9 cells ml^{-1} in 0.85% w/v sodium chloride, that when the copper concentration is no greater than 0.16 m atoms l^{-1} the e.p.r. signals from Cu(II) completely disappear within 30 min of incubation under anoxia. With concentrations greater than this there is an increasingly large residual high-field signal until, at concentrations greater than 0.63 m atoms l^{-1}, the diminution in the signal is proportionately too small to be seen.

PRODUCTION OF Cu(I)

Beswick et al (1976) and Hesslewood et al (1978) reported the disappearance of the Cu(II) e.p.r. signal when cells were treated with copper under anoxia; they concluded that the Cu(II) was being converted to a non-paramagnetic species but they had no direct evidence that this species was Cu(I). However, using the specific colorimetric reagent, bathocuproin (2,9-dimethyl-4,7-diphenyl-1,10-phenanthroline, see Johnson 1964) in aqueous solution, we have demonstrated the appearance of Cu(I) in cell suspensions of *E. coli B* incubated with Cu(II) under anoxia (K. Roser & D.C.H. McBrien, unpublished results).

Using a similar method to that of Kumar et al (1978) we incubated a suspension of *E. coli B* (4×10^8 cells ml^{-1} in 0.85% w/v sodium chloride) with 10 mM bathocuproin and 0.1 m atoms l^{-1} of Cu(II) (cupric sulphate) either in capped tubes under nitrogen (anaerobic) or as a thin film in Erlenmeyer flasks in an orbital incubator (aerobic). At various times samples were removed and the absorbance at 483 nm was measured. Fig. 5 shows that under anoxia the coloured complex between Cu(I) and bathocuproin is formed rapidly whereas under aerobic conditions little change in absorbance is observed.

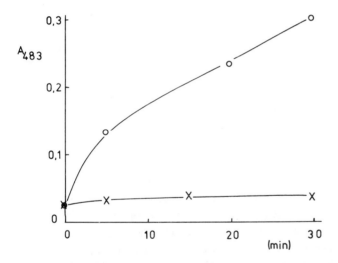

FIG. 5. Absorbance (A) at 483 nm of suspensions of *E. coli B* (4×10^8 cells ml^{-1}) containing bathocuproin (10 mM) and cupric sulphate (0.1 mM). ○—○, anoxic incubation; ×—×, aerobic incubation.

THE REDUCTION OF Cu(II) DOES NOT REQUIRE METABOLISM

The reduction of Cu(II) to Cu(I) can be brought about by heat-killed and by sonicated suspensions of *E. coli*. Suspensions of 10^9 cells ml^{-1} in 0.85% w/v sodium chloride were either held at 100°C for 10 min or sonicated for a total of 2 min in bursts of 30 s, by means of an MSE 60W ultrasonic disintegrator which completely disrupts the cells. The suspensions were then incubated with 0.063 m atoms l^{-1} Cu(II), as cupric sulphate, under aerobic or anoxic conditions. Samples were removed at intervals, and frozen for e.p.r. analysis. From the recorded spectra the magnitude of the e.p.r. Cu(II) signals was assessed semi-quantitatively, by eye, and the results are shown in Table 1. In both types of cell suspension the Cu(II) e.p.r. signal had totally disappeared within 10 min under anoxic conditions but it remained at its original intensity for at least 40 min under aerobic conditions.

TABLE 1

The reduction of copper(II) electron paramagnetic resonance (e.p.r.) signal by heat-treated and by sonicated suspensions of *Escherichia coli*

Treatment	Time (min)	Incubation conditions	
		Aerobic	Anaerobic
Heat	1	+ + +	±
	10	+ + +	−
	20	+ + +	−
	40	+ + +	−
Sonication	1	+ + +	+
	10	+ + +	−
	20	+ + +	−
	40	+ + +	−

Key: intense e.p.r. signal + + +; reduced e.p.r. signal +, ±; no e.p.r. signal −. Suspensions of 10^9 cells ml^{-1} in 0.85% w/v sodium chloride were either held at 100°C for 10 min (heat-treated) or sonicated for 2 min. Suspensions were then incubated with 0.063 m atoms l^{-1} cupric sulphate and samples were removed for e.p.r. analysis at various intervals.

A sonicated cell suspension was separated into 'cell sap' (supernatant) and 'cell coat' (pellet) fractions by centrifugation at 20 000 × g for 30 min and the separate fractions were incubated with copper as described above. The results indicated that the capacity to reduce Cu(II) under anoxia resided principally, but not wholly, in the cell sap fraction. When the cell sap fraction was treated with a protein precipitant (0.6 M perchloric acid), followed by neutralization and centrifugation, the supernatant had lost the capacity to reduce Cu(II).

REDUCTION OF Cu(II) BY SULPHYDRYL GROUPS

The Cu(II) e.p.r. signal disappears completely when 0.06 m atoms l^{-1} Cu(II), as cupric sulphate, is incubated with 0.12 mM cysteine in 0.85% w/v sodium chloride under anoxic conditions. Under aerobic conditions the signal stays at its initial intensity. Similar results are obtained when the same concentration of copper is incubated with 1 mg ml^{-1} yeast alcohol dehydrogenase (EC 1.1.1.1, Boehringer Mannheim) which contains 36 SH groups per molecule. The molecular weight of the enzyme is approximately 145 000 and thus a concentration of 1 mg ml^{-1} is equal to approximately 0.71×10^{-5} M, with an SH concentration of 0.25 mM.

Since the incubation with alcohol dehydrogenase was performed in the absence of substrate or co-factors the reduction of Cu(II) is not being brought about enzymically – the protein is functioning as a supplier of non-metabolic reducing power.

The capacity of the cell sap fraction from *E. coli* and of the 1 mg ml^{-1} solution of alcohol dehydrogenase to reduce the Cu(II) e.p.r. signal under anoxia was abolished when N-ethylmaleimide (10 mM), which reacts with sulphydryl groups (Dawson et al 1969), was added to the reaction mixture immediately before addition of Cu(II).

THE MECHANISM OF COPPER TOXICITY

From the data reported above, one can conclude that cells incubated with Cu(II) under anoxia have a limited capacity for reduction of Cu(II) to Cu(I). This reduction is not dependent upon concomitant metabolism and can be accomplished by cells that have been killed by heat treatment or by sonication as well as by simple solutions of cysteine or alcohol dehydrogenase. The capacity for reduction of Cu(II) to Cu(I) is abolished if the cell extract or protein is pretreated with a reagent that blocks the SH group. Thus, the reduction of cupric to cuprous ions may accomplished by the following reaction.

$$2Cu^{++} + 2\ -SH \rightarrow 2Cu^{+} + -S-S- + 2H^{+} \qquad (1)$$

The Cu(I) produced by reduction within the cell does not have the same potency in killing cells as the Cu(I) applied as a solution. For example, the ED$_{50}$ values for Cu(I) and Cu(II) applied under anoxia were 0.134 m atoms l^{-1} and 1.18 m atoms l^{-1} respectively (Beswick et al 1976), even though, as shown above, the cells have the capacity to produce 0.16 m atoms l^{-1} of Cu(I) from Cu(II). Presumably, the Cu(I) produced intracellularly cannot reach the same sites as the Cu(I) applied to the exterior of the cells, perhaps because it is 'mopped up' by less sensitive binding sites. The nature of the sensitive sites requires elucidation. In two other investigations into copper toxicity (Tröger 1956, Kumar et al 1978), the reduction of Cu(II) to Cu(I) has been observed. In both cases, the conclusions about the significance of this observation were different from those outlined above.

(a) Tröger (1956), using a specific colorimetric reagent, considered that the Cu(II) that disappeared from conidia of *Fusarium decemcellulare* was reduced to Cu(I). He concluded that the reduction was brought about by dehydrogenases and was a detoxification mechanism. Little direct evidence for these conclusions was adduced and the toxicity of Cu(I) was not measured. In the light of our own work it seems unlikely that the reduction of Cu(II) was enzymic or that Cu(I) would have been less toxic than Cu(II).

(b) Kumar et al (1978) have proposed that copper-induced haemolysis, caused by acute copper intoxication in mammals, is brought about by the production of superoxide radicals (O_2^-) during the cyclic oxidation and reduction of copper ions. They proposed that Cu(II) is reduced to Cu(I) by erythrocyte membrane sulphydryl groups (see reaction (1) above) and then that the following reaction occurs.

$$Cu^+ + O_2 \rightarrow Cu^{++} + O_2^- \tag{2}$$

Kumar et al (1978) propose that the superoxide radicals then initiate peroxidation reactions in membrane phospholipids, thus leading to lysis. This process, of course, requires the presence of oxygen. The toxic effects of copper to cells under anoxia include damage to membranes (McBrien & Hassall 1965) but there is little possibility that this damage is mediated by superoxide radicals. Kumar et al (1978) demonstrated that erythrocyte membranes do produce superoxide radicals in the presence of Cu(II) and, with a separate incubation system, they demonstrated that the membranes reduce Cu(II) to Cu(I). It is not clear whether the latter incubation was under aerobic or anoxic conditions. Whatever the role of O_2^- in haemolysis of erythrocytes *in vivo*, the mechanism described by Kumar et al (1978) cannot be invoked as a general mechanism of copper-induced membrane damage since it cannot account for the increase in such damage under anoxia or for the fact that Cu(I) is inherently more toxic than Cu(II). Because we measured the toxicity of Cu(I) (after exposure of the cells to Cu(I) under anoxia) by diluting and plating the cells under aerobic conditions (Beswick et al 1976), superoxide radicals could have been produced during the aerobic stage. However, if this was the principal cause of copper toxicity no difference would have been seen between the effects of Cu(I) and Cu(II).

ANOXIC COPPER TOXICITY IN THE ENVIRONMENT

Anaerobic microorganisms play a significant part in at least two processes of economic importance to humans, in the biological treatment of sewage and in the digestion of ruminant animals. Copper has been observed to interfere with both processes. Since the toxicity of copper is potentiated by anoxia it is interesting to consider these occurrences in more detail.

Municipal effluent treatment plants

Municipal effluent treatment plants receive domestic and industrial waste which can contain substantial amounts of copper. Copper is one of the commonest metals known to cause poor performance in biological waste treatment plants (Kugelman & Chin 1971). The raw sewage received by a treatment plant is usually separated at various stages in the treatment into 'liquid' and 'solid' fractions in settlement tanks (Fair et al 1971). The liquid fraction is usually treated aerobically and the solid sludges are treated by anaerobic digestion, both processes being dependent upon the activity of microorganisms. Most of the metals that contaminate the raw sewage are precipitated in the sludges (National Water Council 1977). The effects of copper on both the aerobic and anaerobic stages of biological sewage treatment have been investigated by McDermott et al (1963a,b). When cupric sulphate at concentrations up to 25 mg l^{-1} was mixed into raw sewage being fed into a pilot plant the reduction in the efficiency of the aerobic ('activated sludge') process was less than 4%. However, the production of methane, a principal product of the anaerobic digestion of sludge, was severely reduced. For example, fifty days after the start of addition of copper at 15 mg l^{-1} to the raw sewage, gas production was only 10% of that for the control.

The interference of heavy metals with anaerobic digestion of sludge can be eliminated by feeding the digesting microorganisms with ferrous sulphate (Kugelman & Chin 1971). The sulphate is reduced to sulphide and the heavy metals precipitate out. Treated sewage sludge equivalent to 1.25×10^6 tonnes dry weight is produced annually in the United Kingdom by municipal treatment plants and 68% is sprayed onto land as a means of disposal. Of the sludge sprayed onto land, 77% is deposited on agricultural or horticultural land (Royal Commission on Environmental Pollution 1979). An example of the concentrations of copper that the sludge might contain is provided by the Glasgow treatment plants which produce sludge containing a mean level of 400 mg kg^{-1} and a maximum of 3000 mg kg^{-1} dry wt. of copper (Schmidt 1978). Thus, with proper management, the contamination of raw sewage by copper need not inhibit its biological treatment. Before I consider the disposal of treated sewage sludge as a source of environmental problems I shall consider another significant source of sewage sludge.

Agricultural waste treatment

In the United Kingdom 60 million tonnes of undiluted animal excreta are produced annually indoors, by intensively reared farm animals. Only a minute proportion is treated or processed in any way (Royal Commission on Environmental Pollution 1979). Most of this is spread, as slurry, onto agricultural land. The Royal Commission recommends that more waste treatment facilities should be provided, primarily to reduce the problem of smell. However, the fact that high levels of copper are commonly added to the diet of pigs poses particular problems for the implementation of this recommendation. For at least 25 years it has been common practice in the United Kingdom to include up to 250 mg/kg dry weight of copper in the food for intensively reared pigs (Braude 1975). At this concentration of copper, the pigs show an accelerated live weight gain and improved food conversion. The reason for this effect of copper is not yet understood, despite intensive investigation (Braude 1975). One of the effects of diets high in copper is to alter the composition of the pig intestinal flora (Fuller et al 1960), an effect which could result from the higher sensitivity of anaerobic, compared with aerobic, microorganisms. Most of the copper administered to pigs is not absorbed across the gut wall but is voided in the faeces. When the waste from pigs that have been fed a high copper diet (250 mg/kg) is digested in anaerobic pits the rate of decomposition is significantly reduced compared with the waste from pigs on a normal diet (10 mg/kg) (Brumm & Sutton 1979). Aerobic treatment of pig waste may also be inhibited by the increased concentrations of copper that are usually found in the excreta of pigs on high copper diets (Robinson et al 1971). The interference with anaerobic, but not aerobic, treatment could presumably be overcome by the addition of ferrous sulphate, as described above for municipal treatment plants. Whether treated or not, most pig slurry is sprayed onto agricultural land (Royal Commission on Environmental Pollution 1979). Wet slurry from piggeries in which 250 mg/kg of copper is included in the diet usually contains 15–35 mg/kg copper, but in extreme cases the concentration reaches ten times that value (Braude 1975). The slurry contains 90% moisture (Royal Commission on Environmental Pollution 1979) and therefore it will usually contain 150–350 mg/kg copper dry wt. or, in extreme cases, 3500 mg/kg dry wt.

Disposal of sewage sludge onto land

The disposal of sewage sludge from municipal treatment plants onto land in the United Kingdom has been considered several times in the recent past by different bodies (Agricultural Development Advisory Service 1971, National Water Council 1977, Royal Commission on Environmental Pollution 1979). Permissible levels of copper, zinc and nickel in sewage used on agricultural land have been recommended (Agricultural Development Advisory Service 1971). The Royal Commission considered that 'the presence of ... metals ... in sewage sludge applied to land is of concern for several reasons'. The reasons listed include possible eventual toxicity to plants (which is especially a problem with copper), the entry of metals into human food chains, the cumulative and irreversible contamination of agricultural soils and the ingestion of contaminated soils by ruminant animals when grass is short. Another effect of applying sludge that is heavily contaminated with copper to the land is that the rate of natural decomposition of organic matter in the soil is reduced (Leeper 1978). The Royal Commission suggests strict controls over the use of sludge from municipal treatment works to ensure that metal levels in soil are monitored and that contaminated grass is not consumed by animals. For reasons that are not fully explained the Royal Commission does not recommend the same controls over the spreading of animal slurry on the land. With regard to the contamination of slurry by copper from pig diets, they say 'we were informed that this does not appear to be a problem'. The figures quoted above for the copper contents of municipal and agricultural sludges suggest otherwise — agricultural slurry can contain similar amounts of copper to sludge from the municipal treatment plants of a heavily industrialized area.

Effects of copper upon ruminant digestion

Ruminants, especially sheep, are more sensitive to copper poisoning than are other animals (Buck 1978, Todd 1969). Pasture herbage does not generally absorb much copper even from heavily contaminated soils. However, cattle and sheep eating grass from land treated with copper-containing sludge may ingest sufficient quantities of copper, on sludge-splashed foliage or in soil swallowed with the grass, to alter the composition of the rumen microflora (Forsberg 1978) or to produce pathological symptoms of copper intoxication (Gracey et al 1976, Dalgarno & Mills 1975). Calves are as susceptible as sheep to copper toxicity (Hill 1975) but older cattle are relatively resistant (Chapman et al 1962). The greater sensitivity of ruminants to copper intoxication does not generally seem to arise from the indirect effect of copper on the rumen microflora but from a direct effect of copper absorbed into the bloodstream (Buck 1978). Although for ruminant animals the anoxic potentiation of copper toxicity, together with dependence of the animals on anaerobic microorganisms and the presence of relatively high concentrations of copper in the environment, poses a potential threat of toxicity, in practice few problems seem to have arisen so far (Braude 1975).

In restricting my discussion to the effects of copper upon anaerobic bacteria I have set aside the possibly more important environmental effects of copper used as an animal-feed supplement in agriculture. More evidence is required before it is possible to decide whether the economic benefits to farmers of using copper in this way outweigh the risks. The United States Food & Drug Administration has already decided to ban the use of animal diets high in copper (Scheinberg & Sternlieb 1976), on grounds that have been criticized as irrational by Braude (1975). The Royal Commission on Environmental Pollution (1979) has called for more research into disposal of sewage sludge. One of the topics that would benefit from more attention is the possibility of long-term environmental damage resulting from the uncontrolled spreading of 'high-copper' pig slurry onto the land.

ACKNOWLEDGEMENTS

I am grateful to P. Williamson for technical assistance and to K.A.K. Lott for his help in obtaining and interpreting the e.p.r. spectra.

References

Agricultural Development Advisory Service 1971 Advisory Paper No 10: Permissible levels of toxic metals in sewage used on agricultural land. Ministry of Agriculture, Fisheries and Food, London

Beswick PH, Hall GH, Hook AJ, Little K, McBrien DCH, Lott KAK 1976 Copper toxicity: evidence for the conversion of cupric to cuprous copper in vivo under anaerobic conditions. Chem Biol Interact 14:347-356

Braude R 1975 Copper as a performance promoter in pigs. In: Copper in farming symposium. Copper Development Association, Potters Bar, Hertfordshire (Roy Zool Soc Symp), p 79-94

Brumm MC, Sutton AL 1979 Effect of copper in swine diets on fresh waste composition and anaerobic decomposition. J Anim Sci 49:20-25

Buck WB 1978 Copper/molybdenum toxicity in animals. In: Oehme FW (ed) Toxicity of heavy metals in the environment. Marcel Dekker Inc, New York vol 1:491-515

Chapman HL, Nelson SL, Kidder RW, Sippel WL, Kidder CW 1962 Toxicity of cupric sulphate for beef cattle. J Anim Sci 21:960-962

Dalgarno AC, Mills CF 1975 Retention by sheep of copper from aerobic digests of pig faecal slurry. J Agric Sci 85:11-18

Dawson RMC, Elliott DC, Elliott WH, Jones KM 1969 Data for biochemical research, 2nd edn. Oxford University Press, p 388

Fair GM, Geyer JC, Okun DA 1971 Elements of water supply and waste water disposal, 2nd edn. Wiley, New York

Forsberg CW 1978 Effects of heavy metals and other trace elements on the fermentative activity of the rumen microflora and growth of functionally important rumen bacteria. Can J Microbiol 24:298-306

Fuller R, Newland LGM, Briggs CAE, Braude R, Mitchell KG 1960 The normal intestinal flora of the pig. IV. The effect of dietary supplements of penicillin, chlortetracycline or copper sulphate on the faecal flora. J Appl Bacteriol 23:195-205

Gracey HI, Stewart TA, Woodside JD 1976 The effect of disposing high rates of copper-rich pig slurry on grassland on the health of grazing sheep. J Agric Sci 87:617-623

Hesslewood IP, Cramp WA, McBrien DCH, Williamson P, Lott KAK 1978 Copper as a hypoxic cell sensitizer of mammalian cells. Br J Cancer (suppl) 37:95-97

Hill R 1975 Copper toxicity. In: Copper in farming symposium. Copper Development Association, Potters Bar, Hertfordshire (Roy Zool Soc Symp), p 43-50

Johnson WC 1964 (ed) Organic reagents for metals, 5th edn. Hopkin and Williams, Chadwell Heath, vol 2:55-57

Kugelman IJ, Chin KK 1971 Toxicity, synergism and antagonism in anaerobic waste treatment processes. In: Anaerobic biological treatment processes. American Chemical Society, Washington DC (Advances in Chemistry Series 105), p 55-90

Kumar KS, Rowse C, Hochstein P 1978 Copper-induced generation of superoxide in human red cell membrane. Biochem Biophys Res Commun 83:587-592

Leeper GW 1978 Managing the heavy metals on the land. Marcel Dekker, New York

McBrien DCH, Hassall KA 1965 Loss of cell potassium by *Chlorella vulgaris* after contact with toxic amounts of copper sulphate. Physiol Plant 18:1059-1065

McBrien DCH, Hassall KA 1967 The effect of toxic doses of copper upon respiration, photosynthesis and growth of *Chlorella vulgaris*. Physiol Plant 20:113-117

McDermott GN, Moore WA, Post MA, Ettinger MB 1963a Effects of copper on aerobic biological sewage treatment. J Water Pollut Control Fed 35:227-241

McDermott GN, Moore WA, Post MA, Ettinger MB 1963b Copper and anaerobic sludge digestion. J Water Pollut Control Fed 35:655-662

Murayama T 1961 Studies on the metabolic pattern of yeast with reference to its copper resistance. I. Respiration and fermentation. Mem Ehime Univ Sect II Nat Sci 4:207-218

National Water Council 1977 Report of the working party on the disposal of sewage sludge to land. Department of the Environment, London

Robinson K, Draper SR, Gelman AL 1971 Biodegradation of pig waste: Breakdown of soluble nitrogen compounds and the effect of copper. Environ Pollut 2:49-56

Royal Commission on Environmental Pollution 1979 Seventh report. Agriculture and pollution. HMSO, London

Scheinberg IH, Sternlieb I 1976 Copper toxicity and Wilson's disease. In: Prasad AS, Oberleas D (eds) Trace elements in human health and disease. Vol 1: Zinc and copper. Academic Press, New York, p 415-438

Schmidt RL 1978 Copper in the marine environment – Part 1. Crit Rev Environ Control 8:101-152

Tröger R 1956 Studien über die fungicide Kupferwirkung bei *Fusarium decemcellulare*. Arch Mikrobiol 25:166-192

Todd JR 1969 Chronic copper toxicity of ruminants. Proc Nutr Soc 28:189-198

Discussion

Hurley: You mentioned, Dr McBrien, that excess copper applied to agricultural fields was unlikely to give rise to copper toxicity in animals, via the food chain, because the grass takes up very little of the applied copper. Does that hold for *all* edible plants that are likely to be grown on such fields, and is there any possibility of toxicity to humans?

McBrien: I believe that most plants that are fed to humans or to animals do not absorb much copper. Some plants *are* able to absorb copper but they are not grown commonly as crop plants for feeding to either animals or humans.

Hassan: What is the effect of excess copper on soil microflora?

McBrien: As I said in my talk, one of the consequences of the build-up of copper in the soil is that the organic matter decays more slowly. All the copper present in sewage sludge is bound to organic matter when the sludge is deposited onto the soil, so the copper is not readily available in large amounts for plants. Eventually, though, the organic matter will decay and then the copper may be released. I do not know whether the copper would be released rapidly or slowly. However, because the organic matter decomposes more slowly when high levels of copper are present, there must be an effect on the microflora of the soil.

Lewis: When the constituents of sewage sludge and slurry are recycled onto the land the right precautions should be used. We are, at present, examining an area of land that belongs to a sewage farm. Sewage sludge has been put onto this land for many years and the adult grazing cattle kept on the area are *copper-deficient* by the end of the grazing season! So there is certainly a problem of poor absorption of copper by grass. Some plants, like the clover family, *are* more likely to take up copper from the soil and the long-term effect of this must be considered if the land use is to be changed. A safety effect of the distribution of sewage sludge onto the land is that the sludge contains nitrogen which encourages the growth of grass and so leads to a dilution effect on the copper. However, when we sampled these pastures during the autumn, it was obvious that the animals had been eating a lot of soil but their blood levels of copper were low.

Mills: Was there molybdenum in the sludge?

Lewis: I have no figures for that but the molybdenum in autumn grass was between 1–2 mg/kg dry matter.

Delves: Was there evidence of an increased uptake of any other trace elements that might antagonize copper?

Lewis: We did not find any evidence of excess absorption of zinc, cadmium or lead.

Dormandy: Dr McBrien, in your original cultures of microorganisms, did the presence of iron have any effect on toxicity?

McBrien: We did not examine the effect of iron.

Österberg: In your study on the anaerobic and aerobic effects of copper, the anaerobic effect was larger than the aerobic one. Do you know the mechanism by which Cu(I) might kill the cell under anaerobic circumstances?

McBrien: I know little about the precise mechanism, but membrane damage occurs. For example, potassium leaks out of algae that have been in contact with copper (McBrien & Hassall 1965).

Dormandy: If lipid peroxidation is responsible for the haemolytic effect it is conceivable that the process is catalysed by iron and therefore that the effect of copper could be influenced by the iron content of the system.

McBrien: We cannot, on the basis of our evidence, rule out the possibility of haemolysis being brought about by superoxide radicals *in vivo*. But that interpretation would imply that haemolysis involves a different kind of membrane damage from the damage that occurs to microorganisms. The idea that there might be two quite different mechanisms may not be acceptable.

Hassan: Perhaps what you have been calling anaerobic conditions are not *absolutely* anaerobic; there may be a small amount of oxygen present which would be enough to generate superoxide radicals anyway, and so cause the damage.

McBrien: But then one would expect that if a small amount of oxygen produced a small amount of damage there would be more damage when there is more oxygen. We do not, in fact, observe that.

Mills: I would like to comment further on the environmental implications of this subject. The European Economic Community is concerned about large amounts of dietary copper given as a growth stimulant to pigs, primarily because of its subsequent effect on the fauna and flora of the soil. There is good evidence that if the zinc and iron balance of the diet is correct the addition of copper to the diet causes no problems to the pigs. However, Britain, Denmark and Holland, which use copper in pig foods, were given about two years to provide evidence that there was no adverse *environmental* impact of the copper supplement. There was very little state-sponsored research to answer the problem, and as a result, the question is now arising again. The hazard to animals of a flux of large amounts of copper through the environment depends not only on the valency state of the copper. If faecal residues, rich in copper, are stored under anaerobic conditions before disposal on the land there is a good chance that the copper sulphide formed during storage will be oxidized when the material is subsequently dispersed. There is a variety of opinions in Northern Ireland, England and Scotland about the hazards of this oxidation to animals such as sheep, which have a low tolerance of copper: some data suggest that there is no risk of excessive retention of copper by such animals; others have found increases in copper storage in the livers of grazing sheep although this has not yet been associated with increased fatalities. The U.S. Environmental Protection Agency has been keeping a watchful eye on these European arguments with the result (see Scheinberg & Sternlieb 1976) that it has recently proposed a recommendation that animal feeds should contain no more than 15 mg copper/kg. This restriction was largely suggested to prevent accumulation of copper in the livers of animals that might subsequently be eaten by patients with Wilson's disease. If one is to rationalize this approach, then perhaps one should also ban the consumption of some calf and sheep livers by the population at large, because it is not unusual for these foods to contain large quantities of copper that might also be dangerous to people with Wilson's disease.

Harris: Dr McBrien, you used *E. coli* which is a facultative anaerobe commonly found in the human intestine. We know that the intestinal flora provides useful micronutrients for consumption by humans. Is it possible that when copper is not absorbed, and when there is therefore too much copper in the intestine, that the balance of these organisms will be damaged?

McBrien: I don't know, but I doubt that the amount of copper that is normally eaten with food would be sufficient to kill the intestinal microflora.

Danks: Some of us believe that the acute haemolytic crisis that occurs in Wilson's disease may have the following sequence of events: a large regenerative liver nodule, common in cirrhosis of this type, exceeds its blood supply, becomes hypoxic and ischaemic, undergoes necrosis, releases a lot of copper, and then sets off a cycle of acute copper poisoning. From what you said, Dr McBrien, it seems that such an episode of hypoxia in such a region of liver might be potentiated by a considerable switch of Cu(II) to Cu(I) within that region. Is that a reasonable suggestion?

McBrien: Yes, it probably is reasonable. We (Hesslewood et al 1978) used mouse leukaemic cells in some of our work because there were some people at the Hammersmith Hospital (London) who were interested in the use of Cu(I) as a radiosensitizer of anoxic cells (Cu(II) is not so useful in this respect). We confirmed the possibility that anoxic centres in tumours may convert Cu(II) to Cu(I) and therefore may make it possible to kill the tumour more effectively by the use of radiation.

The levels of copper required for radiosensitization are less than the levels that cause gross damage to cells, but there are other, more effective radiosensitizers than copper, so the project was not pursued further.

References

Hesslewood IP, Cramp WA, McBrien DCH, Williamson P, Lott KAK 1978 Copper as a hypoxic cell sensitizer of mammalian cells. Br J Cancer (suppl) 37:95-97

McBrien DCH, Hassall KA 1965 Loss of cell potassium by *Chlorella vulgaris* after contact with toxic amounts of copper sulphate. Physiol Plant 18:1059-1065

Scheinberg IH, Sternlieb I 1976 Copper toxicity and Wilson's disease. In: Prasad AS, Oberleas D (eds) Trace elements in human health and disease. Vol 1: Zinc and copper. Academic Press, New York, p 415-438

Final general discussion

Mills: At an earlier stage of the meeting we agreed that we would return to a discussion of the possible role of copper in resistance to viral and bacterial infection.

Danks: It has been said in some single-case reports that marked susceptibility to infection is characteristic of Menkes' syndrome. I would like to deny that this is a marked feature of the condition; we have observed about 10 children throughout the course of the illness and they did not suffer infective illness any more frequently than other children who have a comparable degree of brain damage. In any brain-damaged child, episodes of chest infection occur because, e.g., he inhales his food as saliva and because he coughs inadequately. There has also been a suggestion (Pedroni et al 1975) that T cell function was impaired in Menkes' syndrome. A later study has shown, however, that this is not so (Sullivan & Ochs 1978).

Mills: For a long time the question has been raised about the possible relationship between enzootic ataxia and multiple sclerosis. This idea was stimulated by the fact that four of the seven people who did pathological work on swayback (enzootic ataxia) tissue at Cambridge developed multiple sclerosis (Campbell et al 1947). Some Irish neurologists are again becoming interested in this possible relationship and are re-examining the case-histories. Since the pathology of enzootic ataxia can be reproduced experimentally purely by perinatal depletion of the copper reserves of the lamb (Mills & Fell 1960, Fell et al 1965) in an environment where there has been no chance of previous contact with a causal viral or bacterial agent, there may not be a very strong case for a relationship between the two diseases.

Terlecki: For many years research workers have been seeking a possible aetiological relationship between the disease called scrapie in sheep and multiple sclerosis in humans, but so far, conclusive evidence is lacking.

However, many workers now agree that scrapie is rather similar to Creutzfeld-Jacob's disease in humans. The two diseases have similar clinical features, pathogenesis and neuropathology, and each is caused by 'unconventional viruses' (Kimberlin 1976).

Swayback (enzootic ataxia) in lambs, however, is generally agreed to be unrelated aetiologically to multiple sclerosis in humans. Some diseases of humans, including multiple sclerosis, are now being re-examined in the light of 'slow-virus disease', a concept first introduced by Sigurdsson in 1954, and of which scrapie is one of the animal models.

Shaw: Was the evidence that all four veterinarians did, in fact, die of multiple sclerosis absolutely water-tight? I believed that three of the four case histories did not indicate multiple sclerosis unequivocally (Campbell 1963).

Terlecki: Well, I have personally examined sections of brain from one of the patients and it was clear that he had suffered (and died) from the disease. As for the other three, I was only verbally informed that they too were the victims of the disease.

Hassan: I believe that in chronic granulomatous disease the patients have an impaired ability to generate superoxide radicals, not because of a shortage of copper but because of lack of the enzyme that generates the superoxide radical. Babior (1978) calls this enzyme 'superoxide synthetase'.

Dormandy: Have copper concentrations been measured in people with chronic granulomatosis, who have a neutrophil defect?

Mills: I know that selenium concentrations have been examined in those patients but I don't know of any studies on copper in relation to the disease.

Lewis: I think we should be cautious about interpreting results from *in vitro* experiments on the effects of copper on bacteria. The sheep that we use in our work have been copper-deficient, as a flock, for about 20 years. We have never vaccinated them against the common sheep diseases like pulpy kidney and enterotoxaemia. This is a closed flock, with no introductions since 1969, and the people who look after the animals have contact with other groups of sheep, but the incidence of bacterial disease has been low and the animals are generally very healthy. These copper-deficient sheep are with us from the time they are born up to 13 or 14 years of age; so, if there *were* a disease problem in copper-deficient animals, I would expect that we should be the first to see it.

Sourkes: Several people at this meeting have mentioned bone changes and scurvy-like changes. A shortage of ascorbate is possibly responsible for such changes. Dopamine β-monooxygenase (EC 1.14.17.1), which catalyses the conversion of dopamine to noradrenaline (norepinephrine), is the one enzyme that is known to use ascorbic acid as the coenzyme. I wonder if there is any relationship between the scurvy-like changes in copper deficiency and a shortage of ascorbate.

Harris: Ascorbic acid has also been identified as a cofactor of proline, 2-oxoglutarate dioxygenase (EC 1.14.11.2) and lysyl hydroxylase, which are both involved in collagen synthesis. Without the hydroxylation step there is an impairment in collagen synthesis which is relevant to changes in the collagen matrix of the bone. Incidentally, proline, 2-oxoglutarate dioxygenase also required iron and α-ketoglutarate stoichiometrically.

Sourkes: So the bone changes that one might see in scurvy could be due to lack of ascorbate, although the copper would presumably still be available for the lysyl oxidase?

Harris: I can't quite bring copper into the picture here. Proline, 2-oxoglutarate dioxygenase is not a copper enzyme.

Mills: The origin of the skeletal lesions in copper deficiency is still far from clear. Work going on in several centres (e.g. Moredun Institute, Edinburgh and Rowett Research Institute, Aberdeen) suggests that defects develop in osteoblast activity and that there is an enhancement of osteoclast activity. But for neither of these do we understand what the role of copper might be. So in addition to the defects in the collagen matrix there are at least these two other phenomena taking place. These are roles of copper about which we know very little. These effects develop at a very early stage of the deficiency, long before overt clinical signs appear.

Österberg: I have a question about cytochrome c oxidase (EC 1.9.3.1). This enzyme is used to generate energy for the cell in the oxidative process of phosphorylation that produces ATP in order to compensate for the great entropy losses occurring during protein biosynthesis. If cytochrome c oxidase activity falls in copper deficiency (and there are some indications for that, from Dr Mills' description of the mitochondrial changes in early copper deficiency), then I think there would also be a decrease in protein biosynthesis. It might be advisable to examine cells that are known to reproduce rapidly, like lymphocytes, intestinal cells and osteoblasts, because these cells would be most likely to be affected when protein biosynthesis is reduced as a result of a marked decrease in the concentration of cytochrome c oxidase.

McBrien: Many drugs affect protein synthesis directly, and the side-effects (e.g. loss of hair or sensitive skin) of such anti-tumour agents are not really comparable to those that I have heard described for copper deficiency.

Österberg: Some of those symptoms are related to inborn errors of metabolism. This is somewhat different from what I was discussing, which was plain copper deficiency of the kind described by Dr Mills and Dr Graham at this meeting. It seems as if copper deficiency affects cells that are quite rapidly reproducing and therefore protein biosynthesis and cytochrome *c* oxidase may be involved.

Dormandy: But then you would expect anaemia to be a prominent feature, because bone marrow cells turn over more rapidly than most other cells.

Mills: One problem is that a highly localized lesion can develop. We mentioned earlier that a lysyl oxidase defect could develop at one site long before it appeared at another. Why there should be sensitive and insensitive sites is unknown at present.

Harris: There is a rather disturbing point that we have alluded to during the meeting. Sometimes we understand the biochemical processes clearly and in infinite detail. For example, we can describe how thiamine pyrophosphate (derived from the vitamin, thiamine) is involved in a number of biochemical reactions but we cannot explain the symptoms that occur in beriberi or in thiamine deficiency. So the link between the biochemical understanding at the molecular level and the overt pathological symptoms is what we don't know. I don't like to leave the discussion on a discouraging note, but perhaps it is realistic to point this out.

Mangan: I asked Professor Graham earlier whether he had found any effects on lining cells – e.g. intestinal lesions or peptic ulcerations among copper-deficient children – and he said that he had not, with the possible exception of the one child whose chronic diarrhoea seemingly responded to copper.

Österberg: In contrast to these findings are those of Dr Mills indicating that there are grave cell changes to be seen in the intestine of copper-deficient cattle (Mills et al 1976).

Mills: The rat and the bovine species are somewhat different in that although there is a loss of enterocytic cytochrome *c* oxidase in both species at an early stage, mitochondrial lesions in the enterocytes are evident only in cattle (Mills et al 1976, D. Dinsdale, personal communication). The effects are serious: initially there is a reduction in cytochrome *c* oxidase activity, whether assayed spectrophotometrically or by histochemical techniques, and this reduction is followed by mitochondrial distortion. In only some cattle do these effects progress to loss of the mucosal cells and villus atrophy. The animals that develop villus atrophy appear to suffer much more from the effects of the deficiency, probably because they develop a general defect in nutrient absorption. The extent of mucosal damage is similar to that found in coeliac disease. The great puzzle is why this progression does not occur in all animals.

Dormandy: Dr Hill attempted to unify the role of copper by saying that it must be involved somehow in electron transport and in protection against free radical damage. If that is so, one might expect copper deficiency, and perhaps copper excess, to influence susceptibility to X-ray irradiation and even to ultraviolet light (Dormandy 1980). Is there evidence that any of the syndromes we have mentioned is influenced by exposure to sunlight or to X-rays?

Sourkes: I should imagine that copper would be called into play in the melanin pigmentation that occurs after exposure to ultraviolet light.

Dormandy: There may also be a relationship between copper and susceptibility to certain skin cancers which are associated with exposure to ultraviolet light.

Hurley: Are Menkes' children more susceptible to sunburn?

Danks: Yes; they are extremely susceptible to sunburn but they don't live long enough for us to find out whether they are susceptible to skin carcinogenesis. Some of the heterozygotes develop the intriguing phenomenon of piebald sunburning because of the mosaic X-chromosome inactivation (D.M. Danks, unpublished work). They have some patches of skin that tan well, but intervening patches that sunburn rather badly.

Sourkes: But this would be similar for albinos, who lack the melanin-forming cells and the tyrosinase that such cells contain.

Danks: The piebald sunburning might occur purely because the children with Menkes' syndrome can't produce enough melanin to protect themselves, rather than because the copper deficiency sensitizes them to exposure.

Harris: In all the diseases we have talked about we have not yet mentioned vitiligo. I understand from one report that the disease was treated successfully by exposing the patients to excessive sunlight − the method was termed heliotherapy (Genov et al 1972). I don't know how it might be related to copper status. That is another unanswered question!

References

Babior BM 1978 Oxygen-dependent microbial killing by phagocytes. I and II. N Engl J Med 298:659-668 and 721-725

Campbell AMG 1963 Veterinary workers and disseminated sclerosis. J Neurol Neurosurg Psychiatry 26:514-515

Campbell AMG, Daniel P, Porter RJ, Russell WR, Smith HV, Innes JRM 1947 Disease of the nervous sytem occurring among research workers on swayback in lambs. Brain 70:50-58

Dormandy TL 1980 Free radical reactions in biological systems. Ann R Coll Surg Engl 62:188-194

Fell BF, Mills CF, Boyne R 1965 Cytochrome oxidase deficiency in the motor neurones of copper deficient lambs: a histochemical study. J Comp Pathol 6:170-177

Genov D, Bozhkov B, Zlatkov NB 1972 Copper pathochemistry in vitiligo. Clin Chim Acta 37:207-211

Kimberlin RH (ed) 1976 Slow virus diseases of animals and man. Elsevier, Amsterdam (Frontiers of Biology, vol 44)

Mills CF, Fell BF 1960 Demyelination in lambs born of ewes maintained on high intakes of sulphate and molybdate. Nature (Lond) 185:20-22

Mills CF, Dalgarno AC, Wenham G 1976 Biochemical and pathological changes in tissues of Friesian cattle during the experimental induction of copper deficiency. Br J Nutr 38:309-331

Pedroni E, Bianchi E, Ugazio AG, Burgio GR 1975 Immunodeficiency and steely hair (letter). Lancet 1:1303-1304

Sigurdsson B 1954 Rida, a chronic encephalitis of sheep, with general remarks on infections which develop slowly and of some of their special characteristics. Br Vet J 110:341-354

Sullivan JL, Ochs HD 1978 Copper deficiency and the immune system (letter). Lancet 2:686

Chairman's closing remarks

COLIN F. MILLS

Nutritional Biochemistry Department, Rowett Research Instsitute, Bucksburn, Aberdeen AB2 9SB, UK

Although I don't intend to present an overall summary of this meeting it could be helpful to explore some of the principal features that have emerged and to consider some aspects that have escaped emphasis during our discussions in the last three days.

The meeting started with discussion of the dietary origins of copper deficiency, and the influence of dietary components on copper availability. We considered the range of foods, including dairy products, some cereal products and some therapeutic diets, that are particularly low in copper. Copper deficiency that is inadvertently induced in infants by the use of low-copper preparations for intravenous or intragastric feeding was first described over seven years ago. However, appreciation is dawning only slowly that the methods used for preparation of the amino acid components of these feeds tend also to remove copper and other elements from the feed, and that it is therefore necessary to replace the copper and, often, the zinc. As far as the medical press is concerned, some of these reports are no longer novel, and do not receive the publicity they deserve; but copper deficiency of this type nevertheless continues to occur, as our discussions have emphasized. Similar accidents resulting in deficiencies have been reported also with zinc and with selenium (e.g. Suita et al 1978, Kay et al 1976, Van Rij et al 1979).

One point that has not emerged is that the human species' greatest protection against trace element deficiency is the variety of its diet, except when genetic defects in trace element absorption or metabolism exist, as in Menkes' syndrome or in acrodermatitis enteropathica. Animal nutritionists learnt many years ago that when economic circumstances or the desire to simplify production methods led to the use of diets with fewer staple components, this was when trace element problems began to appear.

The protection afforded by a varied diet is denied to some human populations in developing countries. Indeed, a range of typical diets in the Gambia, recently studied by UK Medical Research Council staff, have copper contents that by standards of the World Health Organization (1973) must be regarded either as frankly deficient or only marginally adequate. Although measures to correct protein–calorie malnutrition merit greater emphasis in many such areas, these data also suggest that some infant populations may be at risk from inadequate dietary supply of copper.

We have not considered at this meeting the geochemical variables that influence the copper content of the staple crops consumed by animals or humans. Such factors certainly influence the geographical distribution of copper deficiency in animals, but little is yet known of their influence on the copper status of human beings, as indicated in a recent Royal Society Discussion Meeting (1980) on environmental geochemistry and health.

Our discussions of copper absorption and transport indicated that we know remarkably little about transport mechanisms, or whether homeostatic regulation of body copper content is normally achieved at the absorptive stage. In addition, Dr George Graham's comments on the link between diarrhoea and copper deficiency highlighted gaps in our knowledge about normal gastrointestinal function in the economy of copper and, conversely, about the effect of deficiency on gastrointestinal function. In some species, diarrhoea is an early manifestation of a gradually developing copper deficiency, particularly of one induced by the action of molybdenum as a copper antagonist (e.g. Underwood 1977) and the diarrhoea can begin when there are no other gross manifestations of deficiency. A strong case can be made, therefore, for investigation of cause-and-effect relationships between copper deficiency and defects in gastrointestinal function.

When discussing inhibitors, we emphasized the great differences in the patterns of inhibitors that affect agricultural animals and humans. Those who are concerned with the incidence of copper deficiency in agricultural animals are well aware that metabolic inhibitors are responsible for the incidence of copper deficiency in many parts of the world. In some instances the inhibitor has a geochemical origin, thus influencing the composition of the diet; in others it originates from metal-rich waste products of heavy industry. The quantitative aspects of such antagonistic interrelationships are still poorly understood and it is difficult to predict how much the copper requirement increases per unit of metabolic inhibitor in the diet. We know little of the mechanisms of inhibition or the variables that influence the biopotency of the antagonist.

We did not discuss in detail the effect on humans of a change in diet, or the effect of naturally occurring inhibitors of copper metabolism upon copper status. We touched only briefly upon the phytate–copper antagonism, and perhaps it is worth emphasizing that phytate seems to be almost as effective an inhibitor of copper metabolism as it is of zinc (Davies & Nightingale 1975). The practical importance of dietary phytate as an inhibitor of zinc metabolism is disputed because there is disagreement about the extent of phytate degradation during cooking. There is, however, plenty of evidence that many of the meat substitutes available for human consumption are rich in phytate. The impact of these high concentrations of phytate on copper absorption and retention has not yet been examined.

We can now determine when copper intake is low, and we emphasized that even when dietary content of copper is below the recommended dietary allowances (RDAs) of the World Health Organization and of the US Academy of Sciences, *overt* signs of copper deficiency may not be apparent. Perhaps those of us who have helped to derive RDAs for copper should therefore point out the scarcity of the data upon which these estimates have to be based. In addition, we have no data from which to assess either the effect of variables on the utilization of dietary copper by humans or the influence of growth rate upon demands for copper. Work with young ruminants indicates that growth may account for as much as half the total requirement for copper in the early stages of development. Thus, if growth is restricted for any reason, a surprisingly low intake of copper may not give rise to any clinical abnormality. Nevertheless, those people who may revise RDAs in the future would certainly be happier if these observations of clinical normality despite a low dietary copper were supported by other data indicating normal copper reserves and the absence of covert pathological changes.

What, then, are the early manifestations of deficient copper status and what are the appropriate ways to detect them? Our discussions on this point covered the role of copper in the activity of superoxide dismutase (EC 1.15.1.1), lysyl oxidase, cytochrome *c* oxidase (EC 1.9.3.1) and caeruloplasmin, and the value of these enzymes as markers for copper status. From these discussions emerged a pragmatic approach to the use, as markers, of changes in caeruloplasmin or superoxide dismutase activity – an approach based on the rapid changes of the former and the slower changes of the latter during copper depletion. Less pragmatic was our discussion of attempts to predict the type of lesion we should look for on the basis of lysyl oxidase function. There are indications that in the young growing animal the failure of lysyl oxidase is quite rapid and that, in contrast to what many of us had thought, its effects on the structure of both collagen and elastin develop quickly. These effects are most marked in rapidly growing species such as the chick, the turkey and the child with Menkes' syndrome. However, the reason why connective tissue lesions should be so strongly evident in the Menkes' child, compared with the child who develops nutritional copper deficiency is not at all clear.

We do not yet know why cytochrome c oxidase activity falls during copper deficiency, nor whether this decline becomes rate-limiting for oxidative reactions. We know that copper is incorporated into the enzyme and has a functional role there, but it is also involved in the synthesis of the α-haem moiety of the enzyme. The degenerative changes that appear in the membranes of copper-deficient mitochondria (Leigh 1975, Dallman & Goodman 1970) could be due to interruption of either of these processes or to the direct or indirect role of copper in modulation of lipid desaturase activity and, thus, of lipid composition (Wahle & Davies 1975). Until these alternatives and their implications are more widely investigated it may remain difficult to link the observed biochemical defects with specific pathological consequences.

Several clinicians at this meeting have complained that our knowledge of the functions of copper does not yet allow either early recognition or reliable prediction of the development of pathological changes. However, differential changes in the activity of caeruloplasmin and superoxide dismutase in plasma and erythrocytes at least offer a means of determining whether copper deficiency has been of a short or long duration.

Perhaps biochemists should remind clinicians that the copper content or copper-enzyme activity in blood — the only tissue usually offered for examination — is seldom an adequate index of the development of copper-responsive metabolic lesions in tissues as diverse as blood vessels, motor neurons, enterocytes, tendons and the mitochondria of heart muscle. Analysis of blood during copper deficiency may indicate that lesions are developing in other tissues, but it will not necessarily reveal the severity, sequence or distribution of these lesions. Nevertheless, a reduced activity of erythrocyte superoxide dismutase can sometimes usefully indicate a low copper status of longer standing and, consequently, a higher risk of a pathological response. Similarly, a low neutrophil count may aid diagnosis in the nutritionally copper-deficient child. However, Dr Graham (pp 20, 222–223) has pointed out the problems associated with interpreting changes in blood copper content and changes in neutrophil count, in this context.

Biochemists are interested in studying single metabolic lesions, and the more pronounced such a lesion is, the greater the likelihood of its detection. Clinicians, veterinarians and pathologists, on the other hand, want to know the relevance of such biochemical defects to progressively developing syndromes. Biochemists can rarely provide these answers because the rate of development of a single lesion or its relationship to others is seldom studied, and its examination may even be regarded as a form of drudgery. However, it is essential that an adequate biochemical picture should be painted of such progressive changes if we are to understand better the relationships between sub-clinical copper deficiency, metabolic lesions and covert pathological lesions.

Adequate biochemical investigation of the fine details of copper metabolism cannot depend only on studies of tissue homogenate, which the histochemist would regard as the 'bucket' approach. However, technical limitations do govern the rate of progress. First, we do not have sufficient sensitivity in analytical electron microscopy or microprobe techniques to determine which cells most rapidly lose copper during deficiency. Histochemical studies on copper-dependent enzymes tell us that some cells are more sensitive to copper depletion than others, but at present we are unable to determine the chemical basis for this difference. Dr Terlecki mentioned that neuronal lesions can develop in one species of cell during copper deficiency while other cells remain normal: e.g. cytochrome *c* oxidase activity can be reduced in motor neurons whereas its activity in the matrix of surrounding cells is unchanged (Fell et al 1965). Probe techniques can be adequate for investigation of local increases in hepatic copper during disease, as in Dr Tanner's studies of copper metabolism in Indian childhood cirrhosis (Tanner et al 1979). Nevertheless, those who investigate normal or copper-deficient tissues find concentrations of copper that unfortunately are far below the sensitivity of existing methods. Thus, we are unable to determine whether differences in the equilibrium distribution of copper between populations of cells influence the pathological outcome of deficiency. A better understanding of these relationships may help us to explain the species differences that we have discussed in response to copper deficiency, and it may also suggest more effective agents for prophylaxis or therapy.

This meeting, perhaps more than many that the Ciba Foundation has organized, has revealed a large number of gaps in our knowledge, which will have to be bridged if we are to define accurately the copper requirements of humans and animals, and thus to reduce the incidence of clinical copper deficiency. We also need to understand the significance for health of the many sub-clinical changes that have been described in copper-deficient subjects.

The collaborative work of biochemists, inorganic chemists, pathologists and clinicians will be needed to fill these gaps. Individuals from all of these disciplines have participated in this meeting and all of us have found the opportunity for joint discussion both interesting and fruitful. I would like to express our gratitude and indebtedness to the Ciba Foundation for allowing us to discuss our common interest in the biological roles of copper, and for giving us such a pleasant environment in which to meet!

References

Dallman PR, Goodman JR 1970 Enlargement of mitochondrial compartment in iron and copper deficiency. Blood 35:496-505
Davies NT, Nightingale R 1975 The effects of phytate on intestinal absorption and secretion of zinc and the whole body retention of zinc, copper, iron and manganese in rats. Br J Nutr 34:243-258
Fell BF, Mills CF, Boyne R 1965 Cytochrome oxidase deficiency in the motor neurones of copper deficient lambs: a histochemical study. J Comp Pathol 6:170-177
Kay RG, Tasman-Jones C, Pybus J 1976 A syndrome of acute zinc deficiency during total parenteral nutrition in man. Ann Surg 183:331-340
Leigh LC 1975 Changes in the ultrastructure of cardiac muscle in steers deprived of copper. Res Vet Sci 18:282-287
Royal Society Discussion Group 1980 Environmental geochemistry and health. Bowie SHU, Webb JS (eds) Royal Society, London
Suita S, Ikeda K, Nagasaki A, Hayashida Y 1978 Zinc deficiency during total parenteral nutrition. J Pediatr Sur 13:5-9
Tanner MS, Portmann B, Mowat AP, Williams R, Pandit AN, Mills CF, Bremner I 1979 Increased hepatic copper concentration in Indian Childhood Cirrhosis. Lancet 1:1203-1205
Underwood EJ 1977 Trace elements in human and animal nutrition. 4th edn. Academic Press, New York
Van Rij AM, Thomson CD, McKenzie J, Robinson MF 1979 Selenium deficiency in total parenteral nutrition. Am J Clin Nutr 32:2076-2085
Wahle KJW, Davies NT 1975 Effect of dietary copper deficiency in the rat on fatty acid composition of adipose tissue and desaturase activity of liver microsomes. Br J Nutr 34:105-112
World Health Organization 1973 Trace elements in human nutrition. WHO Tech Rep Ser 532

Index of contributors

Entries in **bold** *type indicate papers; other entries refer to discussion comments*

Balthrop, J.E. **163**
Bremner, I. 17, 18, **23**, 36, 37, 38, 40, 41, 42, 43, 44, 45, 46, 88, 115, 116, 159, 207, 224, 239, 277, 280

Cass, A.E.G. **71**

Danks, D.M. 5, 14, 18, 20, 40, 42, 43, 44, 118, 119, 120, 137, 158, 159, 177, 181, 182, 206, **209**, 221, 222, 223, 224, 225, 240, 242, 243, 244, 262, 263, 265, 277, 318, 319, 323
Delves, H.T. 13, 14, 16, 316
DiSilvestro, R.A. **163**
Dormandy, T.L. 36, 37, 85, 88, 89, 114, 120, 122, 137, 140, 178, 205, 206, 224, 225, 277, 293, 297, 298, 316, 317, 320, 322, 323

Frieden, E. 14, 15, 17, 37, 38, 43, 45, 65, 66, 68, 86, 87, 88, **93**, 114, 115, 116, 117, 118, 119, 120, 121, 122, 123, 136, 137, 138, 140, 141, 155, 158, 159, 160, 181, 207, 295, 296

Garcia-de-Quevedo, M. **163**
Graham, G.G. 14, 16, 20, 45, 46, 117, 121, 221, 222, 224

Harris, E.D. 14, 20, 40, 41, 86, 87, 113, 118, 119, 123, 139, 141, 154, **163**, 178, 179, 180, 181, 205, 239, 244, 261, 279, 281, 298, 318, 321, 322, 323
Hassan, H.M. 87, 88, 118, **125**, 136, 137, 138, 139, 140, 141, 180, 206, 316, 317, 320

Hill, H.A.O. 18, 38, 39, 45, **71**, 85, 86, 87, 88, 89, 115, 117, 120, 137, 138, 139, 140, 141, 158, 179, 205, 279, 281, 294, 296, 298
Hunt, D.M. 42, 118, 155, 243, **247**, 261, 262, 265
Hurley, L.S. 14, 18, 44, 45, 46, 88, 90, 140, 158, 205, 224, **227**, 239, 240, 242, 243, 244, 262, 264, 294, 295, 315, 323

Keen, C.L. **227**

Lewis, G. 16, 160, 224, 239, 240, 262, 263, 264, 265, 316, 320
Lönnerdal, B. **227**

McBrien, D.C.H. 37, 41, 179, **301**, 316, 317, 318, 321
McMurray, C.H. 15, 37, 46, 68, 122, 138, 140, 161, 180, **183**, 205, 206, 243
Mangan, F.R. 15, 114, 244, 281, 296, 322
Mills, C.F. **1**, 13, 14, 15, 16, 17, 18, 20, 39, 40, 43, 44, 45, 46, **49**, 65, 66, 67, 68, 87, 89, 90, 115, 116, 118, 119, 120, 138, 139, 140, 154, 157, 158, 159, 160, 178, 180, 181, 182, 206, 221, 224, 225, 242, 243, 261, 263, 264, 265, 295, 296, 298, 316, 317, 319, 320, 321, 322, **325**

Österberg, R. 14, 66, 67, 85, 86, 89, 155, 278, **283**, 294, 295, 296, 297, 298, 316, 321, 322
Owen, C.A. 18, 41, 43, 44, 116, 122, **267**, 277, 278, 279, 281, 296, 297

Rayton, J.K. **163**
Riordan, J.R. 39, 140, 180, 225, 244, 261, 265, 280, 281

Shaw, J.C.L. 14, 17, 119, 160, 161, 221, 240, 263, 281, 320

Sourkes, T.L. 16, 42, 46, 118, 119, 120, 122, 123, 137, **143**, 154, 155, 261, 279, 281, 320, 321, 323

Tanner, M.S. 39, 40, 45, 46, 159, 206, 279
Terlecki, S. 121, 240, 243, 262, 263, 265, 319, 320

Indexes compiled by William Hill

Subject index

absorption of copper
7, 24, 43, 49, 237, 239, 267, 326
action of zinc on 27, 53
amino acids affecting 45
by grass 316
cortisone affecting 29, 243
during pregnancy and lactation 27
gut resection and 216
homeostatic control of 25
in crinkled mice 242
in tumour-bearing rats 27
iron deficiency and 46
metallothionein affecting 25
molybdenum and 55
mucosal block to 26
neonatal 44
oestrogens affecting 28
rates of 24, 244
serum copper and 37
species differences 36
through skin 15
accumulation of copper
see also Bedlington terrier disease, biliary cirrhosis, toxicity of copper, Wilson's disease
fetal 17
hepatic 30
perinatal 39
renal 34
achromotrichia
191
acrodermatitis enteropathica
see zinc deficiency

acute-phase reactions
110, 217
adenylate cyclase
123
administration of copper
forms of 19
adrenalectomy
action on diamine oxidases 149
adrenal gland, copper and
261, 265
adrenaline (epinephrine) in copper deficiency
250
agricultural waste treatment, copper toxicity and
312
albumin fraction of plasma, copper in
67, 277
aldehyde dehydrogenase
148, 155
alkaline phosphatase in zinc deficiency
158
amine oxidases
143
caeruloplasmin and 118
classification of 145
copper and 143–156
copper-containing 145
inhibition by semicarbazide 147
metabolism 147
nomenclature 143
amines
copper and 143–156
metabolism 151

regulation of concentrations 106
amino acids
increasing copper absorption 45
neutral 256, 261
sulphur-containing 89
aminoguanidine inhibiting diamine oxidase
149
amodiaquine
152
Anabaena variabilis
96
anaemia
17
copper and iron in 60
Heinz body 193
in brindled and blotchy mice 212
in copper deficiency 106, 192, 211, 218, 222, 322
in crinkled mice 233
in Menkes' syndrome 212
iron deficiency 2
molybdenum and 56
anaphylactic shock, diamine oxidases in
148
anoxia, copper toxicity under
302
antidepressants, caeruloplasmin and
106, 107
anti-inflammatory response to copper
15

antimalarial drugs inhibiting diamine oxidase
152
antioxidant, caeruloplasmin as
108
apocaeruloplasmin in liver
121
apoferritin
102, 173
arginine in catalytic action of copper–zinc-SOD
131
arylamines, oxidation of
104
aryldiamines, oxidation of
98
ascorbate oxidase
95, 97
ascorbic acid
320, 321
azide action on copper–zinc-SOD
129
azurin
97, 141

bacterial disease
319–320
Bedlington terrier disease
272, 280, 281
behavioural changes in copper deficiency
248, 258
benzylamine oxidase
144
benzylamine oxidation
152
bile
  copper excretion in 43, 44, 267, 271
  copper reabsorption into 45
biliary cirrhosis
122, 277
biliary disease
270
binding of copper
239
  cadmium and 27
  cycloheximide inhibiting 32
  mitochondrial 33
  to metallothionein 23

binding sites for copper in albumin
277
blood clotting, caeruloplasmin and
195
blotchy mice
210
  anaemia 212
  hair, pigmentation and skin 214
  myelination 262
  reproduction 243
bone and joint disorders in copper-deficient cattle
191
bone changes
17
  in copper deficiency 11, 213, 320
bone density
19
Border disease
121, 243
brain
  catecholamines 252
  copper concentration 234
  copper concentration in mottled mice 249
  in Menkes' syndrome 214
  in Wilson's disease 271
brain growth in copper deficiency
248
brain lipids in crinkled mice
232
brain phospholipids in crinkled mice
233
breast milk
  copper content 8, 44, 45, 215
  zinc content 215
brindled mice
210
  anaemia 212
  copper concentrations 77
  copper in brain tissue 249
  cultured cells 224
  cytochrome c oxidase in 261
  hair, pigmentation and skin 214, 241
  keratin 241

  myelination 262
  neurological symptoms 248
  reproduction 243

cadmium
  action on caeruloplasmin 54, 65
  action on cytochrome c oxidase 54
  copper absorption and 27
  copper binding and 27
  copper utilization and 54
  dietary concentrations 51
  effect on copper content of liver 51
  effect on testis 51
  in enzootic ataxia 52
  interactions with copper 51, 52, 296
  interaction with copper and zinc 53, 66
  placental copper transport and 53
caeruloplasmin
  2, 29, 81, 93–123, 217
  action of cadmium on 51, 54, 65
  action of molybdenum on 65
  action of silver on 65
  action of zinc on 51, 54, 65
  action on hepatic copper 239
  administration 20
  amino acid sequence 95
  as acute-phase reactant 110
  as copper donor 180, 271
  as scavenger for superoxide anion radicals 110, 117
  as serum antioxidant 108
  as single polypeptide chain 97
  blood clotting and 195
  catalytic activity 98
  controlling iron mobilization 60, 100, 106
  conversion of 267
  copper-binding sites 97
  copper transport and 29, 104, 115
  di-sialated 115
  evolutionary implications 93

ferroxidase activity 101
fetal 119
function 94, 111, 327
homogeneity 118
in amine oxidation 118
in Bedlington terrier
 disease 275
in catecholamine
 metabolism 118
in copper deficiency 106,
 114, 194, 217, 224, 225
incorporation into cyto-
 chrome *c* oxidase 104
in inflammation 158
inhibiting peroxidation 110
inhibition of 118, 120
in Huntington's disease
 108, 119
in Parkinson's disease 108
in plasma and serum 122,
 217
in pregnancy 122
in protein malnutrition 121
in storage of copper 37
in Wilson's disease 108,
 224, 281
kinetics 100
LSD and 108
measurement 7, 11, 122
molecular properties of 96
molecular weight 97
oestrogens affecting 37,
 110, 114, 122, 198, 217
oxidase activity 98, 103
oxygen reduction by 81
pseudoferroxidase activity
 102
regulation of biogenic
 amine concentrations
 106
relation to copper 121, 122
storage in lysosomes 269
SOD and 116
stoichiometry of copper 96
structure 94
substrates 100, 120
subunits 118
uptake in plasma 30
caeruloplasmin synthesis
 270
 copper for 196
 effect of molybdenum on
 56
in polyribosomes 269
cartilage dysplasia
 56
catalase
 127, 136, 137
catechol as caeruloplasmin
 substrate
 100
catecholamines, function in
 copper deficiency
 250
catecholamine metabolism
 amines and 155
 caeruloplasmin in 118
catecholamine synthesis in
 copper deficiency
 250
catechol methyl transferase
 252
cattle
 copper deficiency in 191,
 198
 SOD in 206
central nervous system, copper
 and
 248
cerebrocuprein (SOD)
 127, 257
chelation therapy
 283–299
 in Bedlington terrier disease
 281
 in Wilson's disease 281,
 283, 298
Chlorella sorokiniana
 133
chocolate, copper content
 5
circulatory failure from copper
 sulphate
 20
coeliac disease
 322
collagen
 241
 cross-linking 180, 184,
 185
 ratio to elastin 178
collagen synthesis, copper and
 163–182
connective tissue, role of
 copper
 163–182, 205
contraceptive pill, serum
 copper and
 37
copper acetate
 19, 20
copper-albumin
 20
copper-binding proteins
 257
copper-chelatin
 32
copper-chelating agents
 283–299
copper chloride
 20
copper deficiency
 see deficiency of copper
copper enzymes
 76, 157
 pathological change and
 157
 *see also under individual
 enzyme names*
copper-glycine
 20
copper ions
 302
 distribution of 289, 290
 free concentration 286
 production of 306
 reduction of 307, 308, 309
 sulphydryl reduction 308
copper-metallothionein
 73
 biological properties 38
 concentration in intestine
 26
 degradation 40
 distribution in organs 23
 function 32
 in intestines 23
 in kidney 23, 34
 in liver 17, 23, 31, 39
 in storage 37, 39
 in Wilson's disease 33
 natural occurrence of 31
 storage in liver 39
 turnover rate 40
copper proteins
 23, 72, 83, 86
 *see also metallothionein
 etc.*
 binding sites 87

copper proteins, *continued*
  caeruloplasmin and 105
  catalytic activity 87
  classification 82
  oxidative 76
copper proteins, dioxygen binding sites
  82
copper status, markers of
  194
  see also diagnosis of copper deficiency
copper sulphate
  20
  action on lysyl oxidase 171
  causing peripheral circulatory failure 20
  in parenteral fluids 14
copper-transferrin
  138
copper–zinc-metalloenzyme
  185
copper–zinc-superoxide dismutase (Cu-ZnSOD)
  185
  amino acid composition 128
  catalytic mechanism 130
  in crinkled mice 234
  in fetal copper deficiency 229
  inhibitors 129
  physico-chemical properties of 127
  species differences 128
  stability of 129
  structure of 128
copper (I)-glutathione
  287
copper (I)-penicillamine
  287
cortisone
  affecting copper absorption 29, 243
cow's milk
  copper in 18, 215
  growth and 45
Creutzfeld-Jakob's disease
  320
crinkled mice
  anaemia 233
  brain lipids 232
  copper deficiency 230, 241
  copper–zinc-SOD in 234

  haematological characteristics 233
  hair and pigmentation 241, 242
crustacea
  5, 6
cupreins
  see under copper–zinc-SOD
cyanide
  inhibiting caeruloplasmin 118
  inhibiting superoxide dismutase 129
cycloheximide
  inhibiting copper binding 32
  inhibiting metallothionein synthesis 32
  inhibiting protein synthesis 42, 175, 180
cysteine
  and metallothionein 41
  oxidation of 119
cytochrome c oxidase
  2, 80, 81, 95, 116, 200, 321–322, 327
  caeruloplasmin and 30, 104
  copper function in 207
  effect of cadmium and zinc 54
  hypothermia and 221
  in brindled mice 261
  in copper deficiency 62, 157, 158, 173, 183, 184, 191, 210, 218, 328
  in fetal copper deficiency 229
  in intestinal cells 322
  in intestinal mucosa 15
  in swayback 190, 263
  iron and 61, 63, 65
  oxygen and 85
  relation to erythrocyte SOD 131
cytochrome P-450
  189
cytocuprein (SOD)
  127

Dactylium dendroides
  131
deficiency of copper
  2, 90, 157, 209, 326
  action on amines 151, 152

  adrenal glands and 265
  affecting iron metabolism 116
  anaemia in 192, 211, 218, 222, 322
  behavioural changes and 248, 258
  biochemical signs 228
  bone changes in 213, 320
  brain growth in 248
  caeruloplasmin in 106, 114, 194, 224, 225
  cardiac effects 159, 185, 192, 214
  catecholamine function in 250, 259
  circumstances causing 215
  copper–zinc-SOD in 229
  cytochrome c oxidase and 62, 173, 183, 184, 191, 210, 218, 229, 328
  diagnosis 157–161, 194, 217, 224
  diarrhoea and 15, 46, 159, 191, 194, 212, 216, 221, 222, 223
  dopamine in 255
  dopamine β monoxygenase in 218
  early signs 327
  effect on intestinal mobility 154
  elastin lesions in 205, 214
  fetal 227
  following penicillamine therapy 295
  genetic interactions 229
  growth rate and 192
  hair changes in 214
  in animals 195
  in cattle 191, 198
  in crinkled mice 230, 241
  in humans 209–225
  in infants 17
  in mutants 229, 240
  in parenteral feeding 14
  in populations 198
  in preterm infants 8, 160
  in quaking mice 229
  in ruminants 183–207
  in sheep 189
  intestinal cell changes in 322

lipid synthesis in 189
lysyl oxidase and 167, 169
markers for 194
mechanism 217
molybdenum and 58, 191, 215
myelination in 228, 262, 264
neutropenia in 120, 213
neutrophils in 20, 218, 222, 223, 328
osteoporosis in 18, 19, 190, 191
physiological state and 195
pigmentation affected by 214, 217
prematurity and 215
protein loss and 216
pulmonary tissue in 185
recovery from 121
reproduction affected by 243
skeletal lesions in 321
skin changes in 214
species differences 210
strain differences 229
SOD and 173, 185, 198, 210, 218
vitamin E and 89
desmosine
166
detoxification of copper-metallothionein in
33
developing countries, dietary copper in
20
diagnosis of copper deficiency
157–161, 194, 217, 224
diamine oxidase
alteration in activity 148
copper in 155
endocrine influence on 149
inhibition by antimalarials 152
inhibition of 149
putrescine in study of 148
rate of synthesis 149
resynthesis 150
diarrhoea in copper deficiency
15, 16, 46, 159, 191, 194, 212, 216, 221, 222, 223, 322

dietary intake of copper
17
metallothionein and 25
dietary sources of copper
5–22, 327
diethyldithiocarbamate
129
dioxygen transport
75
disulphide bonding, copper in
182
dopamine
in brain 252
in copper deficiency 250, 255
dopamine-β-hydroxylase
see next entry
dopamine-β-monooxygenase
78, 80, 152, 155, 189, 251, 258, 320
in brain 253
in copper deficiency 218

effluent plants and copper toxicity
311
egg-shells of copper-deficient hens
185
elastin
60, 241
copper deficiency and 205, 214
cross-linking 180, 185, 205
lysyl oxidase in 181, 184
ratio to collagen 178
elastin synthesis
copper and 163–182
electron transport
73, 85, 105, 141, 218, 322
electron trapping
85
enterocytes, copper and zinc-containing metallothionein in
54
enzootic ataxia (swayback)
189, 194, 228, 263, 265
brain effects 240, 262
cadmium in 52
cytochrome c oxidase in 263
incidence 264

myelination in 218, 248
nervous symptoms 247
relation to multiple sclerosis 319
weather conditions and 264
enzymes
71
copper-containing
see copper enzymes
enzyme activity, copper and
224
erythrocuprein (SOD)
127
catalytic function 127
erythrocytes, SOD and copper in
187, 206
Escherichia coli
132, 302, 318
excess copper
see under Bedlington terrier disease, biliary cirrhosis, toxicity of copper, Wilson's disease
exchange of copper
87
excretion of copper
biliary 43, 44, 267, 271
faecal 7
in ruminants 43
storage forms and 39
urinary 7

faecal excretion of copper
7
falling disease
192
ferroxidase
100, 101
fetal caeruloplasmin
119
fetal copper accumulation
17
fetal copper deficiency
240
biochemical signs 228
characteristics 227
cytochrome c oxidase in 229
genetic interactions 229
fetal development, copper in
227–245
fetal liver, copper in
8, 18

fetal resorption in copper-deficient rats
192
flavin copper protein
154
flavoenzymes
144
follicular hyperkeratosis
214
fruit, copper in
5
Fusarium decemcellulare
310

galactose oxidase
80
genetic interactions with copper deficiency
229
globins
76
glucosamine
180
glutathione
284, 294
copper binding 298
glutathione peroxidase
89
granulomatosis
320
grass, copper absorption by
316
growth rate in copper-deficient cattle
192

Haber-Weiss-Fenton reaction
127, 139
haem
cytochrome *c* oxidase and 63
iron release and 62
haemerythrin
76
haemoglobin concentrations
66
haemoglobin synthesis
163
haemocuprein (SOD)
127
haemocyanin
76

haemoglobinuria
193
hair
copper content of 198, 222, 234
in copper deficiency 214
in crinkled mice 241, 242
keratin in 241
heart tissue in copper deficiency
159, 185, 192
heart cells, SOD damage in
140
Heinz body anaemia
193
heparin
148
hepatocuprein (SOD)
30
in storage 37, 38
histaminase
143, 144
histamine
148
histidine
278
hormones, serum copper and
37
human milk, copper content
45
Huntington's chorea, caeruloplasmin and
108, 119
hydrogen peroxide, action on copper–zinc-SOD
129, 130
3-hydroxy-4-methoxy-phenethylamine
108
5-hydroxytryptamine (serotonin)
106
hyperalimentation
14
hypercholesterolaemia in copper deficiency
189
hypercupraemia
272
see also accumulation of copper
hypocupraemia
183–200

see also deficiency of copper
hypothermia in Menkes' disease
221

ileum, copper absorption in
29
imidazolyacetic acid
148
Indian childhood cirrhosis
39, 45, 279, 329
infantile polymeric form of copper
40
infants, copper deficiency in
17
see also diarrhoea in copper deficiency
infection, role of copper in
319–320
inflammation
181, 291
caeruloplasmin in 110, 158
intestine
absorption of copper in 7, 24, 237, 267
cell changes in copper deficiency 322
copper content 236, 239
copper-metallothionein in 23, 26
copper retention in 54
copper transport in 29
cytochrome *c* oxidase activity in 15
effect of copper deficiency on motility 154
intravenous preparations containing copper and zinc
14
ionic copper administration
19
iron
1, 173
absorption 60, 68
albumin concentration and 103
biochemistry 83
caeruloplasmin and 60, 100, 106
copper deficiency affecting 116

SUBJECT INDEX

cytochrome *c* oxidase and 65
deficiency 224
hepatic 60, 65, 116
in biological systems 86
in synthetic diets 9
interaction with copper 60
mobilization 65, 116
oxygen utilization and 103
superoxide dismutase and 65
tissue retention 61
uptake by apoferritin 102
iron deficiency, copper absorption and
46
iron dismutases
88
iron proteins, biological role of
83
iron-superoxide dismutase (FeSOD)
125
isodesmosine
166

jejunum, copper absorption in
29

keratin
in hair of mottled mice 241
structure of 119
kidney
copper in 34, 271, 275
copper-metallothionein in 23, 34

laccase
80, 81, 95, 97
lactation, copper absorption in
27
leucocytes, SOD-deficient
138
leucocytic endogenous mediator (LEM)
110
lipids in brains of crinkled mice
232
lipid formation, copper in
262

lipid metabolism in swayback in sheep
190
lipid peroxidation
109
lipid synthesis in copper deficiency
189
liver
apocaeruloplasmin synthesis in 121
caeruloplasmin action on hepatic copper 239
copper accumulation in 30, 270, 272
copper compartments 268, 279
copper distribution in Wilson's disease 280
copper in 5, 8, 30, 33, 42, 244, 274
copper-metallothionein in 17, 23, 39
copper pools in 41
copper release from 42
copper retention in 18
copper storage in 39
fetal 18
in Indian childhood cirrhosis 40
iron in 60, 65
metallothionein in 30, 33
mitochondrocuprein in 30, 31
subcellular distribution of copper 33
liver cirrhosis
270
liver function, copper and
267–282
lung tissue in copper deficiency
185
LSD
107
lysosomes
caeruloplasmin storage in 269
copper in 33, 34, 274, 279
metallothionein uptake into 40
lysyl oxidase
2, 60, 164, 165, 200, 210, 224, 322, 327

action in elastin and collagen 181
activation 173
activation *in vitro* 168
activation by copper 168
amino acid composition of 179
characteristics 179
copper and 165, 173, 178
copper deficiency and 167, 169
correlation with PPD 171
cross-linking and 205
function of 165
in elastin synthesis 181, 184
occurrence in tissue 166
oestrogen activation of 170
regional differences in tissue 180
serum proteins in activation of 169

magnesium deficiency
221
magnesium-superoxide dismutase (MgSOD)
88
malabsorption syndrome
9
manganese
deficiency 88
dopamine metabolism and 262
in synthetic diets 9
manganese dismutases
88
manganese-superoxide dismutase (MnSOD)
125, 185, 187, 234
maple-syrup-urine disease
9
marasmus
222
markers of copper status
194
maternal diet, copper supplements in
44
melanin
60
copper deficiency and 323
deficient formation 191

melanogenesis
  54, 56
  copper and iron in 60
Menkes' disease
  2, 20, 27, 173, 177, 210,
    211, 219, 229, 261, 263,
    325
  anaemia and 212
  bone changes in 213
  brain disease in 214
  cultured cells in 225
  diarrhoea in 216
  features of 210
  hair content of copper in
    222
  hypothermia in 221
  nervous system in 218, 248
  neutropenia and 213
  noradrenaline and
    caeruloplasmin in 118
  sunburn and 323
  SOD in 248
mescaline
  144
metabolic interactions
  between trace elements
  49–69
metabolism of copper
  molybdenum and
    55, 60
metal ions controlling proteins
  173
metalloenzymes
  158
metalloproteins
  200
metallothionein
  23, 24, 89
  see also copper-metallo-
    thionein
  binding sites for zinc and
    copper 26
  control of copper
    absorption and 25
  copper binding to 23
  cysteine disappearance
    from 40
  degradation 33, 42
  electron transport 73
  in copper detoxification 33
  in enterocytes 54
  in liver 30, 31
  lysosomal uptake 40

metal-binding sites 72
physiological role 40, 72
polymerized forms 40
metallothionein-bound copper
  stores
  17
metallothionein synthesis
  25
  cadmium and zinc affecting
    53
  induction by copper 31
  inhibition by cycloheximide
    32
microsomal enzymes
  189
milk, zinc and copper
  interactions in
  52
mitochondria
  copper binding in 33
  degeneration in mottled
    mice 249
  MAO in 145
mitochondrocuprein
  30, 38
molybdenum
  action on caeruloplasmin
    65
  antagonism with sulphur
    and copper 55, 59
  copper deficiency and 191,
    215
  in plants 68
  in sewage sludge 316
  interaction with copper and
    sulphur 66, 192
monoamine oxidases (MAO)
  144
  in brain 252
  mitochondrial 145
monoamine oxidation
  155
monophenol monooxygenase
  (tyrosinase)
  60, 78, 80
multi-copper proteins
  80
multiple sclerosis, relation to
  enzootic ataxia
  319
myelination
  in copper deficiency 228,
    262, 264

in swayback 190, 218, 248
myocardium in copper
  deficiency
  214
myoglobin
  76
myopathy, selenium and
  90

neonates, copper in
  227, 240
neurological function, copper
  and
  247–266
Neurospora crassa
  72
neutral amino acids
  256, 261
neutropenia
  11, 17, 120, 213
neutrophils
  in copper deficiency 20,
    218, 222–223
  SOD-deficient 138
nickel in sewage
  313
nitrilotriacetic acid
  44
noradrenaline (norepinephrine)
  106, 320
  in brain 252
  in copper deficiency 250
nuts, copper in
  5

oestrogens
  affecting caeruloplasmin
    110, 114, 122, 198, 217
  enhancing lysyl oxidase
    activation 170
  in osteoporosis 114
  inhibiting copper
    absorption 28
  serum copper concentrations
    and 37
osteogenesis
  54, 56
osteoporosis
  114
  in copper deficiency 18,
    19, 114, 190, 191
  in Menkes' syndrome 213

SUBJECT INDEX

oxidases, copper-containing 80
oxidative copper proteins 76
oxygen
 in action of copper and lysyl oxidase 178
oxygen radical
 see also superoxide radicals
 dismutation of 137
 in SOD 139
 in toxicity 310
oxygen toxicity 126, 132
 SOD protecting against 133, 137

paraquat 115, 127
parenteral fluids
 balance studies 11
 containing copper 10, 13, 19
Parkinson's disease 250
 caeruloplasmin in 108
penicillamine 294
 interaction with ions 287, 292
penicillamine therapy
 copper deficiency following 295
 during pregnancy 294
 in rheumatoid arthritis 293, 294, 296, 297
 in Wilson's disease 281, 283, 298
 zinc depletion and 281, 298
perinatal accumulation of copper 39
peroxocarbonates 139
phenylalanine hydroxylase 210
p-phenylenediamine
 correlation of activity with lysyl oxidase 171
p-phenylenediamine oxidase 101, 113, 171

phenylethanolamine-N-methyl-transferase 254, 261
phenylketonuria 9
phospholipids in brain of crinkled mice 233
phosvitin 102
Photobacterium leiognathi 126, 128
phytic acid, copper and 68, 327
pigmentation, copper and 323
 copper deficiency affecting 214, 217
 in crinkled mice 241
pigs, dietary copper 317
placental transport, cadmium, zinc and copper interactions 52, 53
plants, copper in 315
plastocyanin 75, 95, 97, 141
pneumonia, copper deficiency and 18
poliomyelitis 122
polyribosomes, caeruloplasmin synthesis in 269
pregnancy
 caeruloplasmin in 122
 copper absorption in 27
 copper retention in 28
 penicillamine therapy during 294
prematurity, copper deficiency and 160, 215
preterm infants, copper deficiency 8, 160
proteins 71
 copper-binding 41, 42, 257
 metal ions controlling 173

protein loss, copper deficiency and 216
protein malnutrition, caeruloplasmin concentration in 121
protein synthesis, cyclo-heximide inhibiting 175, 180
prothrombin synthesis 294
proton transfer 86
pseudoferroxidase activity of caeruloplasmin 102
Pseudomonas aeruginosa 73, 95
putrescine in diamine oxidase action 148
pyridoxal phosphate 144

quaking mice
 copper binding in 242
 copper deficiency in 229
quercetin dioxygenation 78
quercetin 2,3-dioxygenase 79

radiation damage, SOD protecting against 133
reabsorption of copper into bile 45
recommended dietary allowances (RDA) of copper 5, 6, 17, 18, 216, 327
red deer 263
redox proteins 73
removal of copper 42
retention of copper 18
 see also accumulation of copper
 in intestines 54
 in pregnancy 28

rheumatoid arthritis,
  penicillamine in
  293, 294, 296, 297
routes of administration of
  copper
  15
ruminants
  copper deficiency in
    183–207
  copper excretion in 43
  effect of copper on
    digestion 313

Salmonella typhimurium
  127
scurvy
  320, 321
seborrhoea
  214
selenium in granulomatous
  disease
  320
selenium deficiency,
  vitamin E and
  89
serum concentration of copper
  36, 37
serum proteins in activation
  of lysyl oxidase
  169
sewage sludge
  copper in 316
  disposal 312
  metals in 313
  molybdenum in 316
sheep
  central nervous system in
    240
  copper deficiency in 189
  steely wool 190
  see also enzootic ataxia
shellfish as source of copper
  5
silver, action on caeruloplasmin
  65
skeletal changes
  320
skin, copper absorption
  through
  15
skin cancers
  323

skin changes in copper
  deficiency
  214
slow-virus disease
  320
SOD
  see superoxide dismutase
sodium diethylthiocarbamate
  155
spermine oxidase
  144
stellacyanin
  75, 97
stomach, copper absorption in
  29
storage of copper
  excretion and 39
  forms 37
  in liver 39
  metallothionein-bound 17
Streptococcus faecalis
  126
sulphur, interaction with
  copper and molybdenum
  56, 59, 66, 192
sulphur-containing amino
  acids
  89
sulphydryl groups
  in collagen and elastin
    cross-linking 181
superoxide dismutase (SOD)
  30, 37, 38, 76, 87, 95, 120,
    125–142, 257, 327
  as index of copper
    deficiency 198
  as substrate for caerulo-
    plasmin 120
  biochemical properties 139
  caeruloplasmin as copper
    donor 116
  cellular damage from 192
  classes of 125
  copper binding 89
  copper function in 207
  copper site 77
  crystal structure of 77
  deficiency 138
  effect of copper on
    biosynthesis 131
  formation in liver 269
  function of 131, 141, 187, 205
  in cattle 206

in copper deficiency 173,
  185, 210, 218
induction by substrate 132
in erythrocytes 140, 187, 206
in heart cells 140
in Menkes' disease 248
in mutant *E. coli* 132
in tumour cells 137
iron and 65
membrane-associated 140
protective action of 126,
  133, 137
red cell copper and 206
substrates 140
subunits 137
zinc and 87
superoxide radicals
  haemolysis from 310, 317
  nature and sources of 126
superoxide synthetase
  320
swayback
  see enzootic ataxia
synthetic diets
  copper in 8

testis, effect of zinc and
  cadmium on
  51
tetrathiomolybdate
  60, 67, 120
thiomolybdates, action of
  57, 58
thionein
  38, 72
  polymerization 39
toxicity of copper
  272, 292
  agricultural waste treatment
    and 312, 315
  anaerobic potentiation of
    301–318
  e.p.r. spectra and 302
  in environment 310
  in sheep 34, 37
  intake and 41
  mechanism 309
  municipal effluent
    treatment plants and 311
  renal concentrations 34
  sewage sludge and 312
  under anoxic conditions 302
  zinc and 52

SUBJECT INDEX

trace elements
1
  metabolic interactions with copper 49–69
tranquillizers, caeruloplasmin and
106
transferrin
102
transport of copper
29–30, 49, 71, 173, 240, 326
  caeruloplasmin in 104, 105, 115, 121
  in intestine 29, 267
tryptophan
257
tumour-bearing rats, copper absorption in
27
tumour cells, SOD in
137
tyramine oxidase
143
tyrosinase
210
  see also monophenol monooxygenase
tyrosine aminotransferase
257
tyrosine deficiency
  pigmentation and 217
tyrosine 3-monooxygenase
152, 189, 250, 255, 259

ultraviolet light
323
umecyanin
97
urine
  copper excretion in 7, 217, 297
utilization of copper
  effect of cadmium and zinc on 54

viral disease
319–320

vitamin B_{12} in anaemia of copper deficiency
117
vitamin D, copper and
192
vitamin E
  copper and selenium deficiency and 89
vitiligo
323

weather, swayback and
264
white cells as index of copper function
207
Wilson's disease
173, 219, 270, 273
  atypical 277
  Bedlington terrier disease and 275
  caeruloplasmin in 108, 122, 224, 281
  copper in blood plasma in 297
  copper levels in 271
  copper-metallothionein in 33
  diagnosis of 225
  haemolytic crisis in 318
  hepatic distribution of copper 280
  lysosomes in 280
  mechanism 223
  model for 281
  molybdenum in 60
  monoamines in 152
  pathology 272
  penicillamine therapy in 281, 283, 298

zinc
1, 2, 230
  action on caeruloplasmin 54, 65
  action on copper absorption 27
  action on cytochrome c oxidase 54
  antagonism with cadmium and copper 53
  assessment of 217
  binding sites on metallothionein 26
  copper toxicity and 52
  crossing placenta 52
  dietary concentrations of 51
  effect on copper utilization 54
  effect on hepatic concentration of copper 51
  in breast milk 215
  inhibiting copper absorption 53
  in sewage 313
  in synthetic diets 9
  interaction with copper 51, 52, 296
  interaction with copper and cadmium 53, 66
  penicillamine and 295, 298
  placental copper transport and 53
  superoxide dismutase and 87
  testicular damage from 51
zinc binding
45
zinc deficiency
159
  acrodermatitis enteropathica 325
  alkaline phosphatase in 158
  caused by penicillamine 281
zinc-metallothionein
73, 88
zinc superoxide dismutase (ZnSOD)
125